Water Chemistry

Edited by Murat Eyvaz and Ebubekir Yüksel

Published in London, United Kingdom

IntechOpen

Supporting open minds since 2005

Water Chemistry
http://dx.doi.org/10.5772/intechopen.78128
Edited by Murat Eyvaz and Ebubekir Yüksel

Contributors

Ammar Tiri, Belkhiri Lazhar, Mammeri Asma, Mouni Loutfi, S. Dr. Shanmugan, Omar Cléo Neves Pereira, Altair Bertonha, Rosa Hernández-Soto, Mercedes Salazar-Hernández, José Alfredo Hernámdez-Maldonado, Alba N. Ardilas, Carmen Salazar-Hernández, Majda Breida, Saad Alami Younssi, Mohamed Ouammou, Mohamed Bouhria, Mahmoud Hafsi, Hannachi Yasser, Hafidh Afifa, Ayed Salwa, Jorge Colón, Mario Ramos-Garcés, Joel Sanchez, Isabel Barraza-Alvarez, Yanyu Wu, Dino Villagrán, Thomas Jaramillo, Bahar Ipek, Deniz Uner, Yun Chen, Wenzhong Yang, Sajjad Hussain, Saima Gul, Nadeem Khan, Hammad Khan, Sabir Khan, Hideaki Nakamura, Başak Taşeli

First published in London, United Kingdom, 2020 by IntechOpen
IntechOpen is the global imprint of INTECHOPEN LIMITED, registered in England and Wales, registration number: 11086078, 7th floor, 10 Lower Thames Street, London, EC3R 6AF, United Kingdom
Printed in Croatia

British Library Cataloguing-in-Publication Data
A catalogue record for this book is available from the British Library

Additional hard and PDF copies can be obtained from orders@intechopen.com

Water Chemistry
Edited by Murat Eyvaz and Ebubekir Yüksel
p. cm.
Print ISBN 978-1-78985-557-9
Online ISBN 978-1-78985-558-6
eBook (PDF) ISBN 978-1-78985-429-9

For EU product safety concerns:
IN TECH d.o.o., Prolaz Marije Krucifikse Kozulić 3, 51000 Rijeka, Croatia, info@intechopen.com or visit our website at intechopen.com.

Meet the editors

Editor Dr. Murat Eyvaz is an assistant professor at the Environmental Engineering Department at Gebze Technical University (GTU). He received his BSc degree in Environmental Engineering from Kocaeli University in Turkey in 2004. He completed his MSc and PhD in 2013 at Gebze Institute of Technology (former name of GTU) in Environmental Engineering. He completed his postdoctoral research at the National Research Center on Membrane Technologies in 2015. His research interests are water and wastewater treatment, electrochemical processes, filtration systems/membrane processes, and spectrophotometric and chromatographic analyses. He has coauthored numerous journal articles and conference papers and has taken part in many national projects. He serves as an editor in 45 journals and a reviewer in more than 125 journals indexed in SCI, SCI-E, and other indexes. He has one patent and three patent applications on wastewater treatment systems.

Coeditor Prof. Ebubekir Yüksel is a faculty member at the Environmental Engineering Department in Gebze Technical University. He received his Bachelor degree in Civil Engineering from Istanbul Technical University in 1992. He completed his graduate work (MSc 1995 and PhD 2001) at Istanbul Technical University. His research interests are applications in water and wastewater treatment facilities, electrochemical treatment process and filtration systems at lab and pilot scale, watershed management, flood control, deep-sea discharges, membrane processes, spectrophotometric analyses, chromatographic analyses, and geographic information systems. He has coauthored numerous journal articles and conference papers and has taken part in many national projects. He has produced more than 25 peer-reviewed publications in journals indexed in SCI, SCI-E, and other indexes. He has two patents and three patent applications on energy and wastewater treatment systems.

Contents

Preface

Water, which plays an important role in every aspect of our daily lives, is the most valuable natural resource we have on this planet. Drinking, bathing, cooking, regeneration, cleaning, production, energy, and many other uses of water originate from some of its versatile, useful, basic, and unique features. The access, purification, and reuse of water on our planet, which is of course not endless and not available for direct use, is directly related to the water chemistry that explores its inimitable properties. For a safe drinking water supply, water treatment plants must be operated properly. Appropriate processes should be developed to remove various pollutants from water and wastewater. Surface water quality from rivers and lakes needs to be examined to monitor the adverse environmental effects of water and wastewater treatment plant discharges. Again, the flow and components of groundwater should be monitored to avoid possible contamination of crude oil or gasoline leaks.

This book includes research on water chemistry-related applications in environmental management and sustainable environmental issues such as water and wastewater treatment, water quality management, and other similar topics. The book consists of three sections, namely, water treatment, wastewater treatment, and water splitting, and includes 11 chapters. In these chapters, water-wastewater remediation methods, nanomaterials in water treatment, and water splitting processes are comprehensively reviewed in terms of water chemistry.

Asst. Prof. Murat Eyvaz and Prof. Ebubekir Yüksel
Department of Environmental Engineering,
Gebze Technical University,
Turkey

Section 1

Water Treatment

Chapter 1

Suitability and Assessment of Surface Water for Irrigation Purpose

Ammar Tiri, Lazhar Belkhiri, Mammeri Asma and Lotfi Mouni

Abstract

Surface water is an important resource that can create tensions between different countries sharing the same water sources to know that the agriculture is considered as the last sector that exploits less water compared to the industry which uses very large water quantities. The future strategies of agricultural development in the most of these countries depend on the ability to maintain, improve and expand irrigated agriculture. In this light, this chapter is written in the way to show some steps of the evaluation of surface water for irrigation purpose. The results obtained from this research make it possible to evaluate the suitability of surface water for irrigation and to draw useful recommendations for dam managers and farmers.

Keywords: surface water, hydro chemical analysis, irrigation purpose, water quality index

1. Introduction

Surface water is an essential natural resource that plays a vital role in human life and has an important role in drinking, irrigation and economic sectors. According to FAO statistics, 20% of the land is irrigated but produces 40% of the crops [1]. Irrigation is an effective way to improve productivity significantly. However, there are environmental risks associated with irrigation, especially water stagnation and increased salinity. Agricultural irrigation is a factor of increase and diversification of crops. This is why its development must be encouraged in the world through agricultural, international, national and community policies [2]. For this reason, that successive governments have so far sought to harden the right to water while protecting the interests of irrigators. This strategy, if explained by the social contract that binds the state to farmers, is nonetheless debatable [3]. The priority is no longer today to increase yields among the highest in the world, but to ensure the continuity of drinking water supply services, the preservation of aquatic ecosystems and sufficient water levels to respond industrial needs. The salinity of water has increased in many watersheds around the world and the use of non-traditional resources, bet on the use of available resources, increases efficiency, waste reduction and water maintenance quality establishing surveillance networks, developing and setting standards, and enacting the necessary laws to protect them from pollution [4, 29]. The importance of developing a comprehensive strategy for water demand management appears to be a consequence of the exacerbation of the problem of scarcity of water resources due to drought, which has become more frequent with a longer duration [5]. Therefore, the main goal of this chapter is to evaluate suitability of surface water for irrigation purpose by appropriate parameters and indices in Koudiate Medouar dam in northeast of Algeria.

2. Descriptive of the study area

Koudiate Medouar dam built in 1994 on Oued Reboa is located 7 km north-east of Timgad and 35 km from Batna in Algeria (**Figure 1**). The dam is located at the east longitude 6°30′48″ and north latitude 35°30′57″. It is a reservoir dam that mobilizes the surface waters of Reboa river whose watershed covers 59,000 km^2.

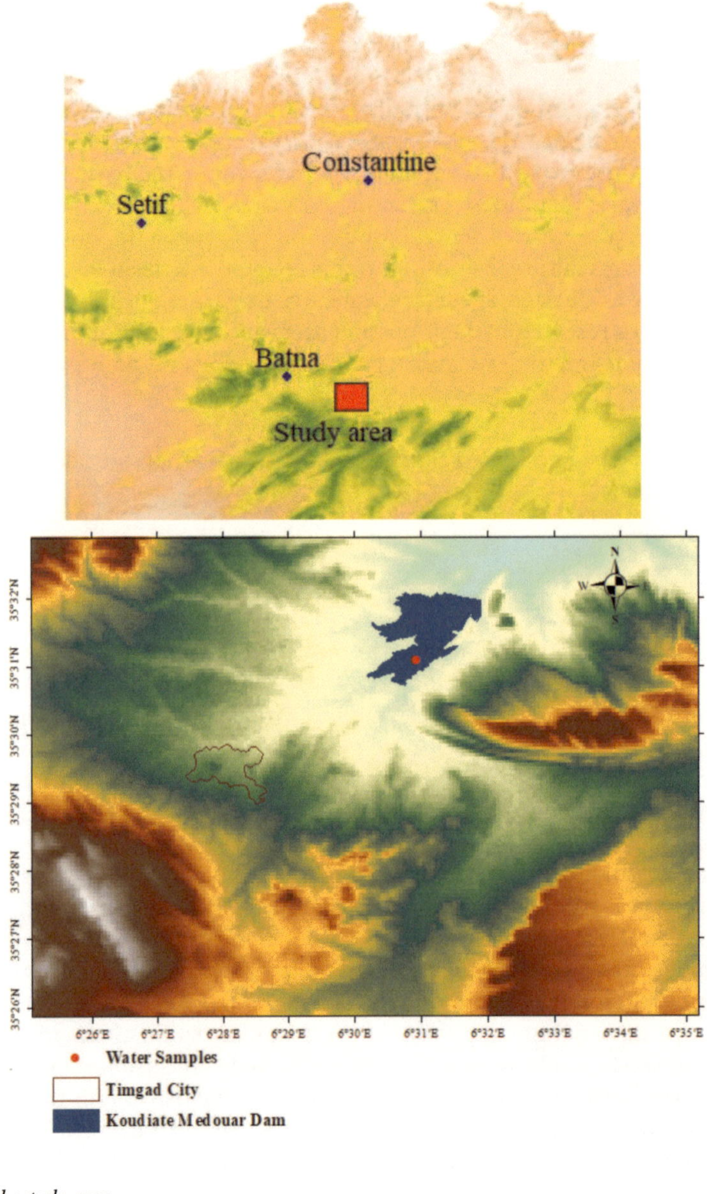

Figure 1.
Location of the study area.

About 48.72% of the population of Batna Wilaya, or 682,000 inhabitants, drink water from this dam that supplies the cities of Batna, Tazoult, Timgad, Ain Touta, Barika, Arris and Ouled Reach in the Wilaya of Khenchela. The climate of the study area is semi-arid, characterized by high temperatures and low rainfall. The average annual rainfall is about 370 mm, while the annual average temperature is around 15°C [6].

3. Methodology

3.1 Water sampling and analysis

Surface water samples collected from the dam basin during a year from February 2017 to January 2018. During water sampling, all samples were filtered on-site by 0.45-μm filter. The water samples were stored in 500-ml high-density polyethylene bottles (HDPE) for laboratory analyses. All water samples were kept at 4°C until they were analyzed with using standard methods APHA [7]. Measurements of pH and electrical conductance (EC) were carried out in field with the use of a portable multi-parameter analyzer Hem [8]. Major cations and anions including calcium (Ca), magnesium (Mg), sodium (Na), potassium (K), sulfate (SO_4), bicarbonate (HCO_3), chloride (Cl) and nitrate (NO_3) were measured in laboratory. Total hardness (TH) and Ca were volumetrically analyzed using standard EDTA. Mg was calculated by taking the difference between TH and Ca. A flame photometer was used to estimate Na and K. HCO_3 and Cl were analyzed by titration with standard HCl and $AgNO_3$, respectively. SO_4 was determined using a turbidimetric procedure. NO_3 was analyzed using the colorimetric method. The reliability of the data set generated was verified through electrical neutrality by the following equation:

$$\text{Error of ion balance} = \frac{\sum\text{Cations} - \sum\text{Anions}}{\sum\text{Cations} + \sum\text{Anions}} \times 100 \tag{1}$$

The analytical data were considered doubtful beyond an error of ±5% [9].

4. Results and discussion

4.1 Descriptive statistics

The statistical summary of the data used in this study is represented in **Table 1**. The results show that pH values varied from 7.4 to 7.8 with a mean of 7.5, indicating that the water is consider as a slightly alkaline water [10]. Electrical conductivity values express the amount of dissolved solids in the water sample. Water samples has EC values that ranged from 1040 to 1800 μS/cm with an average of 1349 μS/cm. EC of the surface water samples was above the fixed value of 1000 μS/cm by WHO [11]. The average values of Ca, Mg, Na and K are 94.04, 42.72, 93.92 and 1.01 mg/L, respectively. The order abundance of the major cations as follows Ca > Na > Mg > K, where the calcium and sodium are the dominate cations in surface water. The calcium values are generally upper than the limits set in WHO guides [11]. The high concentrations of Ca and Na are explained by the ion exchange process between sodium and calcium elements which leads to the precipitation of $CaCO_3$ in the soil profile. The order abundance of the major anion from the highest to the lowest is HCO_3 > Cl > SO_4 > NO_3, indicating that bicarbonate and chloride are the dominants anions in the surface water. The concentration of HCO_3 is varied from 134.2 to

Valid N	pH	EC (μS/cm)	TDS (mg/L)	Na	K	Ca	Mg	HCO_3	Cl	SO_4	NO_3
1	7.50	1080	691.20	103.30	1.26	88.18	45.48	195.22	106.50	90.00	0.6
2	7.40	1060	678.40	98.30	1.29	88.18	43.08	183.00	88.75	92.00	0.2
3	7.50	1040	665.60	95.70	1.01	88.18	45.49	183.00	88.75	91.00	0.3
4	7.40	1060	678.40	87.30	0.99	96.19	45.48	201.30	124.25	90.00	0.2
5	7.50	1100	704.00	99.50	1.00	96.19	43.08	183.00	88.75	90.00	0.3
6	7.80	1500	960.00	92.80	0.90	94.60	41.64	170.80	82.75	82.00	0.5
7	7.50	1500	960.00	92.20	0.81	93.20	42.48	170.80	80.75	78.00	0.3
8	7.40	1300	832.00	93.60	0.92	95.20	41.28	168.80	75.25	80.00	0.6
9	7.60	1800	1152.00	80.00	0.96	96.40	40.56	144.40	81.25	83.00	0.3
10	7.50	1600	1024.00	95.20	0.94	98.40	39.36	134.20	88.25	85.00	1.2
11	7.80	1800	1152.00	95.25	1.00	99.70	42.00	170.80	88.25	82.50	2.5
M	7.54	1300	832.00	95.20	0.99	95.20	42.48	170.80	88.50	85.00	0.3
SD	0.14	302	193.56	6.22	0.15	4.14	2.07	19.87	13.73	4.96	0.681
CV	1.90	22.48	22.41	6.62	14.40	4.40	4.85	11.47	15.21	5.79	107.1

M, mean; SD, standard deviation; CV, coefficient of variation (%).

Table 1.
Summary of the statistical analyses of the physicochemical parameters.

201.3 mg/L with an average of 173.21 mg/L, high concentration of this element related to the water-rock interaction process.

4.2 Determination of the origin of dissolved solids

In order to determine the origin of the highest values of some parameters in surface water the relationships between some parameters are studied.

Figure 2 shows the relationship between total cations and Ca + Mg, we see that all sample points are located below the equilibrium line 1/1 which confirms the alteration process and the exchange of alkaline ions [13].

Total cations versus Na + K is presented in **Figure 3**, we see that the samples are located below the equilibrium line 1/1, indicating that the excessive concentrations of Na and K are due to the accumulated salts in the soil during the evaporation process [13].

Na versus HCO_3 plot shows that there is a distribution of samples below and above the equilibrium line 1/1 indicating the presence of dissolution of the rocks during the infiltrations (**Figure 4**).

The chloride versus sodium plot (**Figure 5**) is employed to verify the relationship and sources of the ions in surface water. A Cl/Na ratio equal one is typically characteristic of halite dissolution, whereas values <1 implies the alkali metal is released from silicate weathering reactions [12, 13].

4.3 Suitability of surface water for irrigation

The evaluation of the water surface suitability of the study area for irrigation was carried out using total dissolved solids (TDS), total hardness (TH), electrical conductivity (EC), the sodium adsorption rate (SAR), the percentage of sodium (%Na), residual sodium carbonate (RSC), the permeability index (PI), the salinity

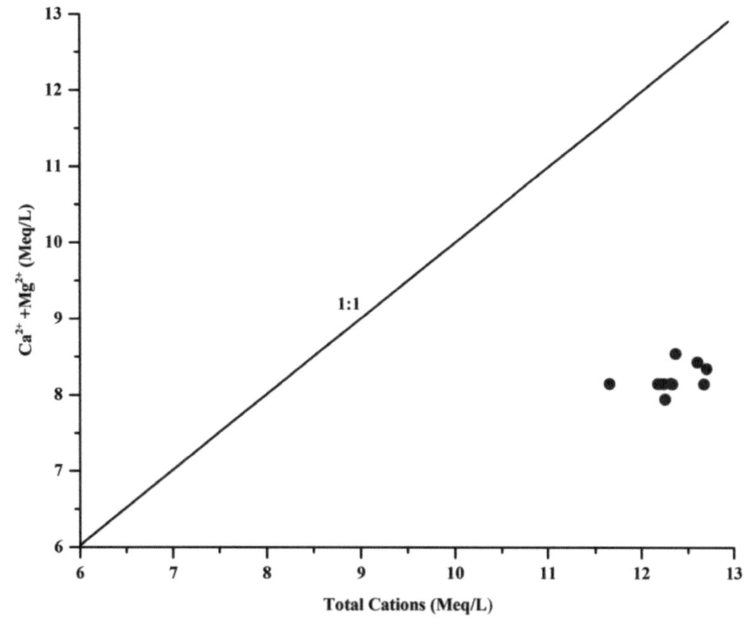

Figure 2.
Relations between the total cations and Ca^{2+} with Mg^{2+}.

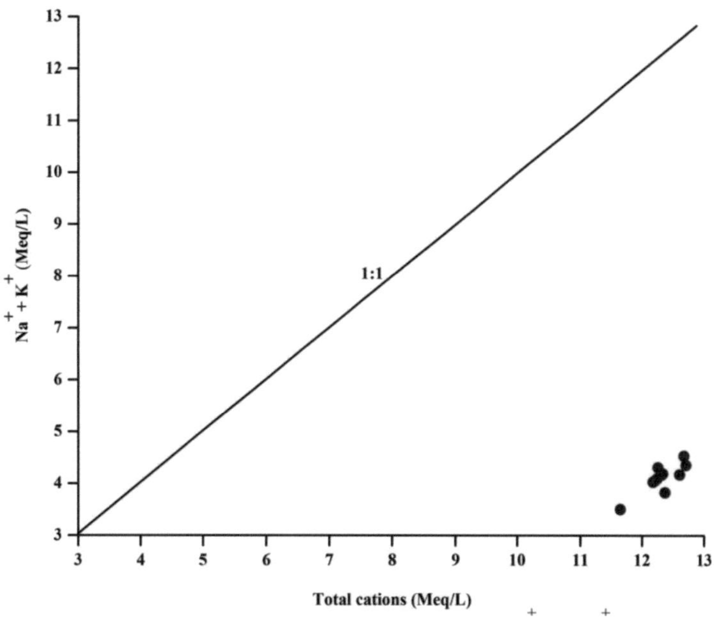

Figure 3.
Relations between the total cations and Na^{+} with K^{+}.

index (PS), the soluble sodium percentage (SSP), the magnesium adsorption rate (MAR), and Kelly's ratio (KR). The results are presented in **Tables 2–4**.

4.3.1 Total dissolved solids (TDS)

Total dissolved solids (TDS) is calculated by the following equations:

$$TDS = 640 \times EC \quad (\text{for } EC < 5\ dS/m) \tag{2}$$

$$TDS = 640 \times EC \quad (\text{for } EC > 5\ dS/m) \tag{3}$$

The results showed that the values of TDS varied from 665 to 1152 mg/L with an average 863 mg/L (**Table 2**). This large variation of TDS values in surface water samples is classified in the high saline water zone and the water samples contain

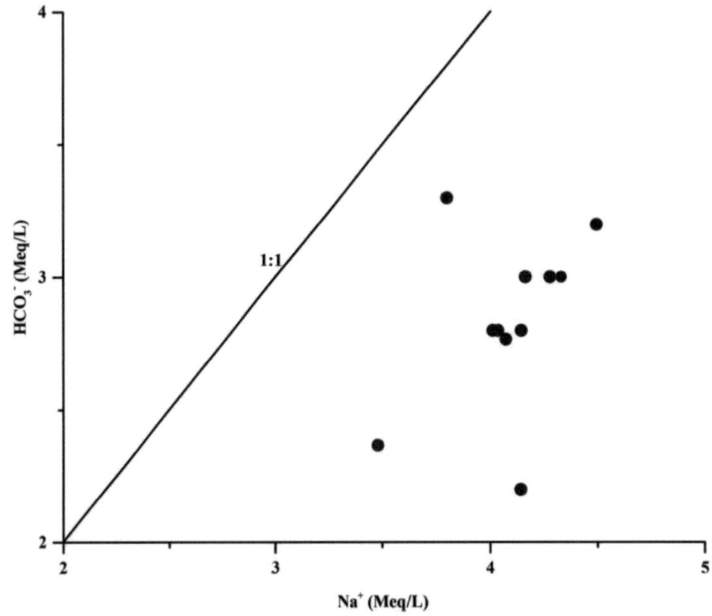

Figure 4.
Relation between: Na^+ versus HCO_3^-.

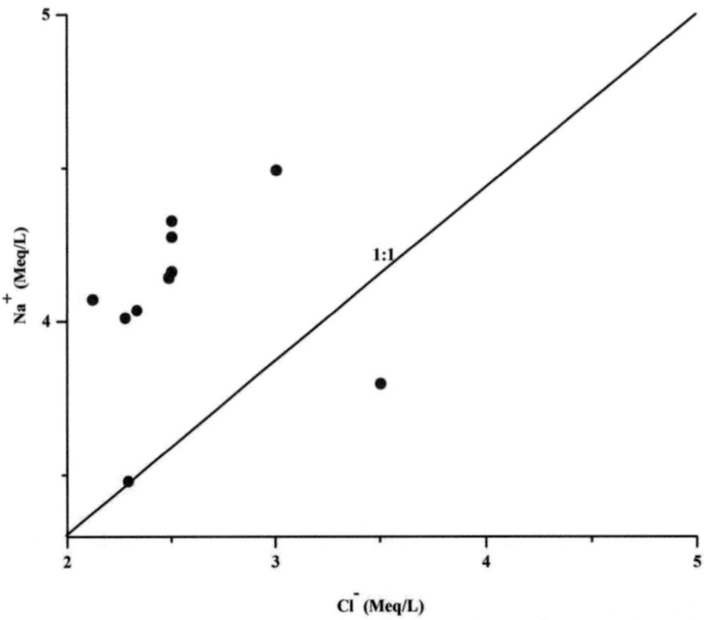

Figure 5.
Relation between: Cl^- versus Na^+.

Parameters	Mean	Median	Minimum	Maximum	Std. dev.	Coef. var.
TDS	863.418	832.0100	665.600	1152.000	193.561	22.418
TH	413.098	410.000	399.950	429.975	8.348	2.021
%Na	40.942	40.917	37.152	43.893	1.918	4.685
RSC	37.889	36.220	−1.560	62.960	20.109	53.075
SAR	4.017	4.006	3.418	4.468	0.286	7.128
KR	0.687	0.686	0.584	0.773	0.054	7.795
MH	31.260	30.933	28.571	34.033	1.807	5.782
PI	46.382	46.223	42.412	49.490	1.958	4.222
PS	99.576	97.470	84.194	133.737	13.901	13.961
MAR	31.260	30.933	28.571	34.033	1.807	5.782
SSP	68.241	68.123	58.003	76.563	5.265	7.715

Table 2.
Summary of the statistical analyses of the irrigation parameters.

Parameters	Desirable permissible [15]	Irrigation [16]
EC	–	1000
TDS	500–2000	–
TH	300–600	712

All major ions and TDS are expressed in mg/l while pH on scale and EC in μS/cm.

Table 3.
Standards used for drinking and irrigation suitability and relative weight for each parameter.

different amounts of major ions [14] (**Table 4**). The large variation of TDS values can be attributed to the variation in the hydrological processes and geological formations in the study area. 36.36% of the total water samples are classified in the good water class while 63.63% of the total samples are represented a permissible water quality with an estimated SD value of 193.53 (**Table 4**). No prescribed value limits the threshold for TDS in irrigation waters (**Table 3**).

4.3.2 Total hardness (TH)

Total hardness is defined as the sum of calcium and magnesium using the following equation:

$$TH = \left[\left(2 \times \frac{Ca^{2+}}{40}\right) + \left(2 \times \frac{Mg^{2+}}{24}\right)\right] \times 50 \quad (4)$$

where Ca and Mg concentrations are in meq/L.

The values of TH varied from 399 to 429.97 mg/L with an average of 413 mg/L (**Table 2**) where the maximum value is below the prescribed limit for irrigation water of 712 mg/L (**Table 3**). The majority of water samples showed TH values are below the standard values used for drinking and irrigation suitability BIS [15] and FAO [16]. The low values of TH are probably due to the presence of alkaline earth ions (Ca and Mg) of weak acids (HCO_3 and CO_3) and strong acids (Cl, SO_4 and NO_3) [30, 31]. Therefore, low alkalinity values reflect immature hydrochemistry of

Classification scheme	Categories	Range (mg/L)	Percentage	Number of samples
Total dissolved solid (TDS)	Excellent	<450	0	0
	Good	450–750	4	36.36
	Permissible	750–2000	7	63.63
	Unsuitable	>2000	0	0
Total handers (TH)	Soft	<75	0	0
	Moderately hard	75–150	0	0
	Hard	150–300	0	0
	Very hard	>300	11	100
Electrical conductivity (EC) in µS/cm	Excellent	<250	0	0
	Good	250–750	0	0
	Permissible	750–2250	11	100
	Unsuitable	>2250	0	0
Permeability index (PI)	Excellent	>75	0	0
	Good	25–75	11	100
	Unsuitable	<25	0	0
Salinity potential (SP)	Excellent to good	<5	0	100
	Good to injurious	5 10	0	0
	Injurious to unsatisfactory	>10	11	0
Magnesium absorption ratio (MAR)	Acceptable	<50	11	100
	Non-acceptable	>50	0	0
Kelly's ratio (KR)	Suitable	<1	11	100
	Unsuitable	>1	0	0
Sodium absorption (SAR)	Excellent	<10	11	100
	Good	10–18	0	0
	Fair	>18–26	0	0
	Poor	>26	0	0
	Hard	>200–300	0	0
	Very hard	>300	0	0
Sodium percentage (%Na)	Excellent	Up to 20	0	0
	Good	>20–40	1	9.09
	Permissible	>40–60	10	90.9
	Doubtful	>60–80	0	0
	Unsuitable	>80	0	0
Residual sodium carbonate (RSC) in meq/L	Good	<1.25	1	100
	Medium	1.25–2.5	0	0
	Bad	>2.5	10	0
Soluble sodium percentage (SSP)	Excellent	0–20	0	0
	Good	20–40	0	0
	Permissible	40–60	0	100
	Doubtful	60–80	11	0
	Unsuitable	>80	0	0

Classification scheme	Categories	Range (mg/L)	Percentage	Number of samples
Irrigation water quality index (IWQI)	Excellent	85–100	0	0
	Good	70–85	11	100
	Poor	55–70	0	0
	Very poor	40–55	0	0
	Unsuitable for irrigation use	0–40	0	0

Table 4.
Classification schemes for surface water quality indicators.

surface water during seepage and hypodermic flow [17]. The water of the Dam is classified as a very hard water (**Table 4**).

4.3.3 Electrical conductivity (EC)

Wilcox [18] proposed a diagram with respect to a combination of EC and %Na for judging suitability of water quality for irrigation. The diagram is divided into five zones, which are excellent to good, good to permissible, permissible to doubtful, doubtful to unsuitable and unsuitable, with increasing salinity hazard and sodium hazard for irrigation.

The results show that EC values of the most water samples represented high saline water for irrigation use which are due to the high concentrations of ions in surface water (**Tables 3** and **4**). Moreover, the EC value which 100% of the surface water in the study area represents permissible water and greater than the value fixed by FOA for irrigation water (EC > 1000 μS/cm).

4.3.4 Sodium adsorption ratio (SAR)

The United States Soil Laboratory Staff (USSLS)'s diagram in Richards [19] illustrates the combined effect of EC, SAR and percent sodium (%Na), residual sodium carbonate (RSC) in the classification of irrigation water quality which, divided into four shared areas between EC and SAR: C1S1 to C1S4, C2S1 to C2S4, C3S1 to C3S4 and C4S1 to C4S4. The salinity hazard classes that have been classified into four classes: low salinity hazard class (C1) with an EC value less than 250 μS/cm; medium salinity risk class (C2) with EC value between 250 and 750 μS/cm; high salinity risk class (C3) with EC value between 750 and 2250 μS/cm; and a very high salinity risk class (C4) with an EC value greater than 2250 μS/cm.

The SAR values varied from 3.41 to 4.46 mg/L with an average of 4.01 mg/L where the water samples are classified in the excellent class of water suitable for irrigation according to the Richard and Wilcox irrigation water quality classification (**Tables 2** and **4**).

The result of the effect of ion exchange processes on soil quality and its capacity in terms of sodium uptake is expressed by the SAR which is calculated by the following equation:

$$SAR = \frac{Na^{+}}{\sqrt{\frac{Ca^{2+}+Mg^{2+}}{2}}} \tag{5}$$

where all ions concentration is in meq/L.

The calculated SAR value expresses that all water samples are classified in the best water range. **Figure 6** shows that all surface water samples are located in the C3S1 zone.

Figure 7 illustrates the Wilcox diagram [18] which highlights the combination of EC and %Na for judging suitability of water quality for irrigation. The diagram is divided into five zones, which are excellent to good, good to permissible, permissible to doubtful, doubtful to unsuitable and unsuitable, with increasing salinity hazard and sodium hazard for irrigation. From **Figure** 7, we see that all water samples are classified in the good to permissible water categories, which is reflected by the EC value which shows more than 100%. Water from the study area is permissible distributed between 750 and 2250 μS/cm (**Table 4**).

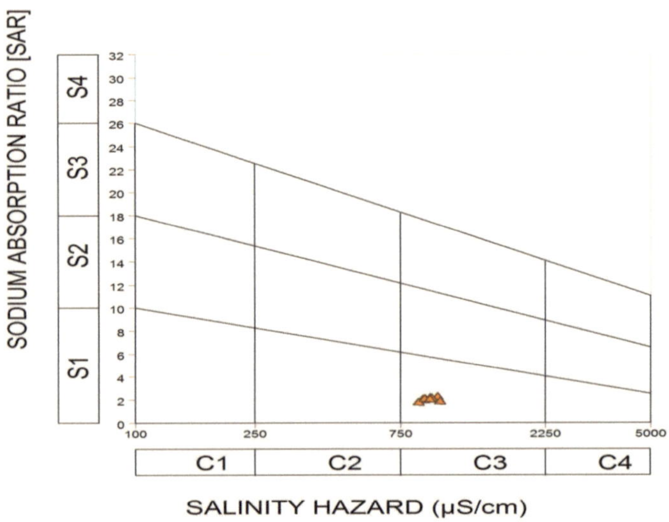

Figure 6.
Classification of Richards [19]. EC versus SAR.

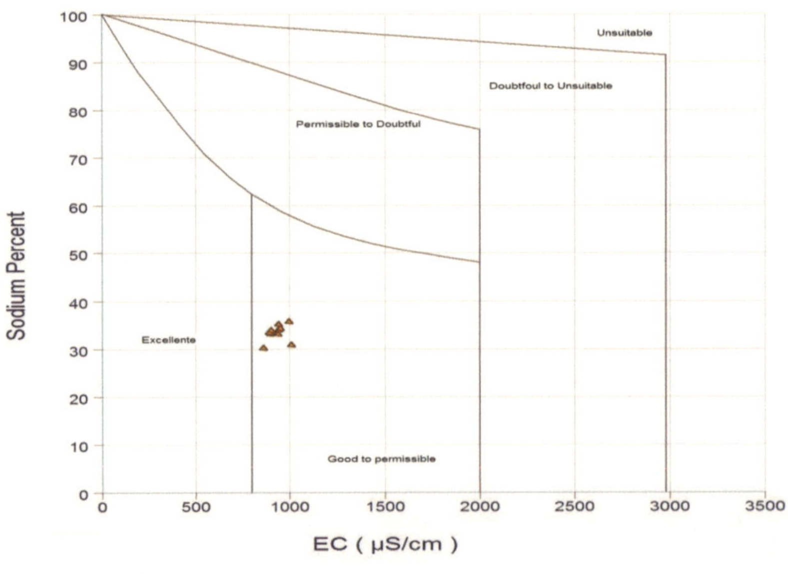

Figure 7.
Classification of Wilcox [18]. EC versus %Na$^+$.

4.3.5 Parentage sodium (%Na)

The values of %Na varied from 37.15 to 43.89 mg/L with an average of 40.94 mg/L (**Table 2**). The water samples are classified in the class good to permissible for irrigation water (**Figure 6**). This variation could be due to the size of the samples, the geological factor, the type of soil, the anthropic activities and the addition of chemical fertilizers, the climatic factor, and the dissolution of the minerals of the lithological composition [20].

Therefore, Eq. (6) shows another way to determine the sodium risk ratio when calculating the sodium ratio (%Na) in order to determine the water quality for irrigation uses.

$$\%Na = \frac{(Na^{+} + K^{+}) \times 100}{(Ca^{2+} + Mg^{2+} + Na^{+} + K^{+})} \tag{6}$$

The values of %Na express that 9.09% of the total water samples are classified in the good water category and 90.9% in the permissible water category (**Table 4**).

4.3.6 Residual sodium carbonate

RSC is a very important parameter in the study of suitability water for irrigation. The RSC is calculated using the equation given below in Eston [21]:

$$RSC = (HCO_3^{-} + CO_3^{2-}) - (Ca^{2+} + Mg^{2+}) \tag{7}$$

where the concentrations are reported in meq/L.

The values of RSC varied from −2.33 to −4.68 meq/L with an average of −3.50 meq/L (**Table 2**). The water samples are classified in the category of good water irrigation according to the recommendations of Eaton [21] and Arslan [22]. In addition, the values of RSC calculated by Eq. (7) for all samples are classified in the good zone (**Table 4**).

4.3.7 Permeability index (PI) and salinity index (PS)

The permeability index (PI) of surface water for irrigation, which in turn is influenced by Na, Ca, Mg and HCO_3 concentration. PI is defined by the following equation:

$$PI = \frac{Na^{+} + \sqrt{HCO_3^{-}}}{Ca^{2+} + Mg^{2+} + Na^{+}} \times 100 \tag{8}$$

The salinity index (PS) is defined as the chloride concentration plus half of the sulfate concentration [33]. PS is computed using the equation bellow:

$$PS = Cl^{-} + \sqrt{SO_4^{2-}} \tag{9}$$

where the concentrations are reported in meq/L.

The values of PI and PS ranged from 42.41 to 49.49 and 84.19 to 133.73 mg/L with an average of 46.38 and 99.57 mg/L, respectively (**Table 2**). The PI and PS for all water samples are classified in the type of water good to excellent and good, respectively (**Table 4**).

4.3.8 Soluble sodium percentage (SSP)

Soluble sodium percentage (SSP) is an important parameter to assess the hazard towards irrigation. SSP is defined by Todd [23] as shown below:

$$SSP = \frac{Na^+}{(Ca^{2+} + Mg^{2+} + K^+)} \times 100 \tag{10}$$

where the concentrations are reported in meq/L.

The values of soluble sodium percentage (SSP) varied between 58 and 76.56 mg/L with an average 68.24 mg/L. The calculation of SSP using Eq. (10) reflects that all water samples are classified in the permissible waters area (**Table 4**).

4.3.9 Magnesium adsorption rate

The magnesium adsorption rate (MAR) is expressed in terms of magnesium hazard (MH), which is computed by Eq. (11) in Raghunath [24], using the values of ions in meq/L.

$$MAR = \frac{Mg^{2+}}{Ca^{2+} + Mg^{2+}} \times 100 \tag{11}$$

The computed values of magnesium hazard from the surface water of the study area are in between 28.57 and 34.03 mg/L (**Table 2**). The majority of the water samples of the study area are less than 5 and hence they are safe for irrigation purpose (**Table 4**).

4.3.10 Kelly ratio

The Kelly's ratio (KR) indicates the degree and the potential effect of sodium on water quality for irrigation.

$$KR = \frac{Na^+}{Ca^{2+} + Mg^{2+}} \tag{12}$$

where the concentrations are reported in meq/L.

The results show that the values of Kelly ratio varied from 0.58 to 0.77 mg/L with an average of 0.68 mg/L (**Table 2**).

Kelly ratio of more than 1 indicates an excess level of Na in water. Kelley [25] suggested that the ratio for irrigation water should not exceed 1. All water samples in the study area fall in suitable water type, indicating that there is no significant excess of sodium in the surface water [29] (**Table 4**).

4.4 Irrigation water quality index (IWQI)

The surface water quality index method for irrigation is a very important tool in determining the overall impact of the various parameters that are used as a single variable. In addition, the method of the surface water quality index for irrigation is considered a very satisfactory way to measure and classify the adequacy of surface water quality for irrigation as unique parameters. Taking into account various water quality variables. In this study, the IWQI model was developed by combining the eight water quality parameters (SAR, RSC, %Na, EC, pH, TDS, Na and Cl), which is based on the recommendations of Amanuel Gidey [26], Meireles et al. [27] and

Parametres	W_{cv}	Q_{rv}	$W_{cv} \times Q_{rv}$
EC	0.000444444	59.95555556	0.026646914
pH	0.117647059	88.58823529	10.42214533
TDS	0.0005	43.1705	0.02158525
%Na	0.011111111	45.48888889	0.505432099
SAR	0.038461538	15.38461538	0.591715976
RSC (meq/L)	0.005882353	21.76470588	0.128027682
Cl	0.01	90.31	0.9031
Na	0.014285714	134.1714286	1.916734694
HCO_3 (meq/L)	0.001639344	28.39508197	0.046549315
Total	0.199971564		14.56193726
IWQI	72.82003972		

Table 5.
The relative weight of hydrochemical parameters in the study area.

Hussain et al. [28]. The irrigation water quality index is calculated by the following equations:

$$Q_{rv} = \frac{C_v}{RS_v} \times 100 \tag{13}$$

Q_{rv} represents the quality rating values, C_v stand for the observed concentration values.

$$W_{cv} = \frac{1}{RS_v} \tag{14}$$

W_{cv} represents the stands for the relative weight coefficient of the parameters, RS_v stands for the recommended standards values of the water quality variable.

$$IWQI = \frac{\sum_{i=1}^{n} W_{cv} \times Q_{rv}}{\sum_{i=1}^{n} W_{cv}} \tag{15}$$

where IWQI represent for water quality index, is a dimensionless parameter ranging from 0 to 100, and *n* stands for the number of water quality variables.

The Irrigation Surface Water Quality Index was calculated by Eqs. (13)–(15) and these values are shown in **Tables 4** and **5** and are compared to the irrigation water quality parameters proposed by University of California Committee of Consultants, Meireles et al. [27], and Mohamed et al. [32]. The values of IWQI ranged between 85 and 100 have no restriction for irrigation water, so values between 70 and 85 have low water, and 55–70 reflects the moderate water area, 40–55 high tolerance crops can grow, and 0–40 this unsuitable for all for all crops.

The IWQI values ranged between 70 and 85 with an average of 78.11 indicating the good waters area.

5. Conclusion

In this chapter, surface water quality and its suitability for irrigation in Koudiate Medouar dam as an example were examined. All samples are suitable for irrigation, and appropriate management measures are suggested to safeguard this resource and improve its quality.

6. Recommendation

Surface water represents a very important source for all consumptions, however it must be preserved, for the moments of crisis especially during the dry years, so it should not be used by farmers and industries, a better management by rational and optimal use must be taken into account by future generations.

In this perspective, the following suggestions have been made:

- Uncontrolled use of agricultural chemicals by farmers where stricter controls are needed to prevent contamination of surface water which in turn affects groundwater during of infiltrations.

- Improved soil texture, soil and water salinity are key factors that need to be controlled and monitored accordingly improvements require deep soil drainage, leaching and drip irrigation.

- Periodic sampling allows taking adequate, appropriate, and consistent measures at the appropriate time to deal with any possible contamination.

- Dam managers must make the public and all farmers aware of water quality forget to implement the appropriate management measures to improve the quality of groundwater.

Author details

Ammar Tiri*, Lazhar Belkhiri, Mammeri Asma and Lotfi Mouni
Research Laboratory in Applied Hydraulics, University of Batna 2 Mostefa Ben Boulaid, Batna, Algeria

*Address all correspondence to: tiri_ammar@yahoo.fr

References

[1] Tiri A, Belkhiri L, Mouni L. Evaluation of surface water quality for drinking purposes using fuzzy inference system. Groundwater for Sustainable Development. 2018;**6**:235-244

[2] USEPA (United States Environmental Protection Agency). Risk Assessment Guidance for Superfund. Human Health Evaluation Manual (Part A). EPA/540/1-89/002 Office of Emergency and Remedial Response. Volume 1. Washington, DC: U.S. Environmental Protection Agency; 1989

[3] Belkhiri L, Mouni L, Tiri A, Narany TS, Nouibet R. Spatial analysis of groundwater quality using self-organizing maps. Groundwater for Sustainable Development. 2018;**7**: 121-132

[4] AGS R. Evaluation of hydrogeochemical characteristics of phreatic alluvial aquifer in southeastern coastal belt of Prakasamdistrict, South India. Environment and Earth Science. 2013;**68**:471-485

[5] Belkhiri L, Tiri A, Mouni L. Chapter 2. Assessment of heavy metals contamination in groundwater: A case study of the South of Setif Area, East Algeria. Books. Achievements and Challenges of Integrated River Basin Management. 2018:17-31. Available from: http://dx.doi.org/10.5772/intechopen.75734

[6] Tiri A, Lahbari N, Boudoukha A. Assessment of the quality of water by hierarchical cluster and variance analyses of the Koudiat Medouar watershed, East Algeria. Applied Water Science. 2017;7:4197-4206. DOI: 10.1007/s13201-014-0261-z

[7] APHA. Standard Methods for Examination of Water and Wastewater. 17th ed. Washington, DC: American Public Health Association; 1989

[8] Hem JD. Study and Interpretation of the Chemical Characteristics of Natural Water: USGS Professional Paper Book 2254. Jodhpur: Scientific Publishers; 1991

[9] Patel P, Raju NJ, BSR R, Suresh U, Gossel W, Wycisk P. Geochemical processes and multivariate statistical analysis for the assessment of groundwater quality in the Swarnamukhi River basin, Andhra Pradesh, India. Environment and Earth Science. 2016;**75**:1-24. DOI: 10.1007/s12665-015-5108-x

[10] Belkhiri L, Mouni L, Narany TS, Tiri A. Evaluation of potential health risk of heavy metals in groundwater using the integration of indicator kriging and multivariate statistical methods. Groundwater for Sustainable Development. 2017;**4**:12-22

[11] WHO (World Health Organization). Guidelines for Drinking-Water Quality. Recommendations. 3rd ed. Geneva: Word Health Organization; 2006

[12] Egbi CD, Anornu G, Appiah-Adjei EK, Ganyaglo SY, Dampare SB. Evaluation of water quality using hydrochemistry, stable isotopes, and water quality indices in the lower Volta River Basin of Ghana. Environment, Development and Sustainability. 2018. DOI: 10.1007/s10668-018-0180-5

[13] Abboud IA. Geochemistry and quality of groundwater of the Yarmouk basin aquifer, North Jordan. Environmental Geochemistry and Health. 2018;**40**(4):1405-1435. Available from: https://doi.org/10.1007/s10653-017-0064-x

[14] Salem WM, El-Sayed M. Hydro-geochemical and isotopic composition of ground water in Helwan area. Egyptian Journal of Petroleum. 2015;**24**(4):

411-421. DOI: 10.1016/j.ejpe.2015.10.004

[15] BIS. Drinking water specications. Bureau of Indian Standards, IS 10500. 2012

[16] FAO (Food and Agriculture Organization). Water quality guidelines for agriculture, surface irrigation and drainage. Vol. 1. Food and Agriculture Organization; 1985. 29 pp

[17] Demetriades A. Understanding the quality of chemical data from the urban environment—Part 2: Measurement uncertainty in the decision-making process. In: Johnson CC, Demetriades A, Locutura J, Ottesen RT, editors. Mapping the Chemical Environment of Urban Areas. Chichester: Wiley-Blackwell; 2011. pp. 77-98

[18] Wilcox LV. Classification and Use of Irrigation Water. Washington, DC: U.S. Department of Agriculture Circular 969; 1955

[19] Richards LA. Diagnosis and improvement of saline and alkali soils. Agricultural Hand Book 60. Washington, D.C: U. S: Department of Agriculture; 1954. p. 160

[20] Sutharsiny A et al. Characterization of irrigation water quality of Chunnakam aquifer in Jaffina Penisula. Tropical Agricultural Research. 2012; **23**(3):237-248

[21] Eaton EM. Significance of carbonate in irrigation water. SoilScience. 1950;**69**: 123-133

[22] Arslan H. Determination of temporal and spatial variability of groundwater irrigation quality using geostatistical techniques on the coastal aquifer of Carsamba plain, Turkey, from 1990 to 2012. Environment and Earth Science. 2017;**76**:38

[23] Todd DK. Salt Water Intrusion of Coastal Aquifers in the United States. Vol. 52. Gentbrugge, Belgium: International Association of Scientific Hydrology; 1960. pp. 452-461

[24] Raghunath HM. Groundwater. 2nd ed. New Delhi: Wiley Eastern Ltd; 1987. 563 p

[25] Kelly WP. Use of saline irrigation water. Soil Science. 1963;**95**:355-439

[26] Gidey A. Geospatial distribution modeling and determining suitability of groundwater quality for irrigation purpose using geospatial methods and water quality index (WQI) in Northern Ethiopia. Applied Water Science. 2018; **8**:82. DOI: 10.1007/s13201-018-0722-x

[27] ACM M et al. A new proposal of the classification of irrigation water. Revista Ciência Agronômica. 2010;**41**:349-357

[28] Hussain HM et al. Evaluation and mapping groundwater suitability for irrigation using GIS in Najaf Governorate, Iraq. Journal of Environmental Hydrology. 2014:**22**

[29] Rao NS. Groundwater quality from a part of Prakasam District, Andhra Pradesh, India. Applied Water Science. 2018;**8**:30. DOI: 10.1007/s13201-018-0665-2

[30] Roy A, Keesari T, Mohokar H, Sinha UK, Bitra S. Assessment of groundwater quality in hard rock aquifer of central Telangana state for drinking and agriculture purposes. Applied Water Science. 2018;**8**(124). DOI: 10.1007/s13201-018-0761-3

[31] Rao NS, Subrahmanyam A, Kumar SR, Srinivasulu N, Rao BG, Rao PS, et al. Geochemistry and quality of groundwater of Gummanampadu sub-basin, Guntur District, Andhra Pradesh, India. Environmental Earth Sciences. 2012;**67**(5):1451-1471.

Available from: https://doi.org/10.1007/s12665-012-1590

[32] Mohamed MM, Murad A, Chowdhury R. Evaluation of ground water quality in the Eastern District of Abu Dhabi Emirate, UAE. Bulletin of Environmental Contamination and Toxicology. 2017;**98**:1-7

[33] Doneen LD. Notes on water quality in agriculture. Water science and engineering paper 4001. California: Department of Water Sciences and Engineering, University of California; 1964

Chapter 2

Water Chemical Remediation for Simultaneous Removal of Phosphate Ion and Blue-Green Algae From Anthropogenically Eutrophied Pond

Hideaki Nakamura

Abstract

Recent organic pollution is caused primarily by the decay of blue-green algae, *Microcystis aeruginosa*, which is seriously multiplied due to phosphorus-based anthropogenic eutrophication. In eutrophic water, the phenomenon of pH rising in the surface water occurs due to photosynthesis by *M. aeruginosa* (rising over pH 10). Such pH value is enough to produce calcium phosphate precipitation. The *M. aeruginosa* cells form colonies and have the outer layer of which is surrounded by a gelatinous sheath. Thus, we considered simultaneous removal of phosphate ion and blue-green algae using calcium chloride from water surface of eutrophic pond. In the present chapter, a simultaneous removal method employing water chemical remediation (WCR) is described. In this method, a flow system was constructed by equipment of a calcium chloride injector and a sand filtration column. As a result, both calcium phosphate and agglutinated algae could be removed from the eutrophic pond water. These water nutrients are removed, phosphorus is concerned about exhaustion as a resource, and the collected algae can be used in various ways as biomass resources. Thus, our system showed the future ability to improve water quality, to remove contaminants, and to recover nutrients from eutrophic water.

Keywords: resource removal, calcium chloride, phosphorus, water bloom, anthropogenic eutrophication, organic pollution

1. Introduction

Anthropogenic eutrophication is caused by the release of nutritive salts, such as phosphates and nitrates from effluents after sewage treatment of domestic and industrial wastewater. The impact of these effluents on the water environment is likely to escalate, especially in closed waters such as ponds, lakes, and dams. In the conventional activated sludge method (ca. 50%) or advanced treatment method (ca. 75%), the nutrient contained in the effluent is not completely removed because only heterotrophic microorganisms, i.e., ecological decomposers, are employed [1].

Excessive influx of nutrient salts into the closed water causes intense growth of blue-green algae (cyanobacteria), and their remains cause organic pollution as

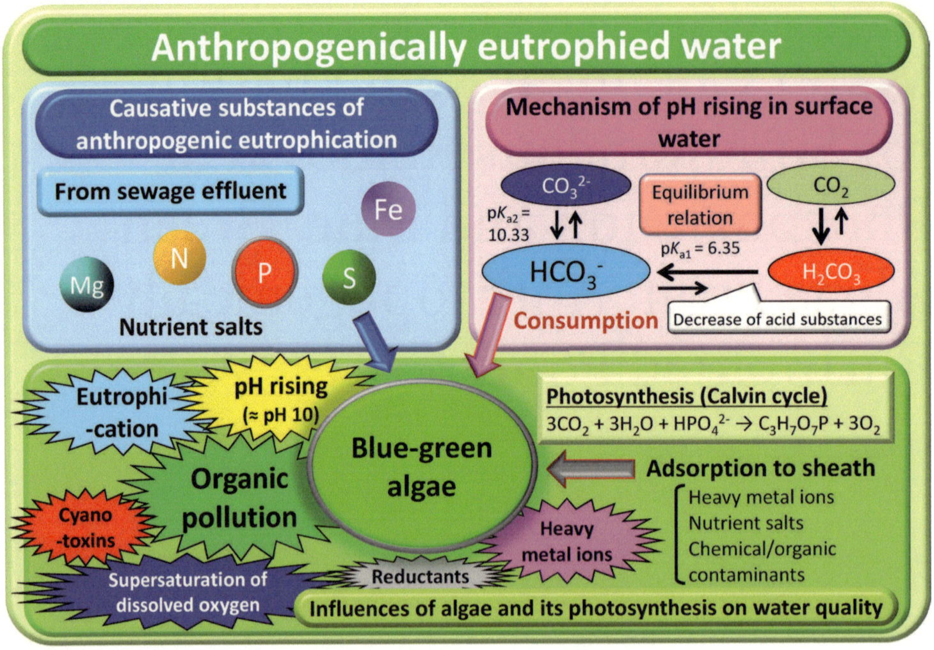

Figure 1.
Influences of blue-green algae on anthropogenically eutrophied water.

excess organic matters [2] (**Figure 1**). When the decomposition of excess organic matter changes from aerobic to anaerobic, anaerobic substances such as hydrogen sulfide and methane are generated, and elution of heavy metal ions occurs. In Japan, it is known that the main species of blue-green algae is *Microcystis aeruginosa* [3, 4]. During daytime, dissolved carbon dioxide (CO_2) is taken by the active photosynthesis of phytoplankton. When a lot of blue-green algae occur, bicarbonate ion (HCO_3^-) and carbonate ion (CO_3^{2-}) take precedence in the afternoon on a clear day. Then, the pH of the surface water rises to around pH 10. Such water environment is more severe for other algae that rely on dissolved CO_2 than algae that consume HCO_3^-. In other words, in both light and carbon environments, blue-green algae are less likely to be limited by environmental changes as a result of growth.

M. aeruginosa that makes the water blooms has a kind of floating bag called gas vesicle inside the cell. *M. aeruginosa* floats on the water surface with the gas vesicles and looks like a green powder called "Aoko" [3]. The inside of the gas vesicles contains a gas with a composition similar to air [5]. The gas vesicle does not always expand, and it collapses when the osmotic pressure in the cell is high. When photosynthesis is actively conducted, there are many low-molecular-weight compounds that are the initial products, and the osmotic pressure in the cell is increased. However, when the initial products are consumed at night or in deep layer for polymer synthesis such as cell structures and stored products, permeation occurs. When the osmotic pressure in the cells decreases, the gas vesicles expand again and increase their buoyancy. Thus, the cells rise to the water surface with light. By the growth cycle, the number of the *M. aeruginosa* cells is increased (**Figure 2**).

When intensely grown, the blue-green algae of *M. aeruginosa* produce cyanotoxins known as microcystins [6]. There are about 50 derivatives of microcystins, which have hepatotoxicity to mammalians. In particular, microcystin-LR is the most toxic substance in microcystins (LD_{50} in mice and rats of 36–122 μg/kg) [7]. The toxic effects of potential human carcinogen microcystin-LR are also investigated [8]. In Japan, the water supply law was defined in 1957, and the concentration of

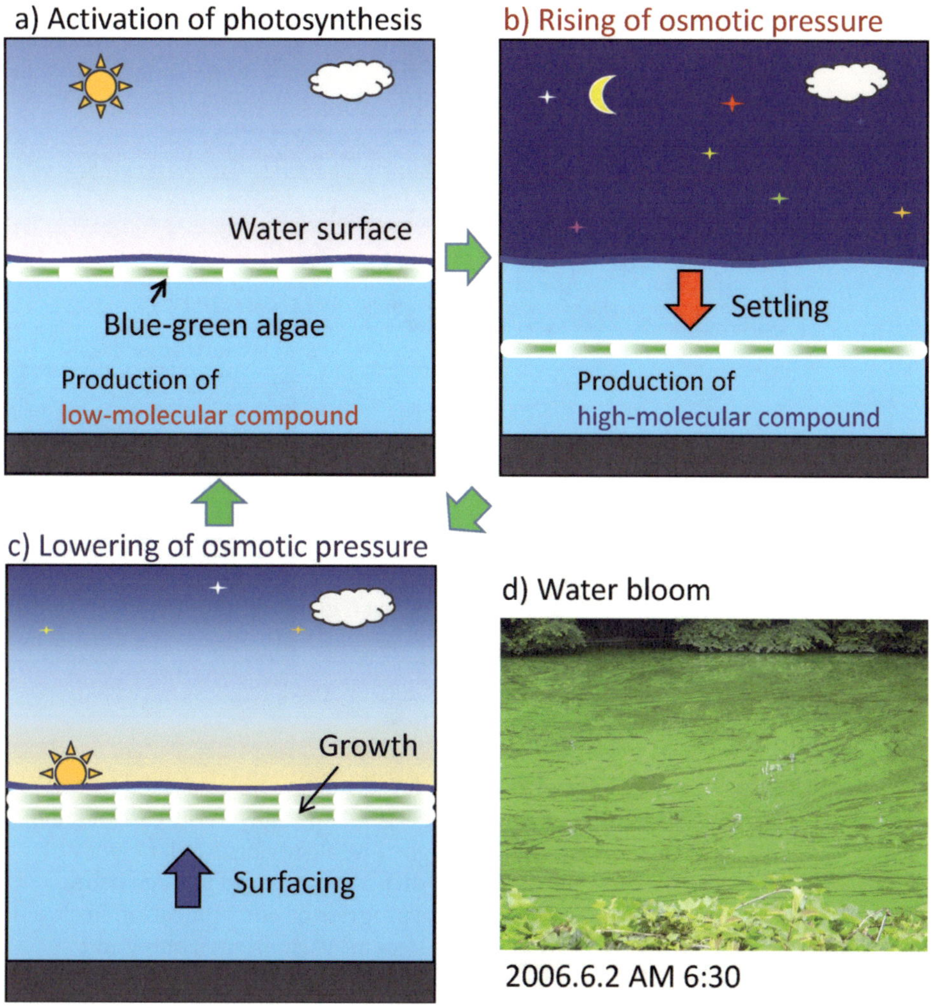

Figure 2.
Circadian rhythm of blue-green algae.

microcystin-LR in raw water and purified drinking water has been determined. Although microcystin-LR has been detected in raw water, it has not been detected in drinking water. This indicates that microcystin-LR has been removed or degraded in the water purification system [9–11]. In fact, in Japanese water purification system, there is no report of health hazard of microcystins caused by drinking water, although problems such as offensive odor of drinking water occasionally occur [12].

The mechanism of anthropogenic eutrophication in closed water body is illustrated in **Figure 3**. By continuous influx of nutrient salts from anthropogenic source to closed water body, its water quality is gradually eutrophied [1]. In the steps from oligotrophic to eutrophic, the aquatic ecosystem continues to increase biomass such as algae, hydrophytes, and fishes. However, in the final step of hypertrophic, its productive ecosystem is rapidly declining, and blue-green algae become dominant species in the water environment. During daytime, the photosynthetic activity by blue-green algae is greatly enhanced, the pH of surface water is increased, and the concentration of dissolved oxygen becomes supersaturated [13]. And since sunlight is absorbed by the thick layer of the water bloom on the surface water, it becomes difficult for the sunlight to reach the water, making it difficult to grow aquatic plants. Similarly, the growth of fish also becomes a difficult situation due to the production

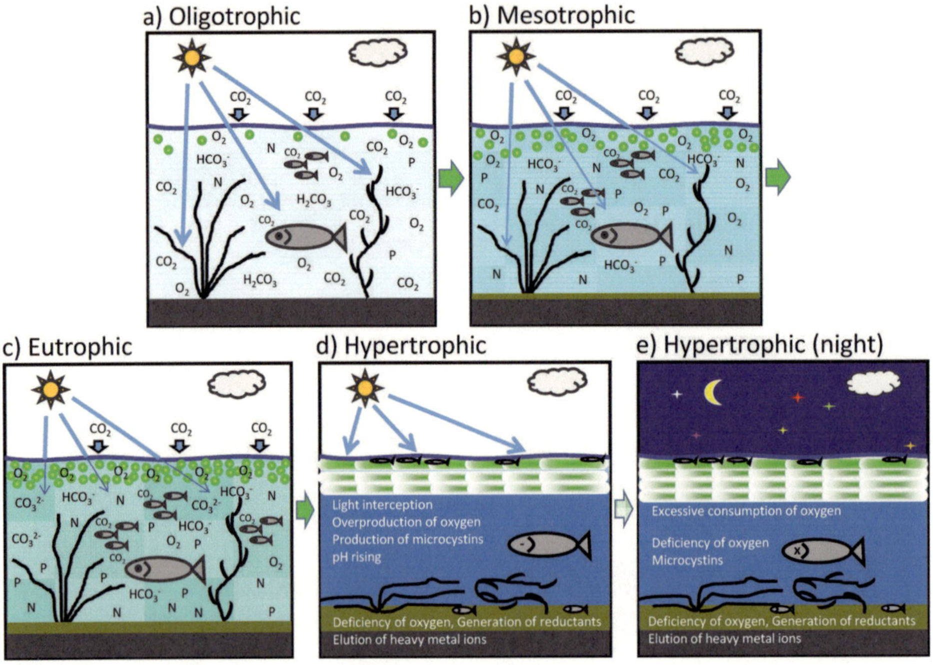

Figure 3.
Mechanism of anthropogenic eutrophication in closed water body.

of microcystins by blue-green algae and the reduced substances generated by the anaerobic decomposition of organic matters [14–16]. Especially in the case of small fishes, excessive consumption of oxygen by nighttime algae causes a deficiency of dissolved oxygen, which makes survival more difficult compared to large fishes.

As mentioned above, it can be seen that the progress of eutrophication in closed waters has a serious impact on the aquatic ecosystem. Various environmental remediation technologies have been developed to improve the situation of such water environment. Among them, one of the surest methods is one of the physical remediation methods. However, this method has problems such as being limited to the place where heavy equipment can be introduced for dredging and high cost of dredging. Therefore, studies on bioremediation that can be performed more easily are actively conducted [17]. As for bioremediation, soil or grand water quality improvement using plants (phytoremediation) or microorganisms (bioaugmentation or biostimulation) has been developed. In particular, the bioaugmentation for sea pollution such as crude oil spill is well-known. But also in water environment, active studies have been made on water quality improvement using plants or algae as a sustainable and environmental friendly method.

Employing plants, the hydroponic phytoremediation using floating raft for water quality improvement has been studied for wastewater treatment [18]. This remediation method has also been developed into an "aquaponics" that applies nutrients contained in aquaculture wastewater to hydroponic cultivation of the plants such as agricultural products [19]. Employing algae, the so-called phycoremediation has also been studied for the improvement of water quality in the eutrophic water body [17, 20]. Algae are highly adaptive and can grow autotrophically, heterotrophically, or mixotrophically in any environment. In the ecosystem, algae as the autotrophic organism belong to the producer; therefore, they intake mineralized substances. The feature can be applied to cost-effective nutrient removal processes [21] and also be applied to a synergistic approach for simultaneous bioremediation and biomass generation [22]. The biomass produced by the algae is energy-rich which can be further processed to make biofuel, biodiesel, and other bio-hydrocarbons. Further, the algae biomass can also be used to obtain a product called as a bio-based product such as bioplastic, fertilizer, animal food, and many more [23].

The phycoremediation employing cyanobacteria calls microalgal remediation or cyanoremediation [24]. The *M. aeruginosa* cells form colonies and have the outer layer of which is surrounded by a gelatinous sheath with a well-defined boundary. The gelatinous sheath adsorbs nutrients [3]. The adsorptive property of the gelatinous sheath has another property as the remediation tools, i.e., it can remove pollutants such as heavy metal ions, nutrients salts, and other chemical and organic contaminants from the water environment and CO_2 from air. In detail, the gelatinous sheath of *M. aeruginosa* is thought to play a role such as maintenance of colony formation, adsorption and concentration of ions, protection against predation, protection against bacterial attack, and relationship with sink-float. The sheath materials were mainly composed of 35.4–47.0% of polysaccharides with uronic acids and 18.2–24.5% of protein [25]. In particular, *M. aeruginosa* is intensively an intake phosphate ion as limiting factor of the water environment and produces polyphosphates and stores them in the cell [3]. Such property of *M. aeruginosa* is also suitable as one of the remediation tools.

Chemical remediation which uses chemicals has been studied especially for soil and grand water. For water quality improvement in the aquatic environment, heavy metal ions, organic contaminants, and radionuclides have been removed by polymeric membranes, microporous solids, and hybrid chemoenzymatic materials [26]. For phosphate removal, calcium compounds such as calcium hydroxide (lime), calcium carbonate, and calcium silicate have been used [27–29]. However, these calcium compounds have low solubility to fresh water; therefore, these compounds are not suitable for effective removal of phosphate ion which dissolved in fresh water. On the other hand, the chemical remediation for blue-green algae has been performed using algaecides (permanganate, aluminum chloride, sodium chloride, acid-soluble cuprous chloride, etc.) [30]. However, the use of the algaecides is concerned about the influences on the aquatic ecosystem. To recover the

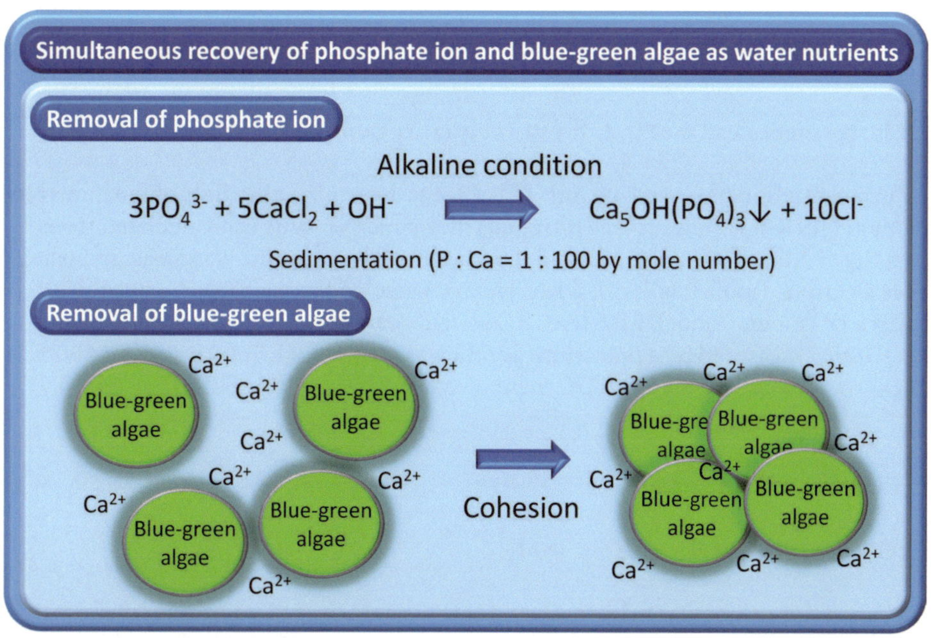

Figure 4.
Principle of our water chemical remediation method.

blue-green algae, iron salt, aluminum salt, permanganate potassium composite, or calcium phosphate precipitation was employed as a coagulation reagent [31–34]. In the calcium phosphate precipitation method, phosphate ion was intensively precipitated with calcium ion by addition of sodium hydroxide [34]. Blue-green algae are coagulated by calcium phosphate precipitation due to that the algae cell surface is negatively charged [35, 36]. The coagulate substance is suitable for algae biomass production due to contain calcium phosphate precipitation as nutrient salt.

As described above, most of chemical remediation methods have problems as the environmentally friendly method. In the present study, we propose a water chemical remediation (WCR) system for simultaneous removal of phosphate ion and blue-green algae from the surface water of the anthropogenically eutrophied pond [37]. The system employs calcium chloride dihydrate for both precipitation with phosphate ion and coagulation with blue-green algae (**Figure 4**). The calcium chloride dihydrate has no ecotoxicity, low toxicity reagent (LD_{50} 2045 mg/kg, rat male/oral), and widely used as food additive and snow-melting agent in Japan. Further, calcium chloride has high water solubility (74.5 g/100 mL at 20°C); the feature is able to reduce the amount of the reagent. For the study, we had been investigating the water quality of the anthropogenically eutrophied pond from 2006 to 2009 [38]. The results are also shown in the present study.

2. Experimental

2.1 Reagent and algae cell culture

Calcium chloride dihydrate was purchased from Wako Chemicals, and the other chemicals were used for analytical grade. The growth of *M. aeruginosa* NIES-87 as a model of blue-green algae: *M. aeruginosa* NIES-87 was purchased from the National Institute for Environmental Studies and grown by two-step incubations. As preincubation, *M. aeruginosa* was grown in a test tube with 10 mL of MA medium until reaching stationary phase (120 rpm, 27°C, light-dark cycle of 12 h) [39]. As present incubation, the growth medium of the preincubation was added to a glass incubation bottle containing 700 mL of MA medium and grown under 27°C for 2 weeks.

2.2 Instruments and methods for water quality monitoring

Concentration of phosphate ion (PO_4-P) was determined by the molybdenum blue method (JIS K-0102). Inorganic nitrogen (inorganic-N) value was calculated from sum the of NH_4^+, NO_2^-, and NO_3^- concentrations. The pH value was measured using a pH electrode (model, IOL-50, DKK, Japan). Dissolved oxygen (DO) was measured using a DO meter (model, MM-60R, TOA-DKK, Japan). Chemical oxygen demand (COD) was measured by a potassium permanganate method. Total organic carbon (TOC) was measured using a TOC meter (model, TOC-5000A, Simadzu, Japan).

3. Results and discussion

3.1 Backgrounds of the present study

When the author was a doctoral student, the author studied on enzyme biosensors for phosphate ion in the reserved water for drinking [40, 41]. Then after, the author had interests to the phenomenon of anthropogenic eutrophication [38] and subsequent organic pollution [1] and also had interests to environmental protection using biosensors [2, 42] and remediation techniques for closed water quality [43]. The present study was carried out based on these backgrounds.

3.2 Investigation of aquatic ecosystem

The pond investigated in this time is on the site of the university and is an adjustment reservoir with an area of 20,000 m^2, an average depth of ca. 1 m, a maximum depth of ca. 3 m, and a storage capacity of about 20,000 kL. Most of the pond water is the inflow of rainwater that has fallen into the university. In addition, the surplus of the treated sewage that is not used for the regeneration of toilet flushing water is discharged to the pond as drainage. In addition, the pond bottom is covered with a rubber sheet to prevent the penetration of the pond water [38].

In this pond, water blooms were often observed by the rise of the water temperature (**Figure 2d**). In 2006, the thick layer of the water bloom on the surface water had been seen at the corner of the pond (**Figure 5**). Using a microscope, two types of blue-green algae in the form of spherical *Microcystis* sp. and filamentous *Planktothrix* sp. were observed from the surface water. Subsequently, the ecology of aquatic animals was investigated. In the case of hypertrophic water body, DO is consumed by aerobic respiration of blue-green algae at night, and then the water body changes to reductive environment [14]. In such reductive condition, survival of small fishes becomes more difficult than that of large fishes. In fact, only large carps and large turtles were observed by visual observation of the water surface. Then, we tried to capture small aquatic organisms such as small fishes using a cell bottle, a four-way net, and casting net. As a result, small fishes such as *Cyprinus carpio* and *Carassius* were observed. Therefore, it was found that the pond maintained the water quality necessary for the small aquatic organisms to survive.

3.3 Periodic monitoring of water quality

Periodic monitoring was conducted twice a month to examine several items related to weather conditions, basic properties of water quality, eutrophication, and organic pollution. Water sampling was conducted by collecting reservoir water with a depth of about 10 cm at 10:30 am.

Figure 5.
Photograph of thick-layered water bloom on the surface water of the pond (2006).

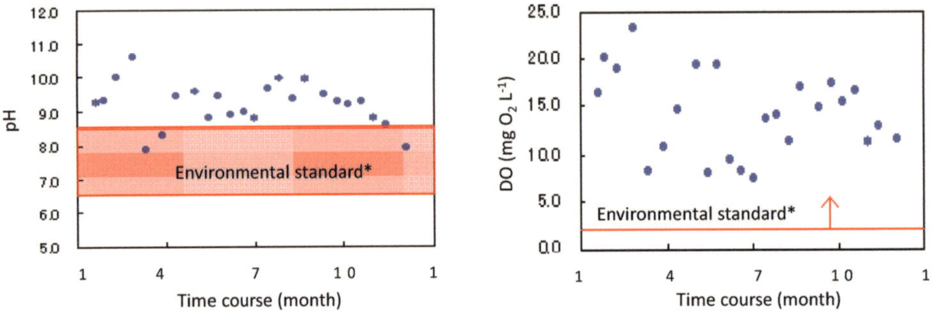

Figure 6.
*Periodic monitoring for indications of basic water quality (2008) [38]. *The Ministry of the Environment (Japan), water-pollution standard, Appendix 2.*

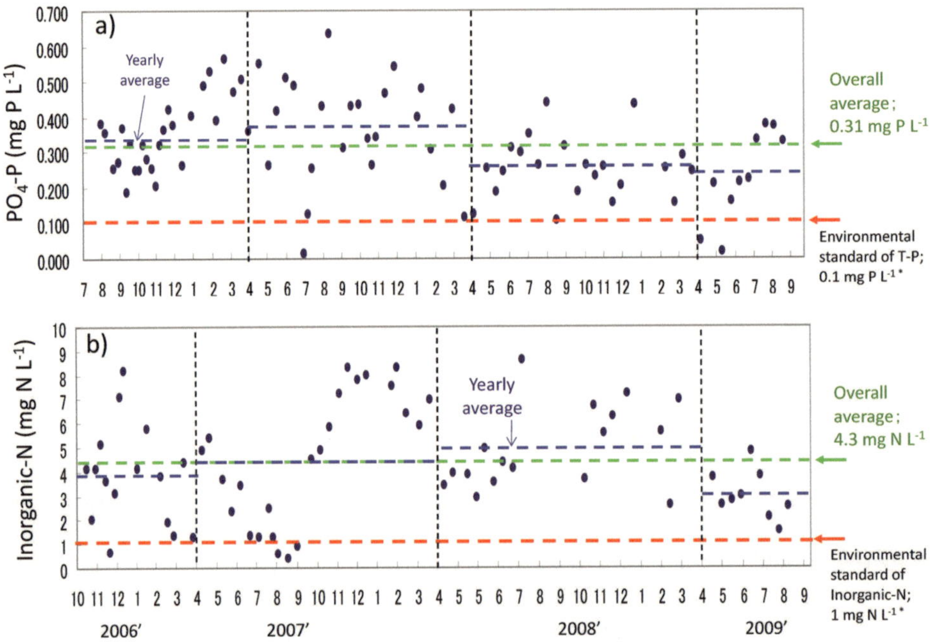

Figure 7.
*Periodic monitoring for indications of eutrophication (fiscal years between 2006 and 2009) [38]. a) Phosphate ion and b) inorganic nitrogen. *The Ministry of the Environment (Japan), water-pollution standard, Appendix 2.*

In the indications of the basic water quality, the values of both pH and DO are influenced by the intensive photosynthesis of blue-green algae (**Figure 6**). The annual average values of both pH and DO were pH 9.2 ± 0.32 and 14.2 ± 4.4 mg O_2/L, respectively. The pH value was above the environmental standard, and the DO value indicated that it was supersaturated. It was speculated that these water quality conditions were caused by intensive photosynthesis of the blue-green algae [13].

In the indications of eutrophication, the concentrations of phosphate ion (PO_4-P) and inorganic nitrogen (inorganic-N) exceeded each environmental standard value in most cases (**Figure 7**). In general, intensive growth of blue-green algae consumes these nutrient salts [44]. Nevertheless, such reduction of nutrient salt concentrations was not observed in this pond. The results supported that the treated sewage flowing into this pond always supplies these nutrients.

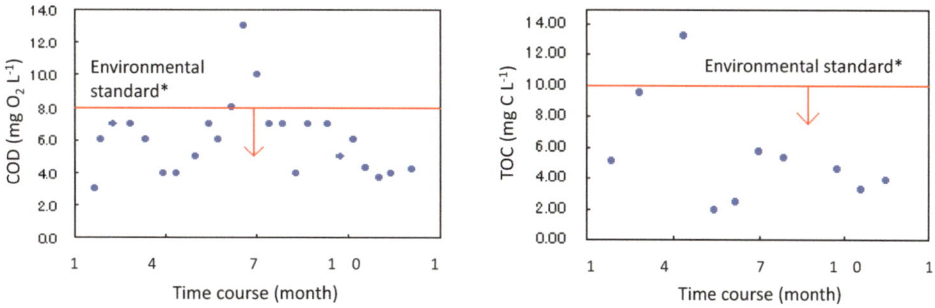

Figure 8.
*Periodic monitoring for indications of organic pollution (2008) [38]. *The Ministry of the Environment (Japan); water-pollution standard, Appendix 2.*

In the indications of organic pollution, the values of COD and TOC did not exceed each environmental standard value in most cases (**Figure 8**). The results showed that the organic substances did not suspend so much in the surface water, although measurement sample was filtrated using 0.45 μm filter as pretreatment. However, it was found that sludge had been accumulated on the bottom of the pond (data not shown).

3.4 Coagulation test and simultaneous removal tests

A coagulation test was performed using growth medium of *M. aeruginosa* NIES-87. $CaCl_2$ solution (0.1 M) of 250 mL and was added to 600 mL of the growth medium, and distilled water was finally added to become 1 L in a measuring

Figure 9.
Constitution of a sand filter for pre-examination. (a) Growth medium of M. aeruginosa NIES-87, (b) coagulation of M. aeruginosa using $CaCl_2$, (c) enlarged view of the coagulation, (d) micrograph of M. aeruginosa coagulated (×20), (e) M. aeruginosa filtered, and (f) filtrate.

cylinder (P:Ca = 1:100 in molar ratio). Then, *M. aeruginosa* was coagulated immediately (**Figure 9**). After leaving for half a day, the medium was used for subsequent simultaneous removal tests.

A medium of 100 mL flowed into a sand filtration column which was prepared for three columns (**Figure 10**). The column was designed for reverse cleaning to recover both phosphate ion and *M. aeruginosa* from the sample. The results showed the possibility as a simultaneous removal of phosphate ion and *M. aeruginosa* (**Table 1**). Next, we examined using the pond water in the same way. As a result, phosphate ion in the pond water (adjusted to pH 10.5) could be removed 65.8% from 0.719 to 0.246 ± 0.0023 mg/L (n = 3). Based on these results, we next constructed a simultaneous removal system.

3.5 Construction of a simultaneous removal system and removal tests

The illustration image of the present system is shown in **Figure 11**. **Figure 12** shows a photograph of the system constructed. This system is a flow type and consists of a surface water inlet, a pump, a photosynthesis enhancer, a calcium chloride injector, a reaction bath for both precipitation and coagulation, a mixing pump, and a sand filtration column. The surface water inlet and the pump were set in a bath that is assumed as pond. **Figure 13** shows a photograph of each part in this system. Using a sand filtration column of the present system, the pond water was treated. As a result, the blue-green algae were removed between 63 and 85%, and phosphate ion was removed between 43 and 69% (**Table 2**). These results were very primitive; however, there must be basic and promising important findings.

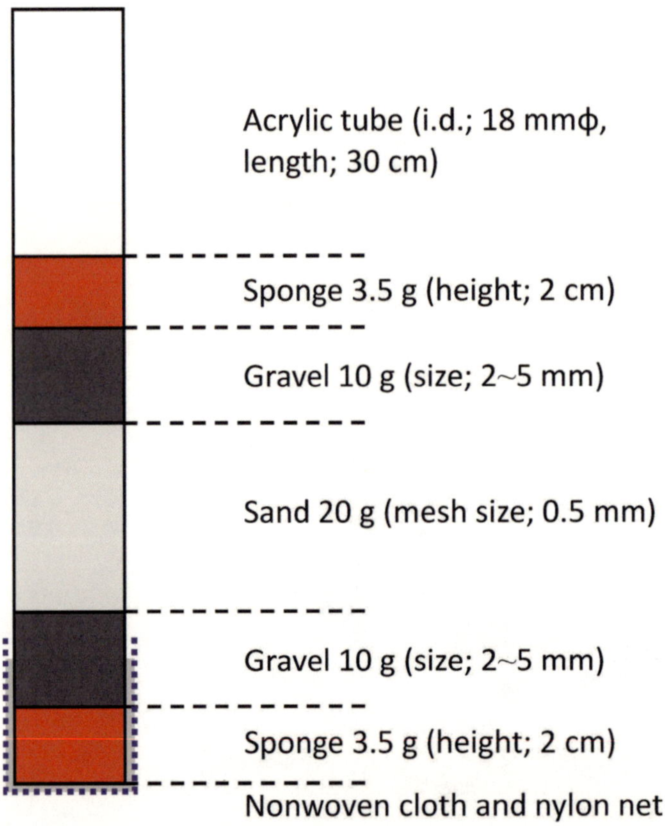

Figure 10.
Constitution of a sand filter for reverse cleaning (not in scale). Sponge is made with nonwoven fabric of nylon and polyester (3M company, USA).

	Cell concentration* (cells/uL)	Removal ratio (%)	Phosphate ion (mg/L)	Removal ratio (%)
Algae-growth medium	4.0×10^4	-	28.7	-
Filtrated	8×10^2	97.4	0.125	99.6

**Cell concentration was calculated from the value obtained by optical density at 600 nm (OD600).*

Table 1.
Sand filtration of algae-growth medium after $CaCl_2$ addition (n = 3).

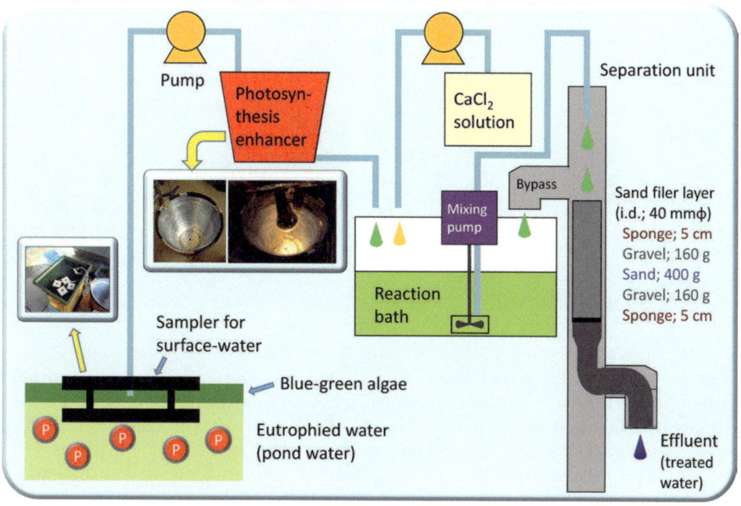

Figure 11.
Schematic design of a simultaneous phosphate ion and blue-green algae removal system.

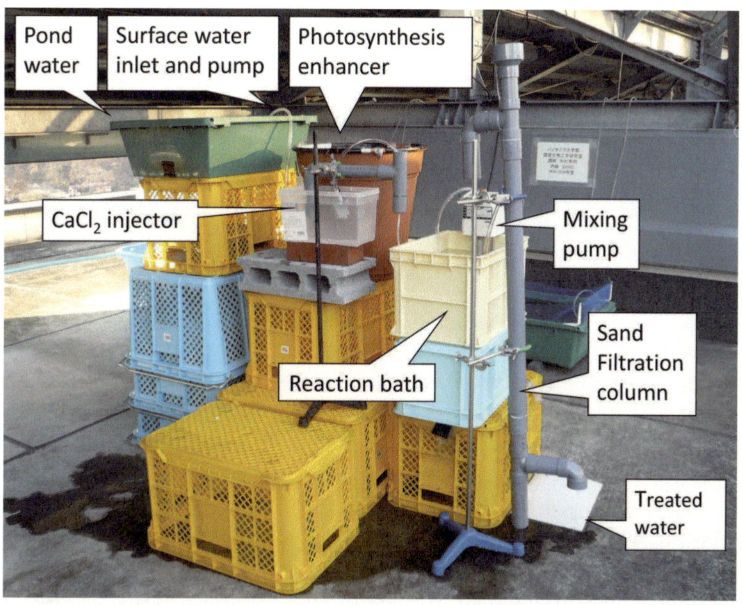

Figure 12.
Photograph of a simultaneous phosphate ion and blue-green algae removal system.

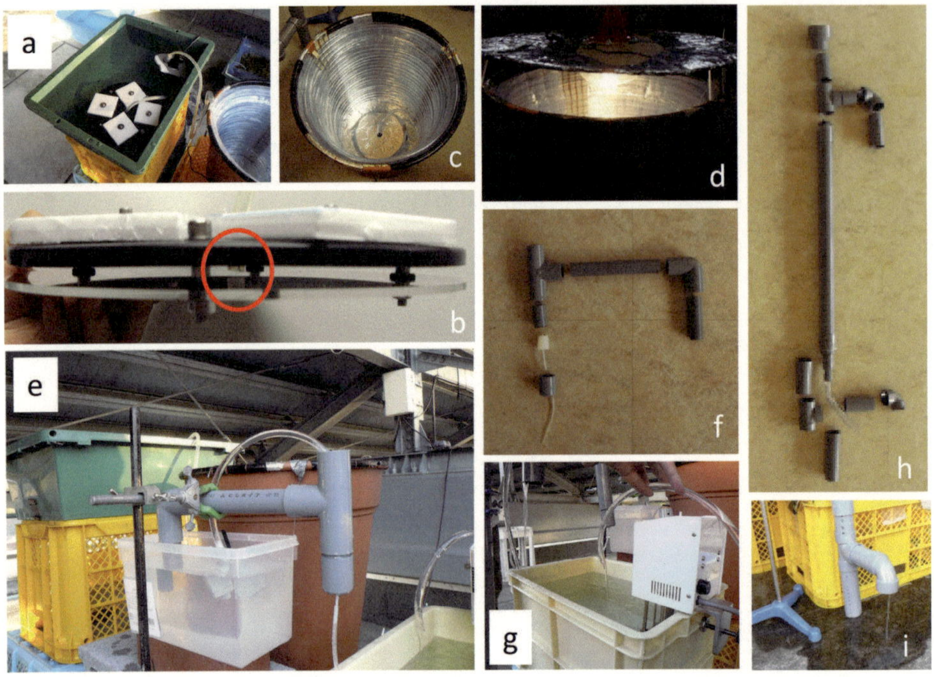

Figure 13.
Parts of a simultaneous phosphate ion and blue-green algae removal system. (a) Surface water inlet and pump in pond water bath, (b) side view of surface water inlet, (c) inside of photosynthesis enhancer, (d) lighting of photosynthesis enhancer, (e) calcium chloride injector, (f) piping structure of calcium chloride injector, (g) reaction bath, (h) structure of sand filtration column, and (i) treated water.

Sample (L)	Cell concentration (cells/uL)	Removal ratio (%)	Phosphate ion (mg/L)	Removal ratio (%)
Before treatment	2.7×10^3	-	1.05	-
1	1.0×10^3	63	0.603	43
3	0.8×10^3	70	0.326	69
5	0.4×10^3	85	0.500	52
7	0.4×10^3	85	0.493	53

Table 2.
Simultaneous removal test of both blue-green algae and phosphate ion from pond water (n = 1).

4. Conclusion

In the present study, we observed the aquatic ecosystem and investigated the water quality of anthropogenically eutrophied pond between 2006 and 2009. As a result, it was found that in this pond, it was easy for blue-green algae to grow intensively, and it was easy for the pH of the surface water to rise by photosynthesis of the blue-green algae (pH 9.2 ± 0.32, n = 24, in 2008). By utilizing these features, we considered to develop a simultaneous recovery system of phosphate ion and blue-green algae using calcium chloride. As the pre-examinations using a growth medium, we observed the phenomenon that the precipitation of calcium phosphate and the coagulation of *M. aeruginosa* occurred simultaneously. The growth medium of 100 mL was filtrated using a prototype sand filtration column which was fabricated to be symmetric for reverse cleaning and to simultaneously recover both phosphate ion and blue-green algae. As a result, from the growth medium, phosphate ion and *M. aeruginosa* could remove 97.4 and 99.6%, respectively. Based on the results, we next constructed a simultaneous phosphate ion and blue-green algae removal system which consisted of a calcium chloride injector and a sand filtration column. Using the sand filtration column of this system, pond water of 7 L was treated, and phosphate ion of 53% and *M. aeruginosa* of 85% were removed, respectively. Our results obtained in the present study were very primitive comparing with the preexamination results using the prototype sand filtration column. However, we believe that our WCR system can be improved to recover both phosphate ion and blue-green algae simultaneously as biomass resources by subsequent studies and applied to remediate hypertrophic water body.

Acknowledgements

The author acknowledges Messrs. Mayu Taguchi, Takuto Toshioka, and Tohru Sasaki for the contributions to their experiments in aquatic ecosystem and the present removal system. The author also appreciates Messrs. Naoya Yasue, Tomoyuki Sakamaki, Norio Suka, Jun Kanno, Yuki Ito, Kohei Taguchi, Kyota Suzuki, Daisuke Tokunaga, and other lab members for their contributions and assistances with the experiments in water monitoring. The present work was financially supported in part by grants of collaborative projects in the Tokyo University of Technology.

Conflict of interest

The author states that there is not conflict of interest.

Abbreviations

CO_2	dissolved carbon dioxide
CO_3^{2-}	carbonate ion
COD	chemical oxygen demand
DO	dissolved oxygen
HCO_3^-	bicarbonate ion
M. aeruginosa	*Microcystis aeruginosa*
TOC	total organic carbon
WCR	water chemical remediation

Author details

Hideaki Nakamura
Department of Liberal Arts, Tokyo University of Technology, Tokyo, Japan

*Address all correspondence to: nakamurahd@stf.teu.ac.jp

References

[1] Nakamura H. Current status of water environment and their microbial biosensor techniques—Part I: Current data of water environment and recent studies on water quality investigations in Japan, and new possibility of microbial biosensor techniques. Analytical and Bioanalytical Chemistry. 2018;**410**(17):3953-3965. DOI: 10.1007/s00216-018-0923-z

[2] Nakamura H. Recent organic pollution and its biosensing methods. Analytical Methods. 2010;**2**(5):430-444. DOI: 10.1039/B9AY00315K

[3] Watanabe MF, Harada K, Fujiki H, editors. Waterbloom (in Japanese). 1st ed. Tokyo: University of Tokyo Press; 1994. 257p. ISBN: 978-4-13-066152-2

[4] Watanabe Y, Watanabe MF, Watanabe M. The distribution and relative abundance of bloom forming *Microcystis* species in several eutrophic waters. Japanese Journal of Limnology. 1986;**47**(1):87-93. DOI: 10.3739/rikusui.47.87

[5] Reynolds CS, Walsby AE. Water-blooms. Biological Reviews. 1975;**50**:437-481. DOI: 10.1111/j.1469-185X.1975.tb01060.x

[6] Watanabe MF, Harada K. Toxic water bloom of blue-green algae: Biological and chemical characteristics. Japanese Journal of Limnology. 1993;**54**(3):225-243. DOI: 10.3739/rikusui.54.225

[7] Dawson RM. The toxicology of microcystins. Toxicon. 1998;**36**(7):953-962. DOI: 10.1016/S0041-0101(97)00102-5

[8] Lone Y, Raj Koiri RK, Bhide M. An overview of the toxic effect of potential human carcinogen Microcystin-LR on testis. Toxicology Reports. 2015;**2**:289-296. DOI: 10.1016/j.toxrep.2015.01.008

[9] Keijola AM, Himberg K, Esala AL, Sivonen K, Hiis-Virt L. Removal of cyanobacterial toxins in water treatment processes: Laboratory and pilot-scale experiments. Environmental Toxicology. 1988;**3**(5):643-656. DOI: 10.1002/tox.2540030516

[10] Himberg K, Keijola AM, Hiisvirta L, Pyysalo H, Sivonen K. The effect of water treatment processes on the removal of hepatotoxins from *Microcystis* and *Oscillatoria*. Cyanobacteria: A laboratory study. Water Research. 1989;**23**(8):979-984. DOI: 10.1016/0043-1354(89)90171-1

[11] Nicholson BC, Rositano J, Burch MD. Destruction of cyanobacterial peptide hepatotoxins by chlorine and chloramine. Water Research. 1994;**28**(6):1297-1303. DOI: 10.1016/0043-1354(94)90294-1

[12] Kimura I. Aquatic pollution problems in Japan. Aquatic Toxicology. 1988;**11**(3-4):287-301. DOI: 10.1016/0166-445X(88)90079-3

[13] Seki H, Takahashi M, Hara Y, Ichimura S. Dynamics of dissolved oxygen during algal bloom in Lake Kasumigaura, Japan. Water Research. 1980;**14**(2):179-183. DOI: 10.1016/0043-1354(80)90235-3

[14] Mackenthun KM, Herman EF. A heavy mortality of fishes resulting from the decomposition of algae in the Yahara River, Wisconsin. Transactions of the American Fisheries Society. 1948;**75**(1):175-180. DOI: 10.1577/1548-8659(1945)75[175:AHMOFR]2.0.CO;2

[15] Ruike K, Kanzo Y, Inamori R, Suzuki R, Xu K, Inamori Y. Effect of microcystin produced from blue-green algae *Microcystis aeruginosa* on animal-plant in agricultural area and its countermeasures during irrigation.

Japanese Journal of Water Treatment Biology. 2014;**50**(2):43-51. DOI: 10.2521/jswtb.50.43

[16] Malbrouck C, Kestemont P. Effects of microcystins on fish. Environmental Toxicology and Chemistry. 2006;**25**(1):72-86. DOI: 10.1897/05-029R.1

[17] Bhatnagar S, Kumari R. Bioremediation: A sustainable tool for environmental management—A review. Annual Research and Review in Biology. 2013;**3**(4):974-993. DOI: 10.9734/arrb/2013/v3i424983

[18] Solanki P, Narayan M, Meena SS, Srivastava RK. Floating raft wastewater treatment system: A review. Journal of Pure and Applied Microbiology. 2017;**11**(2):1113-1116. DOI: 10.22207/JPAM.11.2.55

[19] Diver S, Rinehart L. Aquaponics-Integration of hydroponics with aquaculture [Internet]. ATTRA (The National Sustainable Agriculture Information Service). 2006. p. 28. Available from: https://backyardaquaponics.com/Travis/aquaponic.pdf [Accessed: June 2, 2019]

[20] Phang SM, Chu WL, Rabiei R. Phycoremediation. In: Sahoo D, Seckbach J, editors. The Algae World. Volume 26 of Cellular Origin, Life in Extreme Habitats and Astrobiology. Dordrecht: Springer; 2015. pp. 357-389. DOI: 10.1007/978-94-017-7321-8_13

[21] Olguıin EJ. Phycoremediation: Key issues for cost-effective nutrient removal processes. Biotechnology Advances. 2003;**22**(1-2):81-91. DOI: 10.1016/S0734-9750(03)00130-7

[22] Renuka N, Sood A, Prasanna R, Ahluwalia AS. Phycoremediation of wastewaters: A synergistic approach using microalgae for bioremediation and biomass generation. International Journal of Environmental Science and Technology. 2015;**12**(4):1443-1460. DOI: 10.1007/s13762-014-0700-2

[23] Gani P, Sunar NM, Matias-Peralta H, Latiff AAA, Parjo UK, Razak ARA. Phycoremediation of wastewaters and potential hydrocarbon from microalgae: A review. Advances in Environmental Biology. 2015;**9**(20):1-8. Available from: http://www.researchgate.net/publication/283291000

[24] Gothalwal R, Chillara S. Cyanoremediation: A green clean technology. In: Satyanarayana T, Johri BN, Prakash A, editors. Microorganisms in Environmental Management. Dordrecht: Springer; 2011. pp. 767-786. DOI: 10.1007/978-94-007-2229-3_34

[25] Amemiya Y, Nakayama O. The chemical composition and metal absorption capacity of the sheath materials isolated from *Microcystis*, cyanobacteria. Japanese Journal of Limnology. 1986;**47**(1):87-93. DOI: 10.3739/rikusui.47.87

[26] McPherson G, Pintauro P, O'Connor S. Natural and Active Chemical Remediation of Toxic Metals, Organics, and Radionuclides in the Aquatic Environment. Oak Ridge (State of Tennessee, USA): OSTI Government; 1996. DOI: 10.2172/254373. 141p

[27] Nie M, Deng Y, Nie S, Yan C, Ding M, Dong W, et al. Simultaneous removal of bisphenol A and phosphate from water by peroxymonosulfate combined with calcium hydroxide. Chemical Engineering Journal. 2019;**369**:35-45. DOI: 10.1016/j.cej.2019.03.046

[28] Marani D, Di Pinto AC, Ramadori R, Tomei MC. Phosphate removal from municipal wastewater with low lime dosage. Environmental Technology. 1997;**18**(2):225-230. DOI: 10.1080/09593331808616531

[29] Cairns MJ, Krauss C, Johnston JH. Recovery of phosphate from surface waters using a calcium silicate composite material for potential application in environmental remediation. NSTI-Nanotech. 2013;**3**:746-749

[30] Li X, Zhao L, Yangy GL, Xia L, Ya-li P. Effectiveness and mechanism study of composited algaecide on algae removal. Journal of Beijing University of Technology. 2010;**36**(10):1402-1407. Available from: http://en.cnki.com.cn/Article_en/CJFDTOTAL-BJGD201010018.htm [Accessed: June 2, 2019]

[31] de-Bashan LE, Bashan Y. Recent advances in removing phosphorus from wastewater and its future use as fertilizer (1997-2003). Water Research. 2004;**38**(19):4222-4246. DOI: 10.1016/j.watres.2004.07.014

[32] Huang XL, Chen Y, Shenker M. Solid phosphorus phase in aluminum- and iron-treated biosolids. Journal of Environmental Quality. 2007;**36**(2): 549-556. DOI: 10.2134/jeq2006.0155

[33] Wang L, Fang J, Ma J, Chen Z. Effect of pre-oxidation on algal cell morphology and algal removal by enhanced coagulation process. Journal of Southeast University. 2005;**z1**:182-185. Available from: http://en.cnki.com.cn/Article_en/CJFDTOTAL-DNDX2005S1037.html [Accessed: June 2, 2019]

[34] Phasey J, Vandamme D, Fallowfield HJ. Harvesting of algae in municipal wastewater treatment by calcium phosphate precipitation mediated by photosynthesis, sodium hydroxide and lime. Algal Research. 2017;**27**:115-120. DOI: 10.1016/j.algal.2017.06.015

[35] Sukenik A, Shelef G. Algal autoflocculation-verification and proposed mechanism. Biotechnology and Bioengineering. 1984;**26**(2):142-147. DOI: 10.1002/bit.260260206

[36] Sukenik A, Schröder W, Lauer J, Shelef G, Soeder C. Coprecipitation of microalgal biomass with calcium and phosphate ions. Water Research. 1985;**19**(1):127-129. DOI: 10.1016/0043-1354(85)90333-1

[37] Nakamura H, Kita Y, Sudo K, Yoshioka T, Sasaki T, Nitanai Y, et al. Fundamental study on simultaneous recoveries of phosphate ion and blue-green algae from anthropogenically eutrophied water. In: Proceedings of the International Chemical Congress of Pacific Basin Societies (Pacifichem 2010); 15-20 December 2010; Hawaii, USA; 2010. p. 350

[38] Nakamura H, Suka N, Kita Y, Yasue N, Sakamaki T, Ito M, et al. Investigations of eco-balance and water quality in a holding pond—Organic pollution caused by eutrophication and its effects on water quality (in Japanese). Bulletin of Tokyo University of Technology. 2010;**5**:69-77. ISSN: 18805752

[39] Kasai F, Kawachi M, Erata M, Mori F, Yumoto K, Sato M, et al. NIES-Collection List of Strains. 8th ed. Tsukuba: Environmental Research Center Press; 2009. 350p. ISSN: 0038-1578

[40] Nakamura H. Determination of phosphate ion in fresh water using biosensors (in Japanese) [Doctoral thesis]. Tokyo: The University of Tokyo; 1998

[41] Nakamura H. Digest of doctoral dissertation: Phosphate ion determination in water for drinking using biosensors. Bunseki Kagaku. 2001;**50**(8):581-582. DOI: 10.2116/bunsekikagaku.50.581

[42] Nakamura H. Current status of water environment and their

microbial biosensor techniques—Part II: Recent trends in microbial biosensor development. Analytical and Bioanalytical Chemistry. 2018;**410**(17):3967-3989. DOI: 10.1007/s00216-018-1080-0

[43] Nakamura H, Taguchi M, Yokoyama H, Hirabayashi K, Abiko S, Gotoh M, et al. Study on highly-effective purification of eutrophied freshwater applying hydroponic culture (in Japanese). Bulletin of Tokyo University of Technology. 2010;**5**:43-52. ISSN: 18805752

[44] Okino T. Studies on the blooming of *Microcystis aeruginosa*, characteristics of the water bloom of *Microcystis*. Japanese Journal of Botany. 1973;**20**(6):381-402

Chapter 3

Pollution of Water Sources from Agricultural and Industrial Effluents: Special Attention to NO_3^-, Cr(VI), and Cu(II)

Majda Breida, Saad Alami Younssi, Mohamed Ouammou, Mohamed Bouhria and Mahmoud Hafsi

Abstract

One of the most important challenges facing humanity today is to conserve and sustain water resources (either surface water or groundwater). This challenge became more pronounced with the increase of urban, agricultural, and industrial activities that discharge a considerable amount of wastewater. Therefore, the preservation of water sources from pollutants is a major concern, shared by all, public, industrial, scientific, researchers, and decision-makers. This chapter analyzes in more detail the pollution and pollutants caused by agricultural and industrial activities. Particular attention is given to pollution via nitrogen and heavy metals (NO_3^-, Cr(VI), and Cu(II)) in either international or national level. The effect of these pollutants on human health and environment, their standards/regulations, and the different current methods used for their detection and treatment are all discussed in the chapter.

Keywords: water chemistry, pollution, pollutants, nitrate, copper, chromium, detection methods, treatment methods

1. Introduction

One of the most important challenges facing humanity today is to conserve and sustain natural resources, including water, in order to increase economic and social development along with environment protection. However, urban and industrial sectors not only use the available water but also discharge a considerable amount of wastewater. The impact of these effluents on living organisms may be harmful due to the direct or chronic toxicity of these substances or the products of their degradation. Significant efforts are being made to control the use and spread of these substances. These topics are addressed more thoroughly in this chapter. Particular attention is given to pollution via NO_3^- and heavy metals. Besides the various activities that lead to the pollution of our freshwater bodies and the effect of these pollutants on human health and environment, the chapter also discusses the different methods used for water pollutant detection, standards/regulations, and some of the used treatment methods to protect water bodies.

2. Water quality

Water availability not only depends on the quantity of water, but also on the water quality, the physical structures, laws, regulations, and socioeconomic factors that control its demand and use [1]. The quality of water is defined by a set of general parameters, such as dissolved oxygen (DO), alkalinity, pH, hardness, conductivity, salinity, turbidity, etc. These properties are mostly linked to the geological and ecological characteristics of the water body basin, and although they have a large spatial variability, they are maintained within certain ranges in each specific water body [2, 3]. In fact, degradation in quality of water bodies (surface or underground) has become more apparent because of the increase of economic activity as well as the insufficiency of proper sanitation. This increase is generating a significant number of pollutants.

2.1 Water pollution and pollutants

Water pollution happens when the amounts of pollutants (chemical, physical, or biological) discharged into the water body can no longer be contained by the natural ecosystem. Furthermore, two types of water pollution can be identified: (i) accidental pollution that is punctual and often of large scale and (ii) chronic pollution corresponding to the discharge of permanent pollutants with low doses. The European water framework directive (2000/60/EC) has defined "pollutant" as any artificial substance made by man and dispersed in environment and which is likely able to generate an impact [4]. In addition, water pollutants are those that make water unfit for consumption or degrade some of its properties [5, 6]. However, it seems difficult to define "pollutant" while knowing that the environmental fate of numerous chemicals reveals that most are causing severe to moderate health hazards and are significantly harmful to the environment [26].

2.1.1 Classification of water pollutant

Water pollutants are divided into various categories; each category of pollutants has its own proper means of entering the environment and its own particular hazards. In general, pollutants present in water/wastewater are dissolved or undissolved substances (as illustrated in **Figure 1**). Dissolved pollutants are all substances that are dissociable and transformed into cations and anions. These pollutants are divided in two categories, organic and inorganic substances, while undissolved pollutants are precipitated components, suspended solids (SS), colloidal materials, floating materials, oil, grease, foams, clay minerals, and others.

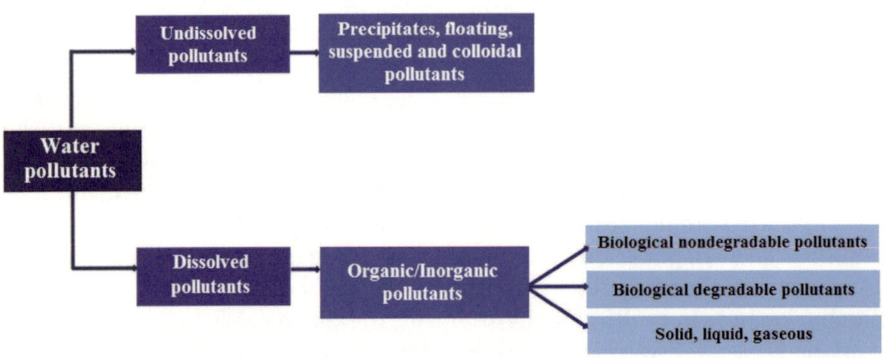

Figure 1.
Classification of water pollutants as dissolved and undissolved substances.

DOI: http://dx.doi.org/10.5772/intechopen.86921

2.1.1.1 Organic pollutants

Organic compounds accommodate carbon and commonly other elements such as hydrogen, oxygen, nitrogen, sulfur, and others. Diverse organic matters occur naturally in water or from anthropogenic sources. These organic compounds may be considered contaminants if their concentrations adversely affect an aquatic system [7]. Furthermore, the organic matter is divided into two types (dissolved and particulate organic matter) depending on their origin and solubility [8]. Organic pollutants can be further divided into (i) oxygen-demanding contaminants, (ii) synthetic organic contaminants, and (iii) crude oil and various petroleum products.

2.1.1.2 Inorganic pollutants

Inorganic contaminants (IOCs) are non-biodegradable pollutants and persist in the environment. IOCs in waters can as well be classified into categories according to nature (as demonstrated in **Figure 2**). The toxicity of a pollutant is directly proportional to its concentration and the presence of other compounds. Metals in high concentration can be toxic which is aggravated when reacting with organic compounds to form organometallic compounds.

2.1.1.3 Suspended solids and sediments

SS appear in the water via the runoff of surface and through municipal sewers. Moreover, the presence of SS in hydrosphere can block the sunlight penetration in water, which is necessary for the vegetation photosynthesis. In addition, if the deposited solids are organic in nature, they increase the growth of anaerobic conditions [10].

2.2 Water pollutants according to activity

As already mentioned, human activities threaten water quality by generating a wild variety of pollutants based on the type of industrial activity. Some of these activities have occurred over several generations, and some have started more recently. In the following section, urban, agricultural, and industrials activities in relation to specific contaminants, their sources, and their effects on water quality in the environment are discussed.

2.2.1 Urban pollution

Population growth and urbanization do have important roles in enhancing water pollution. Urbanization generally leads to higher phosphorus and NO_3^-

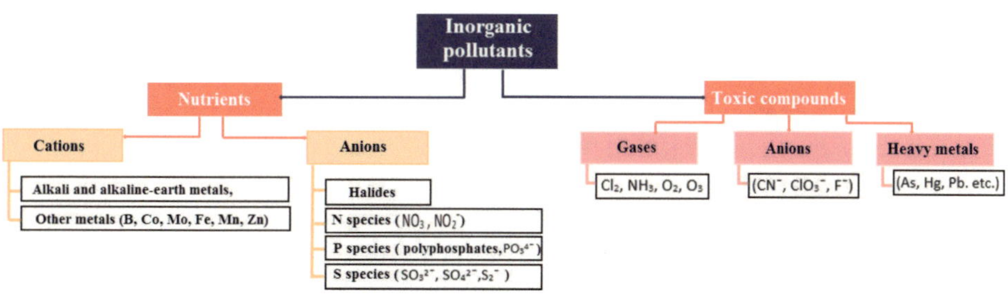

Figure 2.
Classification of inorganic pollutants [9].

concentrations in urban catchments [11, 12]. Urban runoff is the greatest cause of surface water pollutants in numerous parts of the world; according to Singh and Gupta [13], urbanization is second to agriculture as the primary cause of water degradation.

2.2.2 Agricultural pollution

Agriculture is perchance the primeval contribution of man for the survival and prosperity of human kind [14]. Just as agriculture has comprehensively changed the face of the Earth, its impacts have equally profoundly re-wrought the nature of its waters (by degrading both surface and groundwater resources) [15]. These impacts implicate effects on water chemistry, alteration of the hydrological cycles, suspended loads from soil erosion, biocide leaching, and others. Indeed, agriculture can be both cause and victim of water pollution. Since 1990, the connection that exists between land and water use in the agricultural activity is recognized by the Food and Agriculture Organization (FAO) of the United Nations, by its clear requirement "appropriate steps must be taken to ensure that agricultural activities do not adversely affect water quality so that subsequent uses of water for different purposes are not impaired" [16]. The sources of agricultural pollutants could be "point" or "nonpoint" [11, 14, 17, 18]. In fact, water pollution caused by agricultural sources, as nonpoint sources, are hard to supervise and regulate, giving the diffusive nature of agricultural sources. According to Chen et al. [17] the nonpoint source water pollution from agriculture has exceeded that from industry and has become the largest source of nonpoint pollution in China.

2.2.2.1 General information on agricultural pollutants

In 2011, pollutant emissions from the different agricultural sources were firstly considered in the Chinese national wastewater pollutant inventory [6]. These pollutants are organic and/or inorganic coming from large quantities of agrochemicals products, such as insecticides, pesticides, herbicides, fungicides, fertilizers, and veterinary products. According to the Chinese Ministry of Environmental Protection (CMEP) [19], agricultural sources were found to release a total of 11.86 million tons of chemical oxygen demand (COD), accounting for 47.4% of the total COD wastewater from all sources in 2011. Meanwhile, a total of 0.83 million tons of NH_4^+–N was also released from agricultural sources, which represents 31.8% of the total NH_4^+–N wastewater from all sources. In general, agriculture is responsible for the release of four categories of water pollutants into the water environment as follows: (i) nutrients, (ii) pathogens, (iii) pesticides, and (iv) silts.

2.2.2.1.1 Nutrients

Animal wastes and chemical fertilizers are applied to soil to provide the nitrogen, phosphorus, and trace elements necessary for crop growth. When applied, these fertilizers are either taken up by crops, remain in the soil, or enter the aquatic environments. Nitrogen (N) compounds can accumulate in soil crust and vadose zone for years. Furthermore, N in the presence of oxygen is transformed either to N gases, nitrite (NO_2^-), or NO_3^-. The fundamental paths of N cycle include nitrogen fixation, nitrification, denitrification, ammonification, volatilization, and atmospheric deposition. In addition to the natural complexity of the N cycle, N fluxes continue to be substantially modified by human activities especially by agriculture and burning of fossil fuels [20]. **Figure 3** shows a simplified N cycle diagram of nitrogen [21].

2.2.2.1.1.1 Nitrate

NO_3^- is a naturally occurring ion in the nitrogen cycle that is the stable form of N for oxygenated systems; it is of prime concern on a global scale. Nitratation is the conversion of NO_2^- by nitrite-oxidizing bacteria (e.g., *Nitrobacter* spp.) to NO_3^- as in Eq. (1):

$$\text{Nitratation } NO_2^- + \frac{1}{2}O_2 \rightarrow NO_3^- \quad (1)$$

The resulting NO_3^- ion is stable in oxic conditions and can remain in the aquifer for a long time. NO_3^- can be reduced by microbial action into NO_2^- (also called nitritation) or other forms. Nitritation is the ammonia oxidation by ammonia-oxidizing bacteria (*Nitrosomonas* spp.) under aerobic conditions into NO_2^-, according to Eq. (2):

$$\text{Nitritation } NH_4^+ + \frac{1}{2}O_2 \rightarrow NO_2^- + 2H^+ + H_2O \quad (2)$$

According to Rivett et al. [22] conditions under which denitrification will occur require the presence of NO_3^-, denitrifying bacteria, oxygen concentrations (1–2 mg/L), electron donor, favorable conditions of temperature (25–35°C), pH (from 5.5 to 8.0), and other trace nutrients. Chemical and biological processes can further reduce NO_2^- to various compounds or oxidize it to NO_3^-. Due to its high water solubility [23], nitrate is regarded as one of the most widespread groundwater pollutant in the world, imposing a significant hazard to drinking water supplies and thus human health. The movement of NO_3^- into water sources is related to concentration in the soil; this concentration is directly related to the total amount of N [16]. In addition and according to many researches, natural attenuation by denitrification is minimal in aquifers in semiarid areas which allows N to remain as NO_3^- [11].

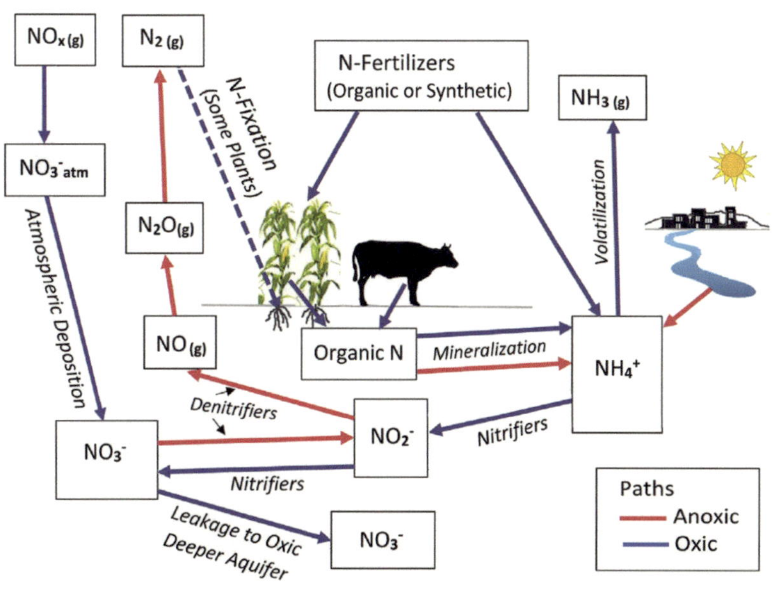

Figure 3.
Simplified nitrogen cycle showing the principal anoxic and oxic paths influencing groundwater N in arid and semiarid areas [21].

2.2.2.1.1.2 Nitrate sources

One of the main sources of NO_3^- is the intensive use of fertilizers in agriculture activity, as already mentioned [23, 24]. It is worth mentioning that in some cases, especially in aquifers of arid and semiarid areas, agricultural and domestic wastes are commonly found mixed together and they both contribute in increasing NO_3^- concentration [25]. According to Paschke et al. [26], an increase in NO_3^- concentration after a 10-year period (1990–2000) is simultaneously related with a decrease of the herbicide atrazine and its degradation product desethylatrazine. NO_3^- has a high mobility and low affinity for adsorbing onto clay particles, which facilities its passage into groundwater [27]. However, despite its high solubility, it is not distributed homogeneously within the aquifer [28]. Indeed, Zhang et al. [29] confirmed the distinctively higher concentration in shallow depths compared with deeper parts of an aquifer. Gutiérrez et al. [21] identified some of the groundwater of aquifers with NO_3^- contamination depending on the type of area, aquifers of semiarid, arid and hyperarid areas (**Table 1**).

The impact of NO_3^- pollution on surface water and groundwater has been the focus of several studies on many sites all over Morocco. According to Menkouchi et al. [30] the pollution of the groundwater by NO_3^- affects nearly all the Moroccan territory with approximately 6% of resources having NO_3^- content more than the national standard. Furthermore, Menkouchi demonstrated the contamination of Boujaad center with NO_3^- (exceeded 80 mg/L). According to the spatial distribution of NO_3^- contents made by Tagma et al. [31], the groundwater in Souss plain is less polluted than Chtouka-Massa plain. In fact, 36% of Chtouka-Massa's wells exceed the regulatory NO_3^- limit while only 7% in Souss plain. Maria Calvache et al. [32] confirmed the highly presence of NO_3^- in Chtouka Ait Baha, Massa, and Tiznit. Recently, Malki et al. [33] has studied and confirmed the NO_3^- contamination of Belfaa and the irrigated area along Massa River. Hydrogeochemical results on groundwater samples collected in 2010 show that the Bou-Areg aquifer is vulnerable to NO_3^- contamination [34]. The contamination of the Loukkos basin, Essaouira, and the basin of Triffa plain in northeast Morocco was confirmed.

2.2.2.1.1.3 Nitrate effects on human health and environment

Results of many studies suggest that ingestion of relatively high levels of NO_3^- could cause health problems. The toxic action of NO_3^- exposure, through diet or drinking water, can be divided into acute (short-term) effects and chronic (long-term) effects [35]. The main concern of an acute toxicity is the capacity of NO_3^- to cause methemoglobinemia (known as blue baby disorder) after oral ingestion. In fact, in the gastrointestinal tract, the ingested NO_3^- is reduced to NO_2^- that binds to hemoglobin to form methemoglobin. In the case of infants, it only takes a 10 mg/L N-NO_3^- to cause methemoglobinemia [21, 36]. For chronic toxicity, the consumption of NO_3^- at levels higher than 50 ppm has been associated with (i) increased thyroid volume and subclinical thyroid disorders, (ii) increased incidence of goiter in children, (iii) carcinogenicity due to the conversion of NO_3^- to NO_2^- and formation of genotoxic/carcinogenic N-nitroso compounds, such as N-nitrosamines and N-nitrosamides, some of which are known carcinogens [23, 35].

Additionally, when N is present in overabundance of the requirements of the biological system, it increase NO_3^- leaching which results in water eutrophication and acidification of sensitive soils [36]. This acidification involves changes in river and lake chemistry that produces secondary changes in the aquatic biota. The United States Environmental Protection Agency (USEPA) identifies eutrophication as the critical problem in those US surface waters with impaired water quality,

Areas	Region	Land use, crops	% >50, N	Range of NO_3^- values
Semiarid	Gaza Strip, Palestine	U, recovery	58%, N = 12	2–185
Semiarid	Rio Grande Valley, USA	A, U	6%, N = 587	0.1–61
Semiarid	Basin and Range, USA	A, U	8%, N = 3539	0.1–130
Semiarid	Andhra Pradesh, India	A, U, sweet lime, cotton	18%, N = 496	0.1–896
Semiarid	North China Plain	A, wheat, maize	9.5%, N = 295	Ave. 45.3
Semiarid	Dakar, Senegal	U	61%, N = 36	0.3–1390
Semiarid	Southwest Niger	Land clearing	25%, N = 28	0.2–176
Semiarid	Seville, Spain	A, cotton, potato	~70%, N = 16	35–630
Semiarid	La Mancha, Spain	A, corn, wheat, barley	18%, N = 684	<0.5–125
	Chihuahua, Mexico	A, U, alfalfa	40%, N = 45	2.2–266
Semiarid	Livermore CA, USA	A, L, vineyard	17%, N = 35	5.2–60.2
Semiarid	Osona, Spain	L, pigs	82%, N = 60	10–529
Semiarid	South Platte River, USA	A, G, corn	47%, N = 30	0.7–229
Semiarid	Grombalia, Tunisia	A, U, citrus trees	Shallow 73%, N = 26 40%, N = 25	0.5–514 2–231
Semiarid	Arava Valley, Israel	A, U, flowers, palms, vine	30%, N = 53	1–330
Arid/hyperarid	Monte Desert, Argentina	G	0%, N = 29	0.3–49.2
Arid/hyperarid	NW China	G, subsistence farming	32%, N = 52	<0.1–113
Arid/hyperarid	Central Iran	U, A, almonds, alfalfa	0%, N = 120	5–40
Arid/hyperarid	Southern Iran	A, wheat	5.9%, N = 34	1.5–70.7

Major land use: A = agricultural, L = livestock, G = grazing, U = urban. With N = number of samples exceeding 50 mg(NO_3^-)/L [21]

Table 1.
NO_3^- concentration (mg(NO_3^-)/L) and % samples exceeding 50 mg(NO_3^-)/L in some aquifers from semiarid, arid, and hyperarid areas.

whereas the United Nations Environment Programme (UNEP) states that eutrophication is "probably the most pervasive water quality problem on a global scale" (UNEP 1991). High NO_3^- concentrations are known to stimulate heavy blue-green algal growth (such as *Cyanobacteria*), which produce poisonous toxins to fish and mammals simultaneously.

2.2.2.1.1.4 Standards and regulations for nitrates

In order to restore the quality of water and deter hazards to humankind, various approaches have been targeted in terms of standards and legislation. In many countries, there are strict limits on the permissible concentration of NO_3^- in drinking water and in many surface waters. The USEPA [37] has set an enforceable standard called maximum contaminant level (MCL) in water for NO_3^- at 10 parts per million (ppm) (10 mg/L) and for NO_2^- at 1 ppm (1 mg/L), [59] and this is for all public water supplies. A summary of water quality guidelines for N-NO_3^-, made by the USEPA, is reported in **Table 2** [39]. Furthermore, the intake limits for NO_3^- in foods were set by the joint expert committee on food additives of the FAO/WHO and the European commission's scientific committee on food, at an acceptable daily intake for NO_3^- of 0–3.7 mg (NO_3^-)/kilogram (kg) body weight [40]. The same goes for Canada that set a MCL at 10 mg/L for NO_3^- as N and 1 mg/L for NO_3^- as N. For Morocco, the quality limits imposed by Order No. 1277-01 of 10 Shaaban 1423 (17 October 2002) setting quality standards for surface water used for the production of drinking water, has also set a MCL at 50 mg/L for NO_3^- [41].

2.2.2.1.1.5 Nitrate detection methods

NO_3^- ions can be detected through laboratory-based methods or in situ sensor-based methods. In general, they are two techniques for NO_3^-detection, direct and indirect methods. While comparing the two methods, the major disadvantage of using a direct method is measurement errors due to interference from other contaminants [38, 42, 43]. Many methods are currently available to the laboratory technicians for the detection and analysis of NO_3^- in a variety of sample matrices. The most commonly used techniques are (i) spectroscopic detection, (ii) electrochemical detection, and (iii) chromatography detection.

2.2.2.1.1.6 Nitrate removal techniques

There are a number of popular and conventional treatment methods available for reduction or removal of NO_3^- in water bodies. In addition, several technologies are being investigated or proposed as denitrification methods. **Figure 4** presents an overview of some of the techniques used for NO_3^- removal from water [23]. Reverse osmosis (RO) is considered as an ex situ and in situ application for the reduction of NO_3^- from water. The efficiency of the process depends on the used pressure; this later should be sufficient to overcome the osmotic pressure [44]. According to Harries et al. [47], RO works well with NO_3^- and NO_2^-. Rejection of 96–98% was

Water use	NO_3^- mg/L as nitrogen	NO_2^- mg/L as nitrogen
Drinking water	10 (maximum)	1 (maximum)
Freshwater aquatic life—acute	32.8 (maximum)	0.06 (maximum) when the chloride is less than or equal to 2
Freshwater aquatic life—chronic	3.0 (30-d average)	0.02 (30-d average) when the chloride is less than or equal to 2
Marine aquatic life—acute	None proposed	None proposed
Marine aquatic life—chronic	3.7 (30-d average)	None proposed
Livestock watering	100 (maximum)	10 (maximum)
Wildlife	100 (maximum)	10 (maximum)
Recreation and esthetics	10 (maximum)	1 (maximum)

Table 2.
Summary of water quality guidelines for nitrogen.

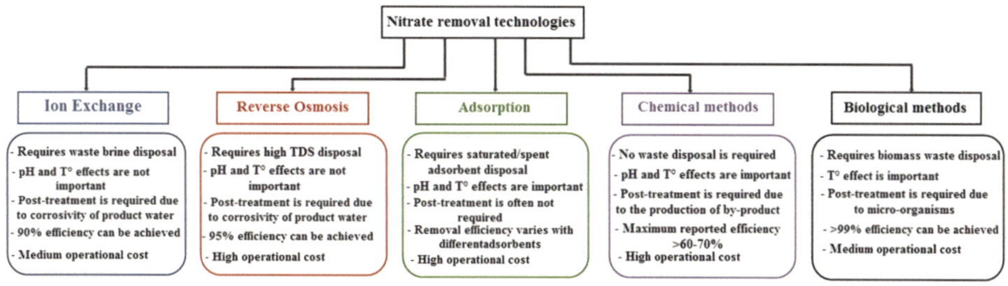

Figure 4.
Nitrate removal techniques.

found for monovalent ions, while 98–99% was found for divalent ions. In addition, ion exchange membranes (IEMs) have great potential in water denitrification. In fact, ion exchange is a reversible reaction between an electrolyte and a complex, where NO_3^- and NO_2^-, which have a negative ionic charge, will bind to the positively charged sites on the anion exchange beads of IEMs [44, 45]. For electrodialysis (ED), according to Mohsenipour et al. [46], the feed water supplied should have a turbidity that is lower than 2.0 nephelometric turbidity units. Furthermore, the concentration of free chlorine in water should be less than 0.5 mg/L, while these of hydrogen sulfide and manganese levels should be lower than 0.3 mg L. With electrodialysis systems about 70–85% of the water that is supplied to the system is available for use as low NO_3^- water [47].

One of the common and promising purification methods is the biological treatment of wastewaters operated by denitrifying bacteria. The idea of using this process to remove NO_3^- from drinking water has gained ground, especially in Europe [44]. Biological denitrification is mostly advised for the removal of relatively low concentration of N components. Furthermore, NO_3^- are efficiently removed when an external organic carbon source, generally methanol, ethanol, or acetic acid, is added [48]. The rate of denitrification also depends on the type and concentration of carbon as well as carbon to nitrogen (C/N) ratio. The reduction of NO_3^- to nitrogen is done in four steps as shown in Eq. (3).

$$NO_3^- \rightarrow NO_2^- \rightarrow NO \rightarrow N_2O \rightarrow N_2 \tag{3}$$

Mohsenipour et al. [46] also reported the efficiency of NO_3^- removal by a biological denitrification method using *Pseudomonas* bacteria and carbon source (starch of 1%) for initial NO_3^- concentrations of 500 and 460 mg/L; the denitrification rates were respectively equal to 86 and 89% [49]. Furthermore, various conventional and nonconventional materials from different origins have been studied and conceived, as adsorbents, to limit the harmful effects of NO_3^- [23]. Huang and Cheng [50] used powdered activated carbon (PAC) and carbon nanotubes (CNTs) for pollution reduction of NO_3^- and demonstrated that the adsorption capacity of CNTs was found to be higher than PAC and decreased for pH higher than 5. Bamboo powder charcoal and nonactivated granular carbon (from coconut shells activated with zinc chloride) demonstrate good removal efficiency in NO_3^- removal [51]. Natural adsorbents such as clay, zeolite, bentonite, and others were also studied [46, 52]. The effect of various variables such as pH, temperature, adsorbent dosage, other ions, and the amount of surfactant was tested on NO_3^- removal, and it been demonstrated that except pH and temperature, the other variables are found to have a marked effect on NO_3^- removal. It is should be noted that the removal of NO_3^- has been conducted using modified and unmodified agricultural waste [74]. Mizuta [51] summarized the various sorbents which have been used so far for the elimination of NO_3^- and demonstrated that hydrotalcite-type compounds/layered double hydroxides and chemically modified adsorbents are found promising sorbents for enhanced removal of NO_3^- from water. Tyagi et al. [53] made a summary of relevant published data with some of the latest important findings on the use of nanomaterials as NO_3^- adsorbents. These nanoparticles can be metallic, semiconductor, or polymeric. **Table 3** reports some of the different nanomaterials used for NO_3^- removal along with their experimental working parameters such as pH, adsorbent dose, initial NO_3^- concentration, and temperature.

2.2.3 Industrial pollution

Industrial wastewater is a generic term involving a wide array of wastewater discharged out of various industries [54]. Indeed, there are many kinds of industrial wastewater, with complex composition, because water fulfills several roles and functions in all types of industries. Measurements of parameters like biochemical oxygen demand (BOD), COD, pH, and alkalinity can allow to classify industrial pollution [55]. Indeed, industrial effluents can be classified according to the dominant nature of pollution, and it may be characterized by a high concentration of organic/inorganic compounds [56]. It should be noted that among the industries generating waste, certain of the most dangerous wastewater comes from sectors such as refineries, mining, tanneries, pharmaceuticals, pulp mills, and sugar production/distillery [57]. The food and agriculture industries generate wastewaters with high BOD, which is estimated to 0.6–20 m^3 wastewater/ton of product (such as bread/butter/milk or fruit juice). For instance, the conventional process in distillery industry generates ~15 L of wastewater per liter of alcohol with a BOD level of about 90,000 mg/L [57]. The most important contributor of wastewater volume (18%), COD (23%), and a major source of NH_4^+–N is the paper and paper products industry, while the raw chemical material and chemical products industry is the dominant source of NH_4^+–N (35.3%) and an important source of COD and petroleum hydrocarbons [59]. Around 80% of the heavy metal discharges into water sources come from four industries, namely, the nonferrous metal manufacturing and processing industry (27.5%), the fur and leather products manufacturing industry (19.4%), the metal product manufacturing industry (17.7%), and the nonferrous metal ore mining industry (14.0%). This is also true for the Moroccan industrial activities and its discharges. According to the Moroccan federation of metallurgical, mechanical, and electromechanical industries, this industrial sector includes more than 1000 companies divided into four main sub-sectors, including iron and steel, metal transformation, coating, surface treatment, and services related to the metallurgical, mechanical, and electromechanical industries.

2.2.3.1 General information on metal pollutants

Although metals are natural components of the Earth's crust and contribute in various physiological processes of living organisms, their elevated concentrations (above the level of homeostatic regulation) produced by industrial activity [58] can be toxic to human and ecological receptors. The majority of elemental metals and their compounds and complexes are extremely stable in the environment [59]. Most metals are extremely soluble in circumneutral waters (as in **Figure 5**) [60].

Furthermore, metals are excellent conductors of electricity and mostly enter chemical reactions as positive ions. These metals are often involved in electron transfer reactions involving oxygen and leads to the formation of toxic oxyradicals [61]. The metals' fate and toxicity are highly dependent on their speciation or the form in which they exist in a given aquatic system [7]. An example of the importance

Nanomaterial	Type	Modification	Initial NO_3^- concentration	pH	Nanomaterial dose	T°	Contact Time	NO_3^- removal efficiency
Nanoparticles	Iron (Fe0) (nanoscale zero valent iron (NZVI))	Coated with graphene oxide (GO)	5 mg/L	6.9	1 g/L	—	8 h	90%
	Iron (Fe0) (NZVI)		100 mg/L		5 g/L	Room	30 min	>96.5%
	Iron (Fe0) (NZVI)	Pd-Cu deposited on NZVI	100 mg/L		0.25 g	Room	1 h	78%
	Iron (Fe0) (NZVI)	Fresh NZVI coated with nickel	1 mg/ml		0.1 g/10 ml	20 °C	2 min	100%
	TiO2/Fe0	—	10 mg/L	3	1:10 ratio of TiO_2/Fe0		30 min	95%
	Nano-alumina	—	40 mg/g	4.4	1 g/L	25 ± 2 °C	70 min	4 mg/g
Nanotubes	CNTs	Oxidized	50–600 mg/L	7	50 mg	25 °C	50 h	70%
	CNTs	Iron oxide nanoparticle immobilized	2.5–100 mg/L		5 mg	25 °C	90 min	91.75%
Nanofibers	Chitosan	Microalgae immobilized	30 mg/L	6.5–7.0	—	—	16 days	87 ± 4%
		Chitosan nanofibers	30 mg/L	6.5–7.0	—	—	16 days	32 ± 3%
Nanoshell	SiO_2-FeOOH-Fe core shell	NZVI	64 mg/L	3	—	—	120 min	99.84%
Nanocluster	Spherical NZVI	—	80 mg/L	3–7	4 mg/L	—	60 min	95%
Nanocomposites	Carbon-silicon	—	50 mg/L		100 mg	—	60 min	44.89%,11.34 mg/g
	Iron/chitosan/zirconium composite	—	1–1000 mg/L	3	—	—		89.3 mg/g
	Magnetic graphene	Nanosized lanthanum hydrous	20 mg/L	6–8	40 mg	293–313 K	30 min	138.88 mg/g

Nanomaterial	Type	Modification	Initial NO_3^- concentration	pH	Nanomaterial dose	T°	Contact Time	NO_3^- removal efficiency
	Nanocomposite	Doped composite	20 mg/L	6–8	50 mg	293–313 K	30 min	83%
	Chitosan/Zeolite/nano ZrO_2	—	20 mg/L	3	0.02 g	35 °C	60 min	23.58 mg/g

Table 3.
Comparison of NO_3^- removal by different nanomaterials [53].

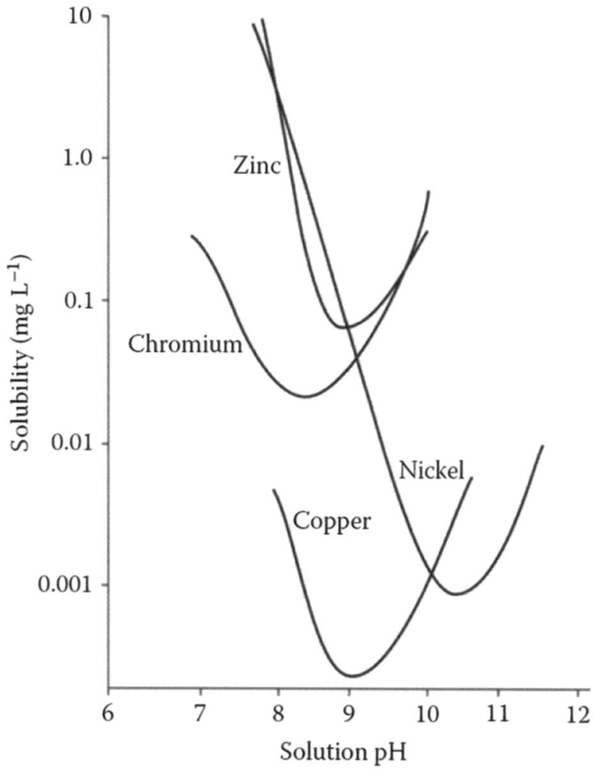

Figure 5.
Solubility curves for common metals in freshwater with pH [60].

of metal speciation and redox conditions is hexavalent chromium (Cr(VI)). Many of metals are transition metals (such as cadmium and Cu) or heavy metals (such as lead and silver) that can form metalloids which bond to organic compounds to form lipophilic substances that are often highly toxic. Metals are also lost from solution by precipitation as the pH changes [7, 60, 61]. **Table 4** shows the ranking of these metals according to their toxicity through biological and carcinogenicity tests [62].

2.2.3.1.1 Heavy metal pollutants

Any metallic element with relatively high density as compared to water and toxic even at low concentrations is termed as "heavy metal" [63, 64]. Among these metals, Cr is one of the top 16 major toxic contaminants that have detrimental effects on human health [64, 65]. Besides Cr, Cu is generally considered as a highly harmful metal at high concentration [66]. Accordingly, the following section discusses and summarizes relevant information about Cr and Cu.

Cr as a metallic element was first discovered and isolated in 1797 by the French chemist Nicolas-Louis Vauquelin [35, 67]. The world resources of Cr exceed 10.9 billion metric tons of shipping-grade chromite [67]. Ferric chromite ($FeCr_2O_4$) is the principal Cr ore, found mostly (with 96% of the world's reserves) in South Africa. Minor common sources include chrome ochre (Cr_2O_3), and crocoite ($PbCrO_4$) are also present [58]. Cr is widely used in engineering and chemical industries because of its durability and esthetic quality. The principal uses for Cr are metallurgical (67%), refractories (18%), and chemical (15%). In addition, Cr exits in three stable forms in the environment with different oxidation states and ionic nature [65]. The physicochemical properties of Cr are presented in **Table 5** [67].

Bioaccumulative character	Chemical elements
Very bioaccumulative	Pb, Zn, Cu
Relatively bioaccumulative	Hg, Ni, Cd
Slightly bioaccumulative	As, Cr
Carcinogenicity	
Strong carcinogenicity	As, Pb
Non-carcinogenic	Cr, Ni, Cd, Zn, Cu
Toxicity	
Group of highly toxic elements	Cd, Hg
Relatively toxic elements	Pb, Ni, As, Cu, Cr(VI)
Slightly toxic elements	Cr(III), Zn

Table 4.
Classification of metals according to their toxicity [62].

Physical and chemical properties of Cr	
Atomic number	24
Atomic mass	51.996
Atomic radius (pm)	185
Main oxidation state(s)	+2, +3, +6
Ionic radius (pm)	87–94 (+2), 75.5 (+3), 55–69 (+4), 48.5–71 (+5), 40–58 (+6)
Electronegativity (pauling)	1.66
Density (g/cm^3)	7.19
Melting/boiling point (°C)	1.907/2.671
Isotopes	4 stable + 17 unstable
Acid/base of oxide	Strong acid
State (st 27 °C, 1 atm)	Solid
Metallic character	Metal
Element group(s)	Transition element
Affinity	Lithophile

Table 5.
Physical and chemical properties of Cr [67].

The oxidation states of Cr can go from −2 to +6; however, only the +6 and +3 oxidation states are commonly encountered in the environment. Cr(III) is highly insoluble, relatively immobile, and most thermodynamically stable [58], while Cr(VI) has high water mobility and solubility and can be easily reduced. Cr(VI) is recognized to be more poisonous (100 times more toxic than Cr(III)), mutagenic, and carcinogenic in nature [68].

2.2.3.1.1.1 Chromium VI sources

To date, naturally occurring Cr(VI) in groundwater has been detected in the following geologic environments: (i) chromite ore bodies, (ii) arid alluvial basins in the southwest United States, (iii) saline brines in evaporate basins, and (iv) serpentinite ultramafic terrains [67], while the major industrial source of Cr(VI) emissions are (i) chemical manufacturing industry, (ii) metal finishing industry, (iii) manufacturers of pharmaceuticals, (iv) electrical and aircraft manufacturers, (v) cement-producing plants (as cement contains Cr), and (vi) production of wood, stone, glass, and clay products. Owlad et al. [69] reported the concentration of Cr(VI) in the wastewater of several industries (as in **Table 6**) and estimated the concentration of Cr(VI) in wastewater caused by industries between 0.1 and 200 mg/L.

2.2.3.1.1.2 Speciation of Cr(VI)

In aqueous solution, Cr(VI) may exist in a variety of oxospecies (viz., dichromates ($Cr_2O_7^{2-}$), chromates ($CrO_4^{\ 2-}$), H_2CrO_4, and acid chromates ($HCrO_4^-$)) depending on pH of the solution, total Cr concentration, the presence of oxidizing and reducing compounds, the redox potential, and kinetics of the redox reactions [64, 70]. At pH < 1, the predominant species is H_2CrO_4, while as the pH is raised from pH 2 to 6, the $HCrO_4^-$ and $Cr_2O_7^{2-}$ anions prevail.

According to Palmer and Puls [71] under pH 6.5, $HCrO_4^-$ dominates when the Cr(VI) concentrations are low (<30 mM), but $Cr_2O_7^{2-}$ becomes significant when concentrations are greater than 1 mM, or it may even dominate when the total Cr(Vl) concentrations are greater than 30 mM. At a pH > 8 only the yellow ion CrO_2^- exists. $Cr_2O_7^{2-}$ dominates in acidic solution at Cr(VI) concentration above 0.01 M, while CrO_4^{2-} dominates in basic solutions independently of Cr(VI) concentration, and H_2CrO_4 is a very strong acid [70]. The chemical equilibria which are involved for Cr(VI) dissociation are as in Eqs. (4)–(6):

$$H_2CrO_4^{+} \leftrightarrow H^{+} + HCrO_4 \tag{4}$$

$$HCrO_4^{-} \leftrightarrow H^{+} + CrO_4^{2-} \tag{5}$$

$$2HCrO_4^{-} \leftrightarrow Cr_2O_7^{2-} + H_2O \tag{6}$$

2.2.3.1.1.3 Cr(VI) effects on human health and environment

Ingesting large quantity of Cr(VI), either by human or animal, can be corrosive to the skin and eyes and causes stomach upsets and ulcers, convulsions, kidney and liver damage [72, 73]. Allergic reactions consisting of severe redness and swelling of the skin have been noted. Some evidence on animal studies show that Cr(VI) compounds can cause cancer in various tissues due to the low water insolubility [35, 67, 72]. Indeed, due to the fact that Cr(VI) is unstable in the body, easy to diffuse across the membrane, and is reduced intracellularly, it can provide very reactive pentavalent Cr and trivalent Cr which can alter DNA. Ingestion of 1.0–5.0 g of Cr(VI) as chromate results in severe acute gastrointestinal disorders, hemorrhagic diathesis, and convulsions [74]. In 1998, Dartsch et al. [75] noticed that 5 mmol/L Cr(VI) resulted in 50% cell death.

Furthermore, Cr(VI) compounds at high concentration are extremely toxic to plants and retard their growth. Cr and Cr(VI) compounds can cause severe phytotoxicity which may result in reduction of seed germination, nutrient imbalance, degradation of pigments, decrease of antioxidant and enzyme concentration, and oxidative stress [76]. Indeed, Rout et al. [77] demonstrated a 25% reduction in seed germination in the presence of 200 μM Cr(VI). According to Davies et al. [78] Cr

Industry	Cr(VI) concentration (mg/L)
Hardware factory	60.0
Chrome tanning plant	3.7
Electroplating plant	1.0
Electropolishing plant	42.8
Tannery plant	3950.0
Tannery plant	100.0
Tannery plant	1770.0
Tannery plant	8.3
Electroplating plant	20.7
Electroplating plant	75.4

Table 6.
Industrial wastewater containing Cr(VI) [69].

is mostly toxic to higher plants at 100 μg/kg dry weights. Sinha et al. [79] reported that Cr is toxic for most agronomic plants at a concentration of about 0.5–5.0 mg/L in nutrient media and of 5–100 mg/g under soil condition. The Cr toxicity in plants affects photosynthesis in terms of CO_2 fixation, photophosphorylation, electron transport, and enzyme activities as reported by H. Oliveira [76, 80].

2.2.3.1.1.4 Cr(VI) detection methods

Given the different chemistries and malignity of Cr(III) and Cr(VI) complexes, the concentration determination of each chemical speciation rather than the total Cr concentration is often desired [81]. Indeed, Cr(VI) is mainly analyzed, and various methods to prevent its reduction have been developed. There are two main groups of Cr speciation methods: off- and online techniques [82]. The off-line methods use pretreatment techniques for separation and concentration of specific Cr species (in the samples) before its insertion into detection instruments. These pretreatment techniques can be (i) colored complex formation methods, (ii) soluble membrane filter techniques, (iii) chromatographic methods, (iv) electrochemical methods, (v) coprecipitation techniques, (vi) ion exchange techniques, (vii) separation using chelating resins, and (viii) solvent extraction. In the online methods, the separation system is coupled with detection system. These methods include (i) flow injection analysis, and (ii) high-performance liquid chromatography (HPLC) that includes ion chromatography (IC), (iii) ion pair chromatography (IPC), and (iv) reversed-phase chromatography.

The globally acknowledged standard methods for selective Cr(VI) detection are spectroscopic techniques using diphenyl carbazide (DPC) method with a limit of detection (LOD) of 0.12 mg/L [83]. In general, the DPC spectrophotometric determination is an inexpensive and sensitive procedure that also permits the speciation of Cr. It is worth mentioning that de Andrade et al. [84, 85] have employed DPC for the flow injection spectrophotometric determination of Cr(VI). In this work, the authors combined the spectrophotometric procedure with the column preconcentration procedure. New reagents are used for Cr(VI) spectrophotometric determination. Andrle and Broekaert [86] suggested the selective determination of Cr(VI) based upon the formation of a complex between ammonium pyrrolidinedithiocarbamate and Cr(VI). To overcome ion interferences, Pyrzynska [83] proposed a new analysis based on the reaction of Cr(VI) with chromotropic acid (in acidic medium) along with the presence of NaF as a masking agent for iron. In fact, Bu et al. [87] indirectly determined Cr(VI) by the use of carbimazole based on the redox reaction of carbimazole with Cr(VI). In the same path, Fan et al. [88] developed a new method for Cr(VI) determination based on the reaction of Cr(VI) and ABTS (2,2′-azino-bis(3-ethylbenzothiazoline-6-sulfonate)). Moreover, the selective detection of Cr(VI) is relatively easier based on electrochemical techniques due to the different reduction potentials of Cr(VI) and Cr(III) [89]. Electrochemically cleaned Au electrode is useful for the detection of Cr. Polymer-based electrochemical sensor are also used for Cr(VI) detection. In 2013, Susan and Aziz reported a screen-printed carbon electrode modified with quercetin to detect Cr(VI) in the water, while Welch et al. [90] studied and confirmed the electrochemical detection of hexavalent Cr species at Au, glassy carbon electrode, and boron-doped diamond electrodes. Chen et al. [91] realized trace detection of Cr(VI) in aqueous mediums using electro-adsorption-assisted laser-induced breakdown spectroscopy.

2.2.3.1.1.5 Standards and regulations for Cr(VI)

According to the USEPA water standards, the maximum limit of Cr in drinking water is 0.1 mg/L which is based on the total Cr (EPA, 1990) [92]. Also, the allowable concentration of dischargeable Cr(VI) into surface water is below 0.05 mg/L; however, the permissible concentration of total Cr (Cr(VI), Cr(III) and other forms) is equal to 2 mg/L [93]. The American conference of governmental industrial hygienists' threshold limit time-weighted averages for Cr(VI) is 0.01 mg/m^3. In addition, the Canadian guidelines has indicated a limit of 0.05 mg/L for total Cr concentration in drinking water [94]. Krishnani et al. [95] reported that the maximum permissible limits of Cr(VI) to discharge into potable water, in land surface water, and industrial wastewater are 0.05, 0.1, and 0.25 mg/L, respectively. According to the Moroccan official bulletin (NM 03.7.001), the Russian federation [73], and the Indian standard [65, 94], the maximum allowable limits for Cr in drinking and domestic water are 0.05 and 0.5 mg/L for Cr(VI) and Cr(III), respectively.

2.2.3.1.2 Copper

From the most common heavy metals that are often present in industrial wastewater, Cu is usually found at high concentrations in industrial discharges (**Figure 6**) [96]. Cu is an abundant trace element found in a variety of rocks and minerals. Over 90% of modern industrial enterprises need Cu products.

The Cu reserves and production (mines and concentrates) in major countries in 2014 are reported in **Table 7** [97]. Cu is involved in reduction, oxidation processes, adsorption/desorption, and dissolution processes which lead to many changes in its speciation. Cu is a transitional metal and appears in nature in four oxidation states: elemental copper Cu(0) (solid metal), Cu(I) cuprous ion, Cu(II) cupric ion, and rarely Cu(III). However in natural water systems, Cu can exist only in two different oxidation states, Cu(I) and Cu(II) [87]. The oxidation states of Cu depend on the concentration of oxygen.

In its oxidation state, Cu forms very stable complexes with both organic and inorganic ligands. The toxicity of serine and citrate is enhanced in case of estuarine bacteria, while Cu-amino acid and Cu-citrate are enhanced in case of *Daphnia* and algae, respectively [99]. In its pure form, Cu has outstanding criteria such as electrical and thermal conductivity, strong chemical stability, good corrosion resistance, and high plasticity and malleability, which makes it so attractive for industries.

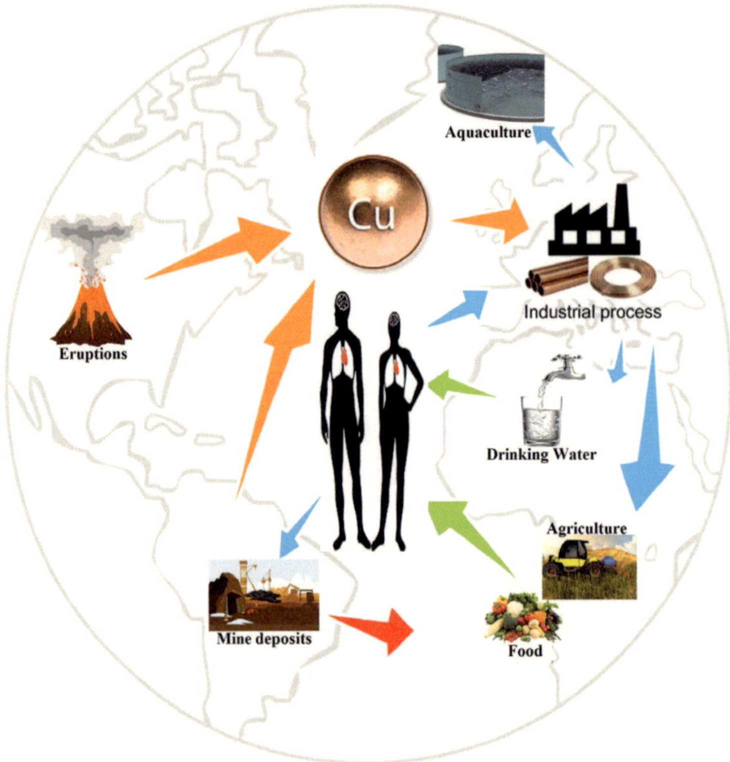

Figure 6.
Ecology of Cu. Colored arrows represent various aspects, namely, orange, elemental Cu fluxes; blue, human Cu interventions; green, human Cu intake; red, contamination [96].

Country	Reserves/ (ten thousand tons)	Percentage of global reserves/(%)	Cu mine production /(ten thousand tons)	Cu concentrate production /(ten thousand tons)
Chile	20,900	29.86	575.0	274.7
Australia	9300	13.29	97.0	
Peru	6800	9.71	138.0	
Mexico	3800	5.43	51.4	
The United States	3500	5.00	138.3	108.6
Russia	3000	5.00	72.0	87.4
China	3000	4.29	192.3	764.9
Poland	2800	4.00		
Indonesia	2500	3.58		
Zambia	2000	2.58	75.6	70.9

Table 7.
Cu reserves and production (mines and concentrates) in major countries [97].

2.2.3.1.2.1 Copper (II)

As stated before, Cu is a redox-active element meaning it can easily go back and forth between the oxidized Cu(II) state and the reduced Cu(I) state [100]. The following Eqs. (7)–(9) represent the reduction processes for Cu ions:

$$Cu^{2+} + 2e^{-} \leftrightarrow Cu^{0} \quad (7)$$

$$Cu^{2+} + e^{-} \leftrightarrow Cu^{+} \quad (8)$$

$$Cu^{+} + e^{-} \leftrightarrow Cu^{0} \quad (9)$$

Cu(II) is suggested to be the primary species in nearly all natural waters. In this oxidation state, Cu forms very stable complexes with both organic and inorganic ligands [101]. $Cu_2(OH)_2^{2+}$ is believed to be the principal cationic hydrolysis product. In contrast, Cu(I) is unstable in aqueous solution and tends to change rapidly to Cu(II). Cu(II) forms strong complexes with the electron donor groups in organic compounds (O, N, and S) [102].

2.2.3.1.2.2 Cu(II) effects on human health and environment

Many studies confirmed that an increased concentration of Cu(II) in the body leads to serious health problems [96, 98, 103]. Cu and Cu(II) is absorbed in the intestine and then transported to the liver and binds to albumin [96]. Due to its catalytic role, Cu can reduce human immunodeficiency and leads to symptoms of anemia, vomiting, neutropenia, hypopigmentation, bone abnormalities, and growth disorders. It was also related to abnormalities in the metabolism of glucose and cholesterol. Brewer et al. [100] also studied the many roles of Cu in human metabolisms and explained properly the different genetic diseases caused by this metal, namely, Wilson's disease, Menkes disease, and Alzheimer's disease. From an environmental point of view, the high concentration of Cu with an extreme acidity inhibits the ordinary development of plants and animals, reduce the biodiversity, contaminate water reservoirs, and even corrode infrastructure. Generally, sheep suffer plenty from Cu poisoning, because of the effects of Cu that manifest at fairly low concentrations [103].

2.2.3.1.2.3 Cu detection methods

The development of analytical methods for the selective detection and visualization of Cu(II) is significant. The common methods for the detection of Cu(II), as for Cr(VI), include liquid chromatography, electrochemical detection, spectrophotometry, solid-phase extraction coupled with atomic absorption spectroscopy, potentiometric techniques, X-ray fluorescence, atomic emission spectroscopy, and inductively coupled plasma mass spectrometry [104, 105]. According to Ramanjaneyulu et al. [106, 107], the existing reagents for the photometric determination of Cu permit the detection of 0.025–30 μg/mL Cu. For instance, the diethyldithiocarbamate reagent is considered one of the most frequently used reagents for Cu(II) determination with a LOD of 0.1 μg/L. Zagurskaya et al. [107] developed a spectrophotometric method of determination of Cu(II) with a ligand for a new coordination compound, the sodium salt of 4-phenylsemicarbazone 1,2-naphthoquinone-4-sulfonic acid (L). The formation of this new complex is accompanied by a color change and LOD of 0.012 mg/L.

2.2.3.1.2.4 Standards and regulations for Cu and its derivative

The maximum permissible limits (MPL) for Cu in water are listed in **Table 8**. Furthermore, the USEPA has set a MPL for Cu discharges either in the soil or in wastewater. The MPL in the soil is equal to 100 mg/L, while the limit for wastewater

	International standard organizations							
	WHO	USEPA	ISI	CPCB	ICMR	BIS	EU	NM
Cu permissible limit (mg/L)	1.0	1.3	0.05	1.5	1.5	1.3	2	2

WHO; USEPA, United States Environmental Protection Agency; ISI, Indian Standard Institution; CPCB, Central Pollution Control Board; ICMR, Indian Council of Medical Research; BIS, Bureau of Indian Standards; EU, European limits via Directive 98/83/EC on the quality of water intended for human consumption; NM, the Moroccan official bulletin (NM 03.7.001).

Table 8.
Permissible limits of drinking water quality [104, 105, 109].

that can be discharged into the public sewage system is 1 and 0.1 mg/L, respectively, for wastewater that can be discharged into public wastewater and which can be discharged into the recipient (surface and groundwater) [108].

2.2.3.1.2.5 Heavy metal (namely Cr and Cu) removal techniques

Faced with more and more stringent regulations, nowadays heavy metals are the environmental priority pollutants. In this regard, a wide range of removal technologies, such as chemical precipitation, ion exchange, adsorption, membrane filtration, and electrochemical, have been the subject of research nowadays.

In precipitation processes, chemicals react with heavy metal ions to form insoluble precipitates (such as hydroxide, sulfide, carbonates, etc.) by pH adjustment. The forming precipitates can be removed by physical means such as sedimentation, flotation, or filtration. **Table 9** reports some of the recent research in Cu(II) and Cr(VI) removal by chemical precipitation. It should be mentioned that Davarnejad and Panahi [110] reported that the adsorption method is considered to be the most common process used to remove different Cu ions from industrial wastewater. The efficiency of adsorption depends on many parameters; high surface area, pore size distribution, functional groups, and the polarity of the adsorbent determines the efficiency of adsorption process [66].

Many different cheap adsorbents have been developed and used for the removal of both Cu and Cr ions from metal-contaminated wastewater. Recently, activated carbons, agriculture byproducts, zeolites, and industrial wastes are widely employed to remove Cr(VI) from waters and industrial wastewaters [113]. Iftikhar et al. [114] investigated the use of rose waste biomass in Cr(III) and Cu(II) removal. In this study, the capacity of adsorption of Cu(II) and Cr(III) by rose biomass varies with temperature, and the maximum adsorption capacities of 55.79 and 67.34 mg/g, respectively, for Cu(II) and Cr(III) were found at 303 ± 1 K, with adsorption over 98% of Cu(II) and Cr(III). Natural zeolites are the most studied natural materials for the adsorption of heavy metals [115]. Ali and Yaşar [116] show the zeolite's high selectivity for Cu ions. In the same context, Barakat [117] studied the effect of pH on Cu adsorption and demonstrate that A4 zeolite is efficiently used to adsorb Cu(II), at natural and alkaline pH. At the same time, Francis et al. [115, 118] studied activated phosphate, zirconium phosphate, and calcined phosphate at 900°C, as net adsorbents to remove Cu(II). In addition, many researchers have reported that electrocoagulation (EC) is a suitable technology for heavy metal removal. The efficiency of this method in removing Cr(VI), Cu(II), and Ni from wastewater of an electroplating plant was investigated by Akbal and Camcidotless [119]. These authors achieved 100% Cu, 100% Cr, and 100% Ni removal at an EC time of 20 min with the use of Fe-Al electrode pair. **Table 10** resumes some of the electrochemical studies on Cr(VI) removal.

Species	Initial metal concentration	Precipitant	Optimum pH	Removal efficiency (%)
Cu(II), Zn(II), Cr(III), Pb(II)	100 mg/L	CaO	7–11	99.37–99.6
Cu(II), Zn(II), Pb(II)	0.018, 1.34, 2.3 mM	H_2S	3.0	100, >94, >92
Cr(III)	5363 mg/L	CaO and MgO	8.0	>99
Cr(VI)	30 mg/L	$FeSO_4$	8.7	>99
Cu(II), ZN	100 mg/L	Lime ($Ca(OH)_2$), Soda ash (Na_2CO_3), Sodium sulfide (Na_2S)	10–11	>99

Table 9.
Heavy metal removal using chemical precipitation [111, 112].

	Concentration	pH	Anode-cathode	Removal	Reference
Electrocoagulation Cr(VI)	5 mg/L	7.0	Al alloy-Fe	98.2%	[113]
	1,000 mg/L	1.2	Carbon steel	100%	[113]
	100 mg/L	2	Fe-Cu	100%	[113]
	1,000 mg/L	4.5	Fe	100%	[113]
	—	4	Al–Al	99%	[113]
Electrochemical Cr(VI)	Concentration	pH	Electrodes	Removal	
	3–50 mg/L	2	Carbon felt	100%	[113]
	12 mg/L	3	CNTs	96%	[113]
	50 mg/L	0.5	Polyaniline	~70%	[113]
	350 mg/L	0.7	Polypyrrole-coated Al3	100%	[113]
	20 mg/L	11	Au nanoparticle-decorated TiO2 nanotube array	100%	[113]
	1470 mg/L	1.84	Stainless steel	100	[112]
	8 mg/L	2.0	Carbon aerogel	98.5	[69]

Table 10.
Cu(II) and Cr(VI) removal electrochemical methods.

3. Conclusion

Widespread pollution of surface and underground water resulted largely from increased pollutant discharges from industrial (specially the heavy metals), municipal, and agricultural sources (nutriments such as N and P), excessive water abstraction from the environment, and poor water resource management and enforcement of pollution control regulations. The preservation of water sources from pollutants, NO_3^-, Cr and Cu, is a major concern, shared by all, public, industrial, scientific, researchers, and decision-makers. Over the past two decades, environmental regulations have become more stringent and require an improved quality of the treated effluents. Consequently, wide ranges of treatment technologies have been developed to remove agricultural and industrial pollutants. The set of methods presented in this chapter is certainly not exhaustive, but its main objective is to show the diversity of techniques used for NO_3^- and heavy metal (Cu(II) and Cr(VI)) removal. The choice of one technique over the others depends on several factors such as the cost of processing, ease of reproducing, and the added value, along with the mode of use of the obtained treated water and the occurrence, or lack thereof, of harmful side products. Consequently, the development of new treatment methods is always a topical research topic.

Acknowledgements

This work was supported by Ministère de l'Enseignement Supérieur et de la Recherche Scientifique et de la Formation des cadres (MESRSFC) (Morocco) and Centre National pour la Recherche Scientifique et Technique (CNRST) (Morocco) (project number PPR/2015/72).

Conflict of interest

The authors declare no conflict of interest.

Author details

Majda Breida[1*], Saad Alami Younssi[1], Mohamed Ouammou[1], Mohamed Bouhria[1] and Mahmoud Hafsi[2]

1 Laboratory of Materials, Membranes and Environment, Faculty of Sciences and Technologies of Mohammedia, University Hassan II, Casablanca, Morocco

2 International Institute for Water and Sanitation, National Office for Electricity and Drinking Water, Rabat, Morocco

*Address all correspondence to: majda.breida-etu@univh2m.ma

References

[1] Reilly TE, editor. Ground-Water Availability in the United States. Reston, VA: U.S. Geological Survey; 2008

[2] Sousa JCG, Ribeiro AR, Barbosa MO, Pereira MFR, Silva AMT. A review on environmental monitoring of water organic pollutants identified by EU guidelines. Journal of Hazardous Materials. 2018;**344**:146-162. DOI: 10.1016/j.jhazmat.2017.09.058

[3] Li D, Liu S. Water quality evaluation. In: Water Quality Monitoring and Management. Elsevier, Basis, Technology and Case Studies, Academic Press; 2019. pp. 113-159. DOI: 10.1016/B978-0-12-811330-1.00004-1

[4] Traceable measurements for monitoring critical pollutants under the European Water Framework Directive (WFD 2000/60/EC). n.d. 3

[5] Zgheib S, Moilleron R, Chebbo G. Flux et sources des polluants prioritaires dans les eaux urbaines en lien avec l'usage du territoire. Techniques Sciences Méthodes. 2011;**4**:50-62. DOI: 10.1051/tsm/201104050

[6] Bouhaouss A, Bchitou R, Hachimi S, Marakchi K, Boulmane M. Apport de la chimiometrie pour l'analyse et l'interpretation de quelques parametres physico-chimiques influençant la repartition des metaux lourds, des elements nutritifs et des anions dans les eaux de l'oued de bouregreg. n.d. 161

[7] Schweitzer L, Noblet J. Water contamination and pollution. In: Green Chemistry. An Inclusive Approach. 2018. pp. 261-290. DOI: 10.1016/B978-0-12-809270-5.00011-X

[8] Khan S, Ali J. Chemical analysis of air and water. In: Bioassays. Advanced Methods and Application. 2018. pp. 21-39. DOI: 10.1016/B978-0-12-811861-0.00002-4

[9] Kowalkowski T, Gadzała-Kopciuch M, Kosobucki P, Krupczyńska K, Ligor T, Buszewski B. Organic and inorganic pollution of the Vistula River basin. Journal of Environmental Science and Health, Part A. 2007;**42**(4):421-426. DOI: 10.1080/10934520601187336

[10] Mareddy AR. Environmental Impact Assessment: Theory and Practice. Oxford, United Kingdom; MA, USA: Elsevier/BH, Butterworth-Heinemann, an imprint of Elsevier; Cambridge; 2017

[11] Zeliger HI. Human Toxicology of Chemical Mixtures: Toxic Consequences Beyond the Impact of One-component Product and Environmental Exposures. 2nd ed. Amsterdam, Boston: William Andrew/Elsevier; 2011

[12] Zhu W, Graney J, Salvage K. Land-use impact on water pollution: Elevated pollutant input and reduced pollutant retention. Journal of Contemporary Water Research & Education. 2008;**138**:15-21. DOI: 10.1111/j.1936-704X.2008.00004.x

[13] Singh MR, Gupta A. Water Pollution-Sources, Effects and Control. n.d. 17

[14] Shulman VL. Trends in waste management. In: Waste, A Handbook for Management. Academic Press; 2011. pp. 3-10. DOI: 10.1016/B978-0-12-381475-3.10001-4

[15] Moss B. Water pollution by agriculture. Philosophical Transactions of the Royal Society B: Biological Sciences. 2008;**363**:659-666. DOI: 10.1098/rstb.2007.2176

[16] World Health Organization. Guidelines for Drinking-water Quality. 3rd ed. Vol. 1. Geneva: WHO Library Cataloguing-in-Publication Data; 2004

[17] Chen Y, Wen X, Wang B, Nie P. Agricultural pollution and regulation: How to subsidize agriculture? Journal of Cleaner Production. 2017;**164**:258-264. DOI: 10.1016/j.jclepro.2017.06.216

[18] Hu Y, Cheng H. Water pollution during China's industrial transition. Environmental Development. 2013;**8**:57-73. DOI: 10.1016/j.envdev.2013.06.001

[19] Ministry of Environmental Protection. 2011 China State of the Environment. Beijing, China: China Environmental Science Press; 2012

[20] Vitousek PM, Aber JD, Howarth RW, et al. Human alteration of the global nitrogen cycle: Sources and consequences. Information Sciences, Ecological Applications; 1997;**7**(3):737-750. DOI: 10.1890/1051-0761(1997)007[0737:HAOTGN]2.0.CO;2

[21] Gutiérrez M, Biagioni RN, Alarcón-Herrera MT, Rivas-Lucero BA. An overview of nitrate sources and operating processes in arid and semiarid aquifer systems. Science of the Total Environment. 2018;**624**:1513-1522. DOI: 10.1016/j.scitotenv.2017.12.252

[22] Rivett MO, Buss SR, Morgan P, Smith JWN, Bemment CD. Nitrate attenuation in groundwater: A review of biogeochemical controlling processes. Water Research. 2008;**42**:4215-4232. DOI: 10.1016/j.watres.2008.07.020

[23] Bhatnagar A, Sillanpää M. A review of emerging adsorbents for nitrate removal from water. Chemical Engineering Journal. 2011;**168**:493-504. DOI: 10.1016/j.cej.2011.01.103

[24] Tellam JH, Rivett MO, Israfilov RG, editors. Urban Groundwater Management and Sustainability. Dordrecht: Springer; 2006

[25] Opsahl SP, Musgrove M, Slattery RN. New insights into nitrate dynamics in a karst groundwater system gained from in situ high-frequency optical sensor measurements. Journal of Hydrology. 2017;**546**:179-188. DOI: 10.1016/j.jhydrol.2016.12.038

[26] Paschke SS, Schaffrath KR, Mashburn SL. Near-decadal changes in nitrate and pesticide concentrations in the South Platte River Alluvial Aquifer, 1993-2004. Journal of Environment Quality. 2008;**37**:S-281. DOI: 10.2134/jeq2007.0656

[27] Schroth G, Sinclair FL, editors. Trees, crops and soil fertility: concepts and research methods. Wallingford: CABI; 2002. DOI: 10.1079/9780851995939.0000

[28] Prieto-Guerrero A, Espinosa-Paredes G. The Distribution of Soil Nutrients With Depth: Global Patterns and the Imprint of Plants. Vol. 53(1). Biogeochemistry, Springer, Kluwer Academic Publishers. 2001. pp. 51-77. DOI: 10.1023/A:1010760720215

[29] Zhang Y, Li F, Zhang Q, Li J, Liu Q. Tracing nitrate pollution sources and transformation in surface- and ground-waters using environmental isotopes. Science of The Total Environment. 2014;**490**:213-222. DOI: 10.1016/j.scitotenv.2014.05.004

[30] Menkouchi Sahli MA, Annouar S, Mountadar M, Soufiane A, Elmidaoui A. Nitrate removal of brackish underground water by chemical adsorption and by electrodialysis. Desalination. 2008;**227**:327-333. DOI: 10.1016/j.desal.2007.07.021

[31] Tagma T, Hsissou Y, Bouchaou L, Bouragba L. Groundwater nitrate

pollution in Souss-Massa basin (south-west Morocco). n.d. 9

[32] Calvache ML, Duque C, Pulido-Velazquez D, editors. Groundwater and Global Change in the Western Mediterranean Area. Cham: Springer International Publishing; 2018. DOI: 10.1007/978-3-319-69356-9

[33] Malki M, Bouchaou L, Hirich A, Ait Brahim Y, Choukr-Allah R. Impact of agricultural practices on groundwater quality in intensive irrigated area of Chtouka-Massa, Morocco. Science of The Total Environment. 2017;**574**:760-770. DOI: 10.1016/j.scitotenv.2016.09.145

[34] Re V, Sacchi E, Allais E. The use of nitrate isotopes to identify contamination sources in the Bou-Areg Aquifer (Morocco). Procedia Earth and Planetary Science. 2013;7:729-732. DOI: 10.1016/j.proeps.2013.03.020

[35] Wexler P, Anderson BD, editors. Encyclopedia of Toxicology. 3rd ed. Amsterdam; Boston: Elsevier/AP, Academic Press is an imprint of Elsevier; 2014

[36] Wilson WS, Ball AS, Hinton R, Royal Society of Chemistry, editors. Managing Risks of Nitrates to Humans and the Environment, University of Essex on 1-2 September, 1997. Vol. 1999. Cambridge: Royal Society of Chemistry;

[37] Powlson DS, Addiscott TM, Benjamin N, Cassman KG, de Kok TM, van Grinsven H, et al. When does nitrate become a risk for humans? Journal of Environment Quality. 2008;**37**:291. DOI: 10.2134/jeq2007.0177

[38] Daniel WL, Han MS, Lee J-S, Mirkin CA. Colorimetric nitrite and nitrate detection with gold nanoparticle probes and kinetic end points. Journal of the American Chemical Society. 2009;**131**(18):6362-6363. DOI: 10.1021/ja901609k

[39] Water Quality Guidelines for Nitrogen (Nitrate, Nitrite, and Ammonia). n.d. 29

[40] https://www.atsdr.cdc.gov/csem/csem.asp?csem=28&po=8

[41] Ghalit M, Gharibi E, Taupin J-D, Lamhamdi A. Mineralization process of spring in carbonate coastal aquifer in the massif Bokkoya (Central Rif, Morocco). 2017. p. 1

[42] Alahi MEE, Mukhopadhyay SC. Detection methods of nitrate in water: A review. Sensors and Actuators A: Physical. 2018;**280**:210-221. DOI: 10.1016/j.sna.2018.07.026

[43] Moorcroft M. Detection and determination of nitrate and nitrite: a review. Talanta. 2001;**54**:785-803. DOI: 10.1016/S0039-9140(01)00323-X

[44] Follett RF, Hatfield JL. Nitrogen in the environment: Sources, problems, and management. The Scientific World Journal. 2001;**1**:920-926. DOI: 10.1100/tsw.2001.269

[45] Luo T, Abdu S, Wessling M. Selectivity of ion exchange membranes: A review. Journal of Membrane Science. 2018;**555**:429-454. DOI: 10.1016/j.memsci.2018.03.051

[46] Mohsenipour M, Shahid S, Ebrahimi K. Removal techniques of nitrate from water. Asian Journal of Chemistry. 2014;**26**:7881-7886. DOI: 10.14233/ajchem.2014.17136

[47] Harries R, Elyanow D, Heshka D, Fischer K. Desalination of brackish groundwater for a prairie community using electrodialysis reversal. Desalination. 1991;**84**:109-121. DOI: 10.1016/0011-9164(91)85123-C

[48] Rossi F, Motta O, Matrella S, Proto A, Vigliotta G. Nitrate Removal from Wastewater through Biological Denitrification with OGA 24 in a Batch

Reactor. Water. 2014;7:51-62. DOI: 10.3390/w7010051

[49] Kim Y-H, Hwang E-D, Shin WS, Choi J-H, Ha TW, Choi SJ. Treatments of stainless steel wastewater containing a high concentration of nitrate using reverse osmosis and nanomembranes. Desalination. 2007;**202**:286-292. DOI: 10.1016/j.desal.2005.12.066

[50] Huang W-J, Cheng Y-L. Effect of characteristics of activated carbon on removal of bromate. Separation and Purification Technology. 2008;**59**:101-107. DOI: 10.1016/j.seppur.2007.05.034

[51] Mizuta K. Removal of nitrate-nitrogen from drinking water using bamboo powder charcoal. Bioresource Technology. 2004;**95**:255-257. DOI: 10.1016/j.biortech.2004.015

[52] Mena-Duran CJ, Sun Kou MR, Lopez T, Azamar-Barrios JA, Aguilar DH, Domínguez MI, et al. Nitrate removal using natural clays modified by acid thermoactivation. Applied Surface Science. 2007;**253**:5762-5766. DOI: 10.1016/j.apsusc.2006.12.103

[53] Tyagi S, Rawtani D, Khatri N, Tharmavaram M. Strategies for nitrate removal from aqueous environment using nanotechnology: A review. Journal of Water Process Engineering. 2018;**21**:84-95. DOI: 10.1016/j.jwpe.2017.12.005

[54] Li D, Liu S. Detection of industrial water quality. In: Water Quality Monitoring and Management. Basis, Technology and Case Studies. Academic Press; 2019. pp. 329-349. DOI: 10.1016/B978-0-12-811330-1.00013-2

[55] Moran S. Industrial effluent characterization and treatment objectives. In: An Applied Guide to Water and Effluent Treatment Plant Design. Butterworth Heinemann; 2018. pp. 221-229. DOI: 10.1016/B978-0-12-811309-7.00016-3

[56] Thomas O, Thomas M-F. Industrial wastewater. In: UV-Visible Spectrophotometry of Water and Wastewater. Elsevier Science; 2017. pp. 317-348. DOI: 10.1016/B978-0-444-63897-7.00010-X

[57] Ranade VV, Bhandari VM, Institution of Chemical Engineers (Great Britain). Industrial Wastewater Treatment, Recycling, and Reuse. Amsterdam: Elsevier; 2014

[58] Youcai Z, Sheng H. Pollution Control and Resource Recovery: Industrial Construction and Demolition Wastes. ISBN: 978-0-12-811754-5

[59] Popek E. Environmental chemical pollutants. In: Sampling and Analysis of Environmental Chemical Pollutants. Elsevier; 2018. pp. 13-69. DOI: 10.1016/B978-0-12-803202-2.00002-1

[60] Gray NF. Water Science and Technology: An Introduction. 4th ed. Boca Raton: CRC Press, Taylor & Francis Group, CRC Press is an Imprint of the Taylor & Francis Group, an Informa Business; 2017

[61] Amjad Z, Demadis KD, editors. Mineral Scales and Deposits: Scientific and Technological Approaches. Amsterdam: Elsevier; 2015

[62] Sancey B. Développement de la bio-adsorption pour décontaminer des effluents de rejets industriels: abattement chimique et gain environnemental. n.d. 324

[63] Lenntech Zater Treatment and Air Purification. Water Treatment. Rotterdalseweg. Netherlands: Lenntech; 2004

[64] Jobby R, Jha P, Yadav AK, Desai N. Biosorption and

biotransformation of hexavalent chromium [Cr(VI)]: A comprehensive review. Chemosphere. 2018;**207**:255-266. DOI: 10.1016/j.chemosphere.2018.05.050

[65] Adam MR, Salleh NM, Othman MHD, Matsuura T, Ali MH, Puteh MH, et al. The adsorptive removal of chromium (VI) in aqueous solution by novel natural zeolite based hollow fibre ceramic membrane. Journal of Environmental Management. 2018;**224**:252-262. DOI: 10.1016/j.jenvman.2018.07.043

[66] Carolin CF, Kumar PS, Saravanan A, Joshiba GJ, Naushad M. Efficient techniques for the removal of toxic heavy metals from aquatic environment: A review. Journal of Environmental Chemical Engineering. 2017;**5**:2782-2799. DOI: 10.1016/j.jece.2017.05.029

[67] Guertin J, Jacobs JA, Avakian CP, Independent Environmental Technical Evaluation Group, editors. Chromium (VI) Handbook. Boca Raton, Fla: CRC Press; 2005

[68] Rojas G, Silva J, Flores JA, Rodriguez A, Ly M, Maldonado H. Adsorption of chromium onto cross-linked chitosan. Separation and Purification Technology. 2005;**44**:31-36. DOI: 10.1016/j.seppur.2004.11.013

[69] Owlad M, Aroua MK, Daud WAW, Baroutian S. Removal of hexavalent chromium-contaminated water and wastewater: A review. Water, Air, and Soil Pollution. 2009;**200**:59-77. DOI: 10.1007/s11270-008-9893-7

[70] Baes CF, Mesmer E. The Hydrolysis of Cation. Berichte der Bunsengesellschaft für physikalische Chemie. John Wiley & Sons, Inc.; 1977;**81**(2). DOI: 10.1002/bbpc.19770810252

[71] Palmer C, Puls R. Ground Water Issue: Natural Attenuation of Hexavalent Chromium in Groundwater and Soils. n.d. 12

[72] Nordberg G, Fowler BA, Nordberg M, editors. Handbook on the Toxicology of Metals, General Considerations. 4th ed. Vol. 1. Amsterdam: Elsevier, Aademic Press; 2015

[73] Kononova ON, Bryuzgina GL, Apchitaeva OV, Kononov YS. Ion exchange recovery of chromium (VI) and manganese (II) from aqueous solutions. Arabian Journal of Chemistry. 2015. DOI: 10.1016/j.arabjc.2015.05.021

[74] Jacobs J, Testa SM. Overview of Chromium(VI) in the Environment: Background and History. n.d. 22

[75] Dartsch PC, Hildenbrand S, Kimmel R, Schmahl FW. Investigations on the nephrotoxicity and hepatotoxicity of trivalent and hexavalent chromium compounds. International Archives of Occupational and Environmental Health. Sep 1998;**71** (Suppl):S40-S45

[76] Mishra S, Bharagava RN. Toxic and genotoxic effects of hexavalent chromium in environment and its bioremediation strategies. Journal of Environmental Science and Health, Part C. 2016;**34**:1-32. DOI: 10.1080/10590501.2015.1096883

[77] Rout GR, Samantaray S, Das P. Effects of chromium and nickel on germination and growth in tolerant and non-tolerant populations of *Echinochloa colona* (L.) Link. Chemosphere. 2000;**40**(8):5

[78] Davies FT, Puryear JD, Newton RJ, Egilla JN, Saraiva Grossi JA. Mycorrhizal fungi increase chromium uptake by sunflower plants: Influence on tissue mineral concentration, growth, and gas

exchange. Journal of Plant Nutrition. 2002;**25**:2389-2407. DOI: 10.1081/PLN-120014702

[79] Sinha V, Pakshirajan K, Chaturvedi R. Chromium tolerance, bioaccumulation and localization in plants: An overview. Journal of Environmental Management. 2018;**206**:715-730. DOI: 10.1016/j.jenvman.2017.10.033

[80] Oliveira H. Chromium as an environmental pollutant: Insights on induced plant toxicity. Journal of Botany. 2012;**2012**:1-8. DOI: 10.1155/2012/375843

[81] Caballero B, Finglas PM, Toldrá F, editors. Encyclopedia of Food and Health. Amsterdam, Boston: Academic Press is an Imprint of Elsevier; 2016

[82] Hamilton EM, Young SD, Bailey EH, Watts MJ. Chromium speciation in foodstuffs: A review. Food Chemistry. 2018;**250**:105-112. DOI: 10.1016/j.food chem. 2018. 01.016

[83] Pyrzynska K. Redox speciation of chromium using sorption-based systems. TrAC Trends in Analytical Chemistry. 2012;**32**:100-112. DOI: 10.1016/j.trac.2011.09.004

[84] de Andrade JC, Rochat JC, Baccan N. Sequential spectrophotometric determination of chromium(III) and chromium(V1) using flow injection analysis. Analyst. 1985;**110**:3

[85] Tunçeli A. Speciation of Cr(III) and Cr(VI) in water after preconcentration of its 1,5-diphenylcarbazone complex on amberlite XAD-16 resin and determination by FAAS. Talanta. 2002;**57**:1199-1204. DOI: 10.1016/S0039-9140(02)00237-0

[86] Andrle CM, Broekaert JAC. Speciation of Cr(III) and Cr(VI) by reversed phase high-performance liquid chromatography using UV-detection. Fresenius' Journal of Analytical Chemistry. 1993;**346**:653-658. DOI: 10.1007/BF00321265

[87] Bu X, Zhang Z, Zhang L, Li P, Wu J, Zhang H, et al. Highly sensitive SERS determination of chromium(VI) in water based on carbimazole functionalized alginate-protected silver nanoparticles. Sensors and Actuators B: Chemical. 2018;**273**:1519-1524. DOI: 10.1016/j.snb.2018.07.058

[88] Fan W, Qiao J, Guan X. Multi-wavelength spectrophotometric determination of Cr(VI) in water with ABTS. Chemosphere. 2017;**171**:460-467. DOI: 10.1016/j.chemosphere.2016.11.153

[89] Ouyang R, Zhang W, Zhou S, Xue Z-L, Xu L, Gu Y, et al. Improved Bi film wrapped single walled carbon nanotubes for ultrasensitive electrochemical detection of trace Cr(VI). Electrochimica Acta. 2013;**113**:686-693. DOI: 10.1016/j.electacta.2013.09.110

[90] Welch C. Reduction of hexavalent chromium at solid electrodes in acidic media: Reaction mechanism and analytical applications. Talanta. 2004;**65**:74-78. DOI: 10.1016/j.talanta.2004.05.017

[91] Chen S-H, Li Y-X, Li P-H, Xiao X-Y, Jiang M, Li S-S, et al. Electrochemical spectral methods for trace detection of heavy metals: A review. TrAC Trends in Analytical Chemistry. 2018;**106**:139-150. DOI: 10.1016/j.trac.2018.07.005

[92] Environmental Protection Agency. Environmental pollution control alternatives. Cincinnati; 1990. EPA/625/5-90/025, EPA/625/4-89/023

[93] Abbasi-Garravand E, Mulligan CN. Using micellar enhanced ultrafiltration and reduction techniques for removal of Cr(VI) and Cr(III) from water. Separation and Purification

Technology. 2014;**132**:505-512. DOI: 10.1016/j.seppur.2014.06.010

[94] Pawlisz AV, Kent RA, Schneider UA, Jefferson C. Canadian water quality guidelines for chromium. Environmental Toxicology and Water Quality. 1997;**12**:123-183. DOI: 10.1002/(SICI)1098-2256(1997)12:2<123::AID-TOX4>3.0.CO;2-A

[95] Krishnani KK, Srinives S, Mohapatra BC, Boddu VM, Hao J, Meng X, et al. Hexavalent chromium removal mechanism using conducting polymers. Journal of Hazardous Materials. 2013;**252-253**:99-106. DOI: 10.1016/j.jhazmat.2013.01.079

[96] Latorre M, Troncoso R, Uauy R. Biological aspects of copper. In: Clinical and Translational Perspectives on Wilson Disease. Academic Press; 2019. pp. 25-31. DOI: 10.1016/B978-0-12-810532-0.00004-5

[97] Li L, Pan D, Li B, Wu Y, Wang H, Gu Y, et al. Patterns and challenges in the copper industry in China. Resources, Conservation and Recycling. 2017;**127**:1-7. DOI: 10.1016/j.resconrec.2017.07.046

[98] Jellouli Ennigrou D, Ben Sik Ali M, Dhahbi M. Copper and zinc removal from aqueous solutions by polyacrylic acid assisted-ultrafiltration. Desalination. 2014;**343**:82-87. DOI: 10.1016/j.desal.2013.11.006

[99] Azlan A, Khoo HE, Idris MA, Ismail A, Razman MR. Evaluation of minerals content of drinking water in Malaysia. The Scientific World Journal. 2012;**2012**:1-10. DOI: 10.1100/2012/403574

[100] Brewer GJ. Background on copper, including why Copper-2 is so specifically neurotoxic. In: Environmental Causes and Prevention Measures for Alzheimer's Disease. Academic Press; 2018. pp. 71-82. DOI: 10.1016/B978-0-12-811162-8.00007-X

[101] Mansilla-Rivera I, Nriagu JO. Copper chemistry in freshwater ecosystems: An overview. Journal of Great Lakes Research. 1999;**25**:599-610. DOI: 10.1016/S0380-1330(99)70765-3

[102] Flemming CA, Trevors JT. Copper toxicity and chemistry in the environment: A review. Water, Air, and Soil Pollution. 1989;**44**:143-158. DOI: 10.1007/BF00228784.103

[103] Thompson LJ. Copper. In: Veterinary Toxicology. Basic and Clinical Principles. Academic Press; 2018. pp. 425-427. DOI: 10.1016/B978-0-12-811410-0.00026-X

[104] Gumpu MB, Sethuraman S, Krishnan UM, Rayappan JBB. A review on detection of heavy metal ions in water—An electrochemical approach. Sensors and Actuators B: Chemical. 2015;**213**:515-533. DOI: 10.1016/j.snb.2015.02.122

[105] Ramdass A, Sathish V, Babu E, Velayudham M, Thanasekaran P, Rajagopal S. Recent developments on optical and electrochemical sensing of copper(II) ion based on transition metal complexes. Coordination Chemistry Reviews. 2017;**343**:278-307. DOI: 10.1016/j.ccr.2017.06.002

[106] Ramanjaneyulu G, Reddy PR, Reddy VK, Reddy TS. Direct and derivative spectrophotometric determination of copper(II) with 5-bromosalicylaldehyde thiosemicarbazone. The Open Analytical Chemistry Journal. 2009;**2**:78-82. DOI: 10.2174/1874065000802010078

[107] Zagurskaya-Sharaevskaya O, Povar I. Determination of Cu (II) ions using sodium salt of 4-phenylsemicarbazone 1,2-naphthoquinone-4-sulfonic acid in natural and industrial environments.

Ecological Processes. 2015;**4**:16. DOI: 10.1186/s13717-015-0042-0

[108] Bradberry SM. Metals (cobalt, copper, lead, mercury). Medicine. 2016;**44**(3):182-184. DOI: 10.1016/j.mpmed.2015.12.008

[109] Puri A, Kumar M. A review of permissible limits of drinking water. Indian Journal of Occupational and Environmental Medicine. 2012;**16**:40. DOI: 10.4103/0019-5278.99696

[110] Davarnejad R, Panahi P. Cu (II) removal from aqueous wastewaters by adsorption on the modified Henna with Fe_3O_4 nanoparticles using response surface methodology. Separation and Purification Technology. 2016;**158**:286-292. DOI: 10.1016/j.seppur.2015.12.018

[111] Chen Q, Yao Y, Li X, Lu J, Zhou J, Huang Z. Comparison of heavy metal removals from aqueous solutions by chemical precipitation and characteristics of precipitates. Journal of Water Process Engineering. 2018;**26**:289-300. DOI: 10.1016/j.jwpe.2018.11.003

[112] Fu F, Wang Q. Removal of heavy metal ions from wastewaters: A review. Journal of Environmental Management. 2011;**92**:407-418. DOI: 10.1016/j.jenvman.2010.11.011

[113] Jin W, Du H, Zheng S, Zhang Y. Electrochemical processes for the environmental remediation of toxic Cr(VI): A review. Electrochimica Acta. 2016;**191**:1044-1055. DOI: 10.1016/j.electacta.2016.01.130

[114] Iftikhar AR, Bhatti HN, Hanif MA, Nadeem R. Kinetic and thermodynamic aspects of Cu(II) and Cr(III) removal from aqueous solutions using rose waste biomass. Journal of Hazardous Materials. 2009;**161**:941-947. DOI: 10.1016/j.jhazmat.2008.04.040

[115] Al-Saydeh SA, El-Naas MH, Zaidi SJ. Copper removal from industrial wastewater: A comprehensive review. Journal of Industrial and Engineering Chemistry. 2017;**56**:35-44. DOI: 10.1016/j.jiec.2017.07.026

[116] Korkmaz M, Özmetin C, Fil BA, Yaşar Y. Determination of parameters affecting copper removal from solutions by clinoptilolite: Adsorption isotherm and thermodynamic. 2013;**3**:8

[117] Barakat MA. New trends in removing heavy metals from industrial wastewater. Arabian Journal of Chemistry. 2011;**4**:361-377. DOI: 10.1016/j.arabjc.2010.07.019

[118] Francis AA, Abdel Rahman MK. The environmental sustainability of calcined calcium phosphates production from the milling of eggshell wastes and phosphoric acid. Journal of Cleaner Production. 2016;**137**:1432-1438. DOI: 10.1016/j.jclepro.2016.08.029

[119] Akbal F, Camcidotless S. Copper, chromium and nickel removal from metal plating wastewater by electrocoagulation. Desalination. 2011;**269**(1-3):214-222. DOI: 10.1016/j.desal.2010.11.001

Chapter 4

Modeling Accumulated Evapotranspiration Over Time

Omar Cléo Neves Pereira and Altair Bertonha

Abstract

The knowledge of accumulated evapotranspiration by seasonal vegetation crops throughout their life cycle can be an important tool in decision-making when considering the economic viability of the crop. This knowledge can help understand how much the plants, subject to specific management, can evapotranspirate at the end of their cycle. This information assists in estimating the quantity of a production variable, for example, the mass of shoot fresh matter, besides indicating a more interesting period for its harvest. The objective of this chapter is, from the daily evapotranspiration estimative throughout the cycle, to model the accumulated evapotranspiration over the entire growth period of the crop. In order to do so, we must understand that the behavior of the response variable, i.e., the accumulated evapotranspiration, over time is not linear and keep in mind that the several observations performed in the same experimental unit have correlations and these correlations are more intense the closer temporally the measurements are. This understanding leads us to the analysis of longitudinal data from the nonlinear mixed effect models perspective.

Keywords: longitudinal data, nonlinear mixed effect model, growth curve, correlation structure, irrigation

1. Introduction

With available water in the soil, the water flow through the plants depends only on atmospheric demand. Therefore, physical variables as temperature, relative air humidity, and wind and solar radiation affect directly the evapotranspiration (*ET*) of a vegetated surface [1]. Besides that, the plants development state may also affect the *ET*.

Seasonal vegetation crops present a small demand for water, while their root system is small, reaching maximum rates in full growths and decreasing in the final stages of development. For these species, the accumulated *ET* over the plant cycle is directly related to its productivity. In other words, the greater the accumulated amount of evapotranspirated water, the greater the quantity of the production variables as shoot fresh and dry matter masses, number of leaves, and leaf area [2, 3].

The objective of this chapter is to illustrate the use of nonlinear mixed effect models to fit accumulated *ET* over time in seasonal vegetation crops.

IntechOpen

2. Response profile

Figure 1 presents the accumulated *ET* from a single lettuce plant over 23 consecutive days, being the first day equivalent to the 35th day after seeding. Throughout this period, the plant's daily *ET* was measured. Hence, the first day in the graph presents the *ET* value from the last 24 h, the second day refers to the last 48 h, and so on. The daily *ET* must be understood as the rate at which the accumulated *ET* occurs over time. In mathematical terms, the daily *ET* can be understood as an approximation of the derivative (rate) of the accumulated *ET* with respect to time.

We wish to describe the behavior of our variable of interest or *response variable*, the accumulated *ET* over time for seasonal vegetation crops. As the amount of evapotranspirated water in 1 day is added to the accumulated *ET* from previous days, its values over time are equal (in case *ET* from a whole day is null) or greater than the immediately preceding value.

Observations from the response variable in more than one moment in the same experimental unity constitute what we call as *response profile*. Therefore, **Figure 1** presents the response profile of a lettuce plant over time. This profile, apparently, presents an *S*-shaped format.

At the first days of observations (**Figure 1**), the accumulated *ET* is small because the plant is at the beginning of its growth. As the days go by, the daily *ET* values increase successively approximately until the 20th day. After the 20th day, the daily *ET*, i.e., the rate of change of the accumulated *ET*, begins to decrease. This causes the accumulated *ET*, which has been growing exponentially, to have a less vigorous growth and, therefore, tending slowly to a maximum value.

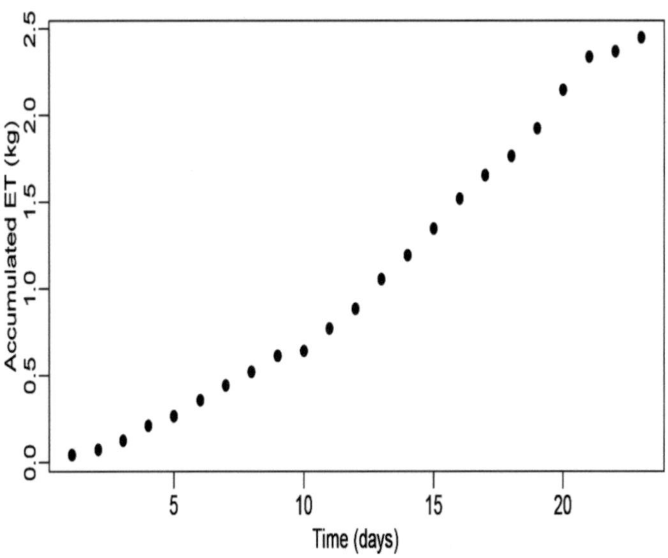

Figure 1.
Response profile of the accumulated ET for a single lettuce plant over 23 consecutive days.

3. A model for growth data

In order to describe the behavior of the accumulated *ET* over time, we also need an *S*-shaped function. In addition, it is expected that the chosen model will be interpretable and parsimonious in the parameters. An empirical option is the polynomial model, which has linear parameters. This type of model can promote good statistical adjustments and be computationally simpler but does not add any theoretical consideration to the physical and/or biological mechanisms that generated the data. On the other hand, a nonlinear model is associated with some theoretical knowledge regarding the studied phenomenon. Besides the interpretability, these models use few parameters when compared to linear models, thus configuring a more parsimonious description of the data [4].

Regarding the accumulated *ET* over time, we have some physical and biological aspects that we can use to choose a proper model. A function that describes this response variable needs parameters that delimit it between a minimum and a maximum value. In other words, the minimum value should be really close to zero and indicate the beginning of the plant's growth, and the maximum should be a value to which the accumulated *ET* tends asymptotically as the end of the life cycle approaches. Another aspect to highlight is that the model should present an inflection point, which indicates the day that the accumulated *ET* rate (daily *ET*) reaches its maximum value.

There are several functions capable of characterizing the accumulated *ET* over time in the sense we just described. For example, we will use the four-parameter logistic function (4*PL*) to describe our response variable. This function is widely used to fit growth or decay data. There are some parametrizations for this function in the literature, but we are using the one given by [4]

$$ET(t) = \phi_1 + \frac{\phi_2 - \phi_1}{1 + \exp\left[(\phi_3 - t)/\phi_4\right]} \tag{1}$$

with $ET(t)$ being the accumulated *ET* over time and ϕ_{1-4} the model parameters (**Figure 2**). ϕ_1 is the inferior horizontal asymptote which gives the accumulated *ET* value when $t \to -\infty$. Biologically, this parameter does not have a consistent interpretation, but it is important in the fitting because it ensures the accumulated *ET*, in times close to zero, to be very small [2, 3]. ϕ_2 is the superior horizontal asymptote and gives the accumulated *ET* value when $t \to \infty$. This parameter can be

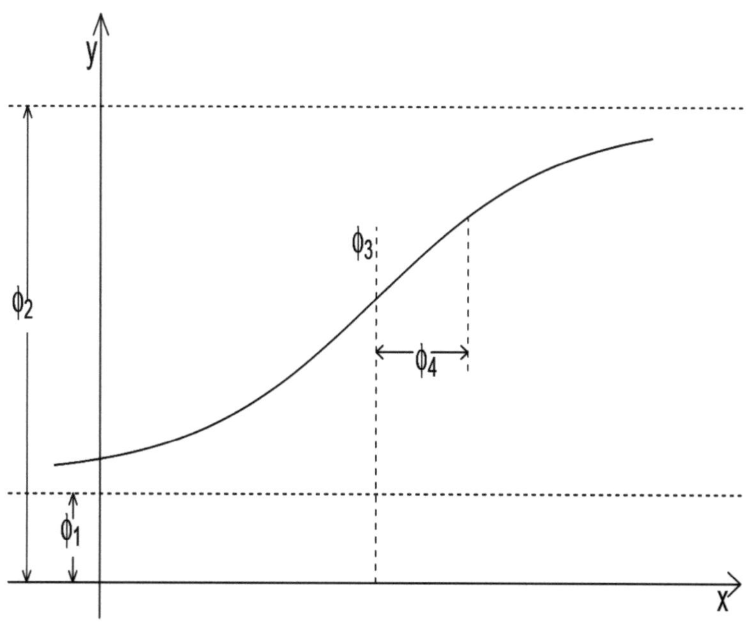

Figure 2.
Graphical representation of the 4PL parameters (Eq. (1)). Figure adapted from [4].

interpreted as the maximum accumulated ET estimative that the plant can reach in its final life cycle. ϕ_3 is the curve's inflection point and indicates the day (time) in which the daily ET reaches a maximum value. The corresponding time to ϕ_3 results in an accumulated ET between ϕ_1 and ϕ_2. More precisely, the accumulated ET until time ϕ_3 is the mean value of both asymptotes, i.e., $(\phi_1 + \phi_2)/2$. ϕ_4 is the scale parameter. The day (time) corresponding to $\phi_3 + \phi_4$ gives $\sim 0.75(\phi_2 - \phi_1)$ of the accumulated ET. Therefore, ϕ_4 indicates how quickly the accumulated ET leaves the proximity of ϕ_1 until it reaches values close to ϕ_2. The greater the ϕ_4, the slower this occurs.

To better understand what we have said, let's observe **Figure 3**. It brings the observed data shown in **Figure 1** represented by the black dots and two other curves representing two different fits made from this data. The first is a fifth-degree polynomial model, and the second is given by Eq. (1). The graph in this figure was extended until the 40th day in order to present the behavior of these fittings over time.

Both the fifth-degree polynomial and the nonlinear *4PL* models have fitted well with the data. However, the polynomial model goes to zero after the 30th day, which physically is impossible. Besides that, this model presents five parameters without any physical and/or biological explanation for the phenomenon in this study. The *4PL* nonlinear model is a strictly increasing function, being compatible with a variable which is accumulated over time. Nevertheless, this fitting does not allow the infinite growth of the accumulated ET as time grows larger and larger. It limits the accumulated ET to a value which can be understood as a maximum amount in which this plant can evapotranspirate throughout its life cycle.

We also see that, from approximately the 20th day, the accumulation of evapotranspiration water is decreasing. This indicates that this day is the inflection point of the nonlinear *4PL* model. For a practical example of the model's inflection point regarding seasonal vegetation crops, suppose that the commercial product of a given crop plant is its leaves. It is known that the production variables present a positive correlation with the amount of total evapotranspired water throughout the

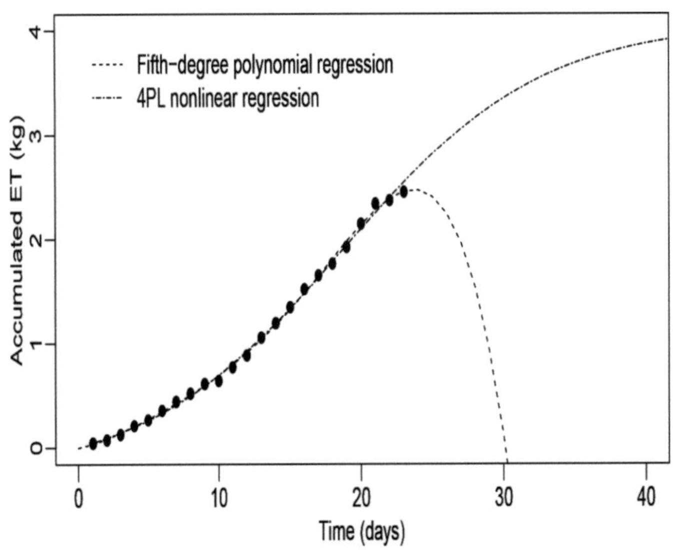

Figure 3.
Accumulated ET response profile for a single lettuce plant over 23 consecutive days. The graph was extended until the 40th day to present the behavior of two fittings made for this data, a fifth-degree polynomial and a nonlinear model given by Eq. (1).

plants' life cycle. That is, the larger the total evapotranspiration water, the higher the values of the production variables [2, 3]. Hence, the inflection point can indicate the proximity of the harvest day. From this point, the *ET* rates decrease, and therefore, the plants start to accumulate less quantities of shoot fresh matter over time. Thus, this point can be considered when thinking in terms of the economic viability of the crop.

Another important aspect, besides the parameter interpretability, is that the nonlinear 4*PL* model is more parsimonious than the linear model. In general, nonlinear models use a small number of parameters than the linear ones, which grant them more parsimony. Besides that, as can be observed in **Figure 3**, in regions outside the data interval, the nonlinear model gives a more trustworthy prediction for the response variable [4].

4. Longitudinal data

Clearly, in order to make sense, an experiment must provide data from more than one experimental unit. In our case, more than one plant should be observed over time, i.e., we must obtain more than one response profile. Studies in which the response variable is observed repeatedly throughout time in the experimental units are called *longitudinal studies*. This kind of work is common in agriculture when analyzing the increase or decrease of the response variable over time [2, 3, 5–8].

Measurements performed in the same experimental unit are most likely to be correlated. Suppose two plants which its *ET* is registered daily. If all covariables (fertilization, cultivate, planting season, soil water, and so on) were kept constant over time, plants with high rates of *ET* in a given day will most likely also have high rates of *ET* in the next day, the same for plants with smaller *ET* rates. If in a day the rate is small, probably in the next day it will also be small. In other words, measurements performed in the same experimental unit tend to be similar over time. It's the individual expression of each plant.

Besides the correlation between the observations within the same experimental unit, we must consider that, most likely, these correlations are greater for observations performed between neighboring times than those performed between more distant times.

5. Mixed effect model

In a longitudinal study, the monitoring of the experimental units over time generates correlated dataset. As mentioned, these correlations within the same experimental unit are stronger among neighboring observations. The greater the time distance between two measures, the weaker the correlation between them. Besides that, when we observe experimental units which received the same conditions regarding growth over time and are part of the same treatment, we have a variability among them that we attribute to chance. The treatment effects, the correlations, and the variabilities in a longitudinal study indicate that we need a tool that, in addition to being flexible in specifying a mathematical model, also emphasizes each experimental unit.

In mixed effect models, we select an ordinary function to describe the response variable regarding the studied covariables, that is, the responses of the experimental units in a population. Besides that, specific coefficients of this function can be unique for each experimental unit. In a mixed effect model, we assume that the experimental units of a population have the same functional form, but the function parameters may vary among the units.

The name *mixed model* comes from the fact that this model combines *fixed effects* and *random effects*. A mixed effect model is a parametric model which describes the relations between the response variable and the covariables (fixed effects) and takes into account the individual responses of each experimental unit (random effects). In other words, the fixed effects parameters describe the relations of the response variable and the covariables in an entire population, and the random effects specify the contribution of each individual within the population [4, 9–12].

To illustrate how to write a 4*PL* nonlinear model with mixed effect, let's assume that we are studying the accumulated *ET* regarding four levels of water in the soil, W_1, W_2, W_3, and W_4 (there could be more levels of water in the soil or less, but to exemplify, let's assume four treatments). We will consider that the response variable, the accumulated *ET*, has a normal probability distribution. Consider ET_{ij} as the accumulated *ET* at the *j* situation, for plant *i*, with $j = 1, 2, \cdots, n_i$ e $i = 1, 2, \cdots, N$, where n_i is the number of observations for the *i*-th plant and *N* is the total number of plants. The nonlinear 4*PL* mixed effect model can be expressed by [2, 3]

$$ET_{ij} = \phi_{1i} + \frac{\phi_{2i} - \phi_{1i}}{1 + \exp\left[(\phi_{3i} - t_j)/\phi_{4i}\right]} + \varepsilon_{ij} \tag{2}$$

being

$$\boldsymbol{\phi}_i = \begin{bmatrix} \phi_{1i} \\ \phi_{2i} \\ \phi_{3i} \\ \phi_{4i} \end{bmatrix} = \begin{bmatrix} \beta_1 + \gamma_1 x_{1i} + \delta_1 x_{2i} + \zeta_1 x_{3i} \\ \beta_2 + \gamma_2 x_{1i} + \delta_2 x_{2i} + \zeta_2 x_{3i} \\ \beta_3 + \gamma_3 x_{1i} + \delta_3 x_{2i} + \zeta_3 x_{3i} \\ \beta_4 + \gamma_4 x_{1i} + \delta_4 x_{2i} + \zeta_4 x_{3i} \end{bmatrix} + \begin{bmatrix} b_{1i} \\ b_{2i} \\ b_{3i} \\ b_{4i} \end{bmatrix} \tag{3}$$

$$= \boldsymbol{\beta} + \boldsymbol{\gamma} x_{1i} + \boldsymbol{\delta} x_{2i} + \boldsymbol{\zeta} x_{3i} + \boldsymbol{b}_i, \tag{4}$$

with the parameters $\boldsymbol{\beta}$, $\boldsymbol{\gamma}$, $\boldsymbol{\delta}$, $\boldsymbol{\zeta}$ representing the fixed effects and $\boldsymbol{b}_i$ the random effects in the model. x_{ki} with $k = 1, 2, 3$ are indicative covariables of treatments or groups and may have values zero or one. The parameter $\boldsymbol{\beta}$ is the reference level in the study. When $x_{1i} = x_{2i} = x_{3i} = 0$, thus, $\boldsymbol{\phi}_i = \boldsymbol{\beta} + \boldsymbol{b}_\mathbf{i}$ and the *i*-th plant belongs to the treatment W_1. When $x_{1i} = 1$ and $x_{2i} = x_3 i = 0$, $\boldsymbol{\phi}_i = \boldsymbol{\beta} + \boldsymbol{\gamma} + \boldsymbol{b}_\mathbf{i}$ and the *i*-th plant belongs to the treatment W_2. If $x_{2i} = 1$ and $x_{1i} = x_3 i = 0$, $\boldsymbol{\phi}_i = \boldsymbol{\beta} + \boldsymbol{\delta} + \boldsymbol{b}_\mathbf{i}$ and the *i*-th plant belongs to the treatment W_3. And lastly, if $x_{3i} = 1$ and $x_{1i} = x_{2i} = 0$, $\boldsymbol{\phi}_i = \boldsymbol{\beta} + \boldsymbol{\zeta} + \boldsymbol{b}_\mathbf{i}$ and the *i*-th plant belong to the treatment W_4. The random effects $\boldsymbol{b}_\mathbf{i}$ are considered independent among the plants and are normally distributed with mean zero and covariance matrix $\boldsymbol{\Psi}$ ($\boldsymbol{b}_\mathbf{i} \sim N(\mathbf{0}, \boldsymbol{\Psi})$). In this case, the covariance matrix is given by

$$\boldsymbol{\Psi} = \begin{bmatrix} \sigma_1^2 & \sigma_{12} & \sigma_{13} & \sigma_{14} \\ \sigma_{12} & \sigma_2^2 & \sigma_{23} & \sigma_{24} \\ \sigma_{13} & \sigma_{23} & \sigma_3^2 & \sigma_{34} \\ \sigma_{14} & \sigma_{24} & \sigma_{34} & \sigma_4^2 \end{bmatrix} \tag{5}$$

with σ_1^2, σ_2^2, σ_3^2, and σ_4^2 being the variances of random effects b_{1i}, b_{2i}, b_{3i}, and b_{4i}, respectively, and σ_{12}, σ_{13}, σ_{14}, σ_{23}, σ_{24}, and σ_{34} are the covariances between them. The variances of the random effects indicate how a model parameter varies between the experimental units. Frequently, we suppose that the errors within the groups ε_{ij} are independent between the observations of the same experimental unit and that they are distributed the same way among the experimental units. For a given sample unit i, we can describe the error ε_{ij} as the time j indexed vector, i.e.,

$$\boldsymbol{\varepsilon}_\mathbf{i} = \begin{bmatrix} \varepsilon_{1i} \\ \varepsilon_{2i} \\ \vdots \\ \varepsilon_{n_i i} \end{bmatrix},$$

where each vector $\boldsymbol{\varepsilon}_\mathbf{i}$ has n_i observations throughout time and we assume that they follow a normal multivariate distribution with mean zero and covariance matrix $\boldsymbol{\Lambda}_\mathbf{i}$, i.e.,

$$\boldsymbol{\varepsilon}_\mathbf{i} \sim N(\mathbf{0}, \boldsymbol{\Lambda}_\mathbf{i}). \tag{6}$$

Most of the times, we consider $\boldsymbol{\Lambda}_\mathbf{i} = \sigma^2 \boldsymbol{I}$, being σ^2 a constant variance for all j times.

By using nonlinear mixed effect models, we must consider the technical difficulty in the parameter estimation. In a mixed effects linear model, the derivative of the logarithmic of the likelihood function allows, in a simple way, the algorithm implementation, like Newton-Raphson, to obtain the estimative of the models parameters. Nonlinear models can, however, present nonlinear random coefficients which make it impossible to directly explain the parameters from the likelihood function. Methods that depend on linear approximations such as the first-order Taylor approximation can be used to estimate the model.

Nonlinear mixed effect model analysis can be, preferably, made by the R software [13] with the package name [14], also, at the SAS software using PROC NLMIXED. An excellent text to learn how to use these skills is given in [4].

6. Covariance structure of $\boldsymbol{\Lambda}_\mathbf{i}$

Mixed effect model allows the dependence between the observations to be specified in the model parameters through random effects. In other words, the experimental unit responses from a population tend to follow a nonlinear growth path; however, each experimental unit has its own growth path, and the mixed effect model allows the inclusion of specific coefficients to obtain fitted growth curves that align better with the individual responses of these experimental units.

Thus, mixed models allow relevant flexibility for the specification of the random effects correlation structure. However, the dependence structure of the observations within the experimental units $\boldsymbol{\Lambda}_\mathbf{i}$ until now has been considered independent, identically distributed with mean zero and constant variance. Depending on the chosen model, the growth responses can be explained just by including specific coefficients for the experimental units. However, this may not be enough, and, in this case, modeling the residual dependence of the data becomes important.

There are cases where dependence on observations not accommodated by the growth function is not well understood or, sometimes, additional covariables that could explain this dependences are absent from the model. Thus, an important resource to model this dependence is to identify the covariance structure that allows correlation between the residuals in different occasions. Then, let us relax on the assumption that the errors are independent and allow them to have heteroscedasticity and/or are correlated within the experimental units.

There are several covariance structures for the residues available in the software to help model longitudinal data. However, in our text, we will highlight only two that we consider more important for these studies, the covariance structure with heterogeneous variance and the first-order auto-regressive.

6.1 Heterogeneous variance

The first covariance structure we will consider for $\mathbf{\Lambda_i}$ is the one that admits heterogeneity of the n_i residual variances. In this structure, which has n_i parameters, we assume that the residuals associated with the observed values at the n_i occasions for the i-th experimental unit are independent:

$$\mathbf{\Lambda_i} = \begin{bmatrix} \sigma_1^2 & 0 & \cdots & 0 \\ 0 & \sigma_2^2 & \cdots & 0 \\ \vdots & \vdots & \ddots & \vdots \\ 0 & 0 & \cdots & \sigma_{n_i}^2 \end{bmatrix}. \tag{7}$$

Other variables, besides time, can also be considered with heterogeneous variance in the model. For example, there are cases in which it is important to model the heterogeneity of the treatments, and we can do it by using mixed models.

6.2 First-order auto-regressive

Another covariance structure for $\mathbf{\Lambda_i}$ which is widely used for longitudinal data is the first-order auto-regressive, also called $AR(1)$:

$$\mathbf{\Lambda_i} = \sigma^2 \begin{bmatrix} 1 & \rho & \cdots & \rho^{n_i-1} \\ \rho & 1 & \cdots & \rho^{n_i-2} \\ \vdots & \vdots & \ddots & \vdots \\ \rho^{n_i-1} & \rho^{n_i-2} & \cdots & 1 \end{bmatrix}. \tag{8}$$

This structure has only two parameters, the variance parameter σ^2, always positive, and the covariance parameter ρ, which may vary between -1 and 1. This kind of structure allows the residues associated with the observations in neighboring occasions to be more correlated than those whose observations are further apart. The $AR(1)$ is preferred for datasets in which the longitudinal observations are equally spaced.

7. Real data example

To exemplify what we have done so far, let's work with some real data of the ET from lettuce plants grown in pots. A total of $N = 12$ were completely randomized into three levels of water in the soil. At the first treatment, W_1, the water level for the plant were kept between 50.0 and 75.0% of the substrate's retention capacity. In the other two treatments, W_2 and W_3, the water level in the substrate was kept between 50.0 and 87.5% and between 50.0 and 100.0%, respectively. When the retention capacity of the substrate reached 50.0%, the pots were irrigated until their maximum level regarding each treatment.

The profile graphs from the accumulated ET for all pots in each treatment are shown in **Figure 4**. Note that the inferior asymptote, when $t \to -\infty$, is apparently

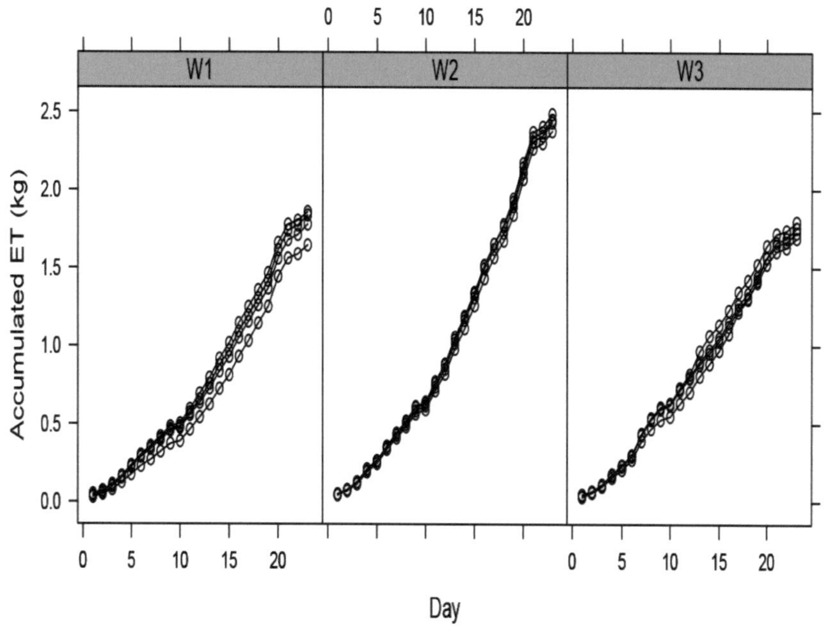

Figure 4.
Response profile for the accumulated ET of the lettuce plants over 23 consecutive days for the soil water levels W_1, W_2, and W_3.

Parameters	$\hat{\beta}$ (W_1)	$\hat{\gamma}$ (W_2)	$\hat{\delta}$ (W_3)
ϕ_1	−0.133363	—	—
ϕ_2	2.439746	0.935210	−0.272847
ϕ_3	16.648272	0.405882^{NS}	−2.302124
ϕ_4	5.765065	—	—

The only nonsignificant parameter (^{NS}) was ϕ_{3i} for the treatment W_2 with p-value >0.34. The other parameters presented p-value < 0.001.

Table 1.
Estimative for the models fixed effects parameters.

the same for all plants, i.e., the individual and treatment effects do not seem to be important for this parameter.

We model this data using Eq. (2) and considering treatment W_1 as baseline. The random effects were added to the parameters ϕ_{2i}, ϕ_{3i}, and ϕ_{4i}, and the treatments were important to explain the parameters ϕ_{2i} e ϕ_{3i}. Besides that, we consider the heterogeneity in the treatments and an AR(1) correlation structure for $\boldsymbol{\Lambda}_\mathbf{i}$. The parameter estimative of the model is presented in **Table 1**.

The parameter ϕ_{2i} seems to be influenced by the other treatments, and its value was estimated in ~ 2.44 kg for W_1, $\sim 2.44 + 0.94 = 3.38$ kg for W_2, and $\sim 2.44 + (-0.27) = 2.17$ kg for W_3. The inflection point, i.e., the day the accumulated ET rate was maximum, also seems to be influenced by the treatments. W_1 and W_2 were not statistically different for the parameter ϕ_{3i}, but W_3, with estimative of ~ 14 days, appears to be statistically different from W_1 (~ 17 days).

The first graph presented in **Figure 5** brings the accumulated ET mean in each day of the four plants in each treatment. The solid lines are fitting for the treatments

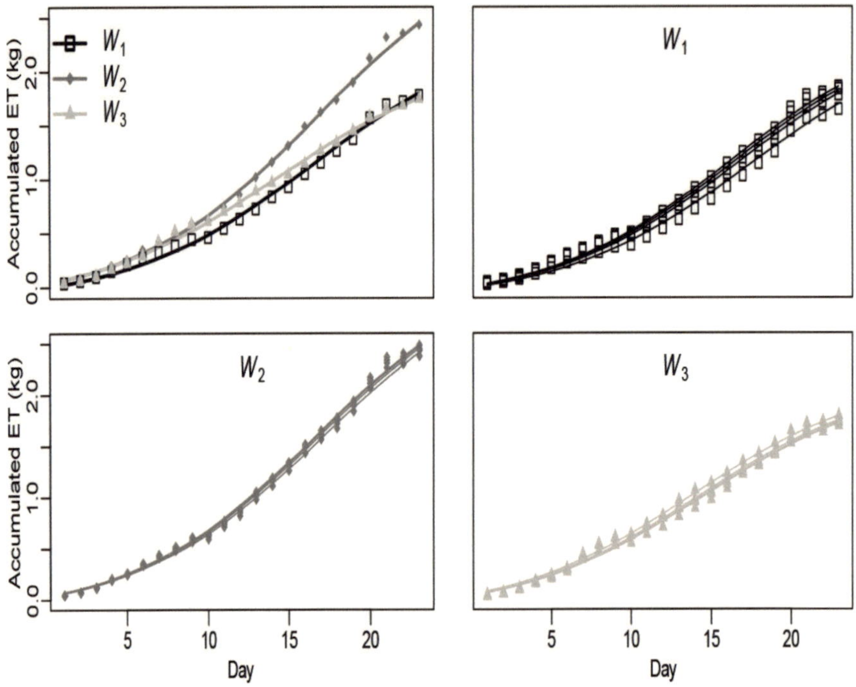

Figure 5.
Response profile of the accumulated ET by the lettuce plants over 23 consecutive days for the water soil levels W_1, W_2, and W_3.

made by Eq. (2). The other graphs present all values observed for the four plants in each treatment, and the solid lines indicate the individual model for each plant.

Author details

Omar Cléo Neves Pereira[1*†] and Altair Bertonha[2†]

1 Department of Statistics, State University of Maringá, Maringá, Brazil

2 Department of Agronomy, State University of Maringá, Maringá, Brazil

*Address all correspondence to: omarcnpereira@gmail.com

† These authors contributed equally.

References

[1] Novk V. Evapotranspiration in the Soil-Plant-Atmosphere System. New York: Springer Science & Business Media; 2012

[2] Pereira OCN, Pereira PVC, Bertonha A, Previdelli ITS. Longitudinal data analysis of stevia rebaudiana evapotranspiration according to water levels. Revista Brasileira de Biometria. 2018;**36**(4):791-801

[3] Pereira OCN, Suguiura TPDS, Pereira AP, Bertonha A, Previdelli I. Analysis of lettuce evapotranspiration across soil water. Natural Resource Modeling. 2019;**32**(2):e12197

[4] Pinheiro JC, Bates DM. Mixed-Effects Models in S and S-PLUS. New York: Springer Science & Business Media; 2000

[5] Brouwer R, De Wit C. A Simulation Model of Plant Growth with Special Attention to Root Growth and its Consequences. In: Proceedings of the 15th Easter School in Agricultural Science. University of Nottingham. Technical Report. IBS; 1968. pp. 224-242

[6] Hansen LD, Hopkin MS, Rank DR, Anekonda TS, Breidenbach RW, Criddle RS. The relation between plant growth and respiration: A thermodynamic model. Planta. 1994;**194**(1):77-85

[7] Paine CT, Marthews TR, Vogt DR, Purves D, Rees M, Hector A, et al. How to nonlinear plant growth models and calculate growth rates: An update for ecologists. Methods in Ecology and Evolution. 2012;**3**(2):245-256

[8] Yan H-P, Kang MZ, De Reffye P, Dingkuhn M. A dynamic, architectural plant model simulating resource-dependent growth. Annals of Botany. 2004;**93**(5):591-602

[9] Brown H, Prescott R. Applied Mixed Models in Medicine. New York: John Wiley & Sons; 2014

[10] Demidenko E. Mixed Models: Theory and Applications with R. New York: John Wiley & Sons; 2013

[11] Fitzmaurice G, Davidian M, Verbeke G, Molenberghs G. Longitudinal Data Analysis. London: CRC Press; 2008

[12] West BT, Welch KB, Galecki AT. Linear Mixed Models: A Practical Guide Using Statistical Software. New York: Chapman and Hall/CRC; 2014

[13] R Core Team. R: A language and environment for statistical computing. In: R Foundation for Statistical Computing. Vienna, Austria; 2018

[14] Pinheiro J, Bates D, DebRoy S, Sarkar D, and R Core Team. Nlme: Linear and nonlinear mixed effects models. R Package Version 3. CRAN Repository. 2017:1-131

Section 2

Wastewater Treatment

Chapter 5

Sustainability Assessment of Wastewater Treatment Plants

Başak Kiliç Taşeli

Abstract

It is thought that this chapter will make a significant contribution to the literature or at least will fill the space on the wastewater treatment plant's effect on climate change. It demonstrates the potential climate change impact of a sequential batch reactor (SBR) and constructed wetland on treating domestic wastewater by giving methods for calculation of their greenhouse gas emissions in terms of N_2O and CH_4. Are wastewater treatment plants sustainable? What aspects determine sustainability? Do tertiary wastewater treatment plants and constructed wetlands (CWs) have less global warming potential (CO_2 emissions) and less energy use than conventional treatment? In accordance with the literature, greenhouse gas calculations of this study showed that CWs and SBR WWTPs do not contribute to global warming negatively.

Keywords: wastewater treatment, sequential batch reactor, greenhouse gas, constructed wetlands, methane, nitrous oxide, sustainability

1. Introduction

Wastewater treatment plants are generally capable of reaching hygienic and environmental standards; however, these were not designed for zero discharge principle in which nutrients, organic matter, and water are recycled and nutrient, organic and water cycles are closed. Are wastewater treatment plants (WWTPs) sustainable? What aspects determine sustainability? Do tertiary wastewater treatment plants and constructed wetlands (CWs) have less global warming potential (CO_2 emissions) and less energy use than conventional treatment?

Since sequential batch reactor (SBR) system sequentially removes carbon, nitrogen, and phosphorous in a single reactor by maintaining anoxic and aerobic stages, it recently has attracted a great deal of interest. High nitrogen and phosphorus removal are achieved by a series of steps, namely, fill, react, settle, draw, and idle steps, as shown in **Figure 1**. Denitrification occurs at the beginning of the fill step taking usually 25% of the total cycling time where raw wastewater is added to the reactor. The step taking up 35% of the total cycle time is called react step where the reactions were finalized. The main purpose of the third step (settle) is to allow solid separation and provide a supernatant ready to be discharged as effluent. The purpose of the fourth step (draw step) ranging from 5 to 30% of the total cycle time is to remove clarified treated water from the reactor. The purpose of last step, "idle," is to provide time for one reactor to complete its fill cycle before switching to another unit.

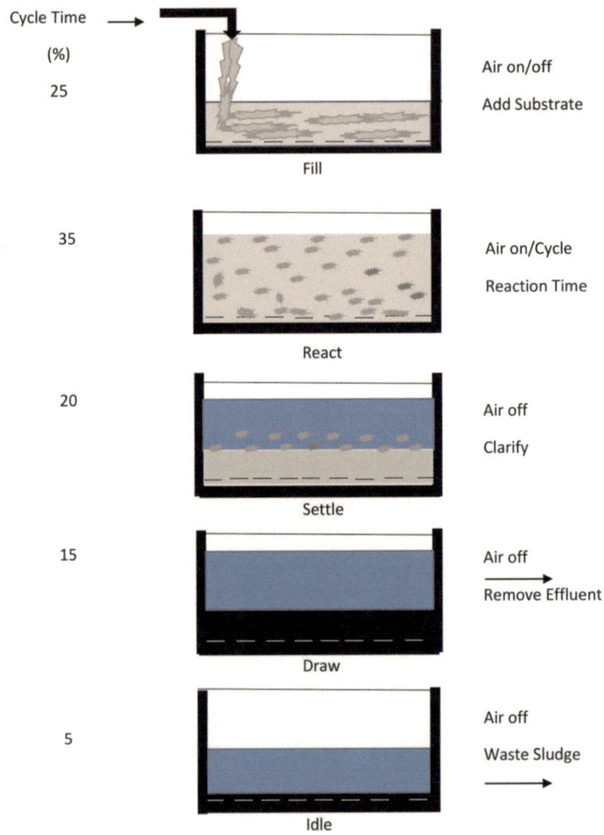

Figure 1.
Operation sequence for sequential batch reactor [1].

In wastewater treatment technologies like activated sludge, membrane methods are not feasible enough for widespread application in rural areas [2]. Constructed wetlands however are attracting great concern due to lower cost, easy operation, and less maintenance requirements as a reasonable option for treating wastewater in rural areas. They are designed and constructed to mimic natural wetland systems for removing contaminants which are basically composed of vegetation, substrates, soils, microorganisms, and water, utilizing complex processes involving physical, chemical, and biological mechanisms (e.g., sedimentation, filtration, precipitation, volatilization, adsorption, plant uptake, and various microbial processes) [3].

While the treatment performance of CWs is critically dependent on the optimal operating parameters (water depth, hydraulic retention time and load, feeding mode and design of setups, etc.) which could result in variations in the removal efficiency of contaminants, plant species and media types are crucial influencing factors for the treatment in CWs as they are considered to be the main biological component of CWs. Emergent, submerged, floating-leaved, and free-floating plants are commonly planted among 150 macrophyte species. The most common used emergent species reported are *Phragmites* spp. (Poaceae), *Typha* spp. (Typhaceae), *Scirpus* spp. (Cyperaceae), *Iris* spp. (Iridaceae), *Juncus* spp. (Juncaceae), and *Eleocharis* spp. (Spikerush) [3].

The greenhouse effect of major greenhouse gases, carbon dioxide (CO_2), methane (CH_4), and nitrous oxide (N_2O) all produced in wastewater treatment operations, is weighted by their global warming potentials (GWP). Over a period of 100 years, 1 ton of methane and nitrous oxide will have a warming effect equivalent to 25 and 298 ton of CO_2, respectively [4]. In the same direction, it is stated that nitrous oxide is a significant greenhouse gas with a lifetime of 114 years with a 298-fold stronger effect of global warming than carbon dioxide and is also responsible for ozone depletion in the stratosphere [5].

In SBR processes, ammonium is transformed into N_2 gas via nitrification and denitrification. N_2O is generated as a by-product or an intermediate due to insufficient oxygen during nitrification in the aeration step and due to insufficient carbon during denitrification in settling and decanting steps [6]. Wastewater treatment facilities are anthropogenic sources of N_2O to the atmosphere, taking account of 3.2–10% of the total emission [7]. Practically, only methane and nitrous oxide are calculated since carbon that is present in wastewater is biogenic and it is assumed that it is returning the carbon to the atmosphere as CO_2 representing no net flux to the system [8].

Based on field measurements, the maximum methane flux occurred in sludge screw conveyor with 823 $g/m^2/d$, and CH_4 emission occurred in every processing unit [9]. Methane is produced by methanogens due to low O_2 and nitrate/nitrite concentration during the anaerobic and anoxic processes. In the same direction, more than 50% of global methane emissions are related to human-related activities like landfill, wastewater treatment, agriculture, and certain industrial process [10].

Are constructed wetlands sustainable? It is reported that constructed wetlands have less global warming potential (CO_2 emissions) and less energy use than conventional treatment [11]. Wetlands also reduced aquatic toxicity and eutrophication compared to conventional activated sludge wastewater treatment [12]. A sustainable solution means minimized costs; minimized energy use; minimized land area required; minimized loss of nutrients; minimized waste production; maximized products like clean water, biogas, biomass, fertilizers, and compost; and maximized qualitative sustainability indicators like social acceptance, institutional requirements, etc. But it is not always possible to design a wastewater treatment that minimizes cost, energy use, and land area, while maximizing performance.

The greenhouse gas emissions measured in N_2O, CO_2, and CH_4 for horizontal flow constructed wetlands (HFCWs) were 3, 1400 and 5 $mg/m^2/d$, respectively [13]. Moreover, vertical flow constructed wetlands (VFCWs) had significantly higher areal gaseous emissions than HFCWs, and gas emissions were correlated to temperature, substrate supply (influent N and C concentrations), and degree of oxidation in the wetland [14]. The quantity and impact of CH_4 and N_2O are important since CH_4 has 25 times and N_2O has 298 times the global warming potential of CO_2 [4].

The CO_2, CH_4, N_2, and N_2O fluxes in both horizontal and vertical subsurface flow constructed wetlands in Estonia were measured and reported that the global influence of constructed wetlands is not significant, that is, even if all global domestic wastewater were treated by constructed wetlands, the emitted GHG would be <1% of total anthropogenic emissions [15]. They also reported the averaged experimental data of 788.33 mg CO_2/m^2 h, 4 mg CH_4/m^2 h, and 0.79 mg N_2O/m^2 h.

The following section will represent CH_4, N_2O, and CO_2 emission calculating principles for both SBR and CW WWTPs.

2. Emission calculating principles for wastewater treatment plants

2.1 Methane (CH_4) emission calculating principles

Estimation of organically degradable material in domestic wastewater, estimation of methane emission factor (EF) for domestic wastewater, and estimation of CH_4 emissions from domestic wastewater are steps for calculating CH_4 emissions.

The direct methane emissions are a function of the amount of degradable carbon in the wastewater and sludge and an emission factor. As can be seen from Eq. (1) (Eq. (1): Eq. 6.2 of IPCC, 2006:CH_4 emission factor for each domestic wastewater treatment/discharge pathway or system), the emission factor is a function of the maximum CH_4 producing potential (B_o) and the methane correction factor (MCF). The B_o value of 0.6 kg CH_4/kg BOD removal, the uncertainty range of ±30%, and the MCF value of 0.05 are recommended [4].

Total organics in wastewater (TOW) (in inventory year, kg BOD/year) is a function of human population and BOD generation per person, and it is expressed in terms of biochemical oxygen demand (kg BOD/year). TOW was calculated by using Eq. (2) (Eq. (2): Eq. 6.3 of IPCC, 2006: total organically degradable material in domestic wastewater) [4].

$$\text{CH}_4 \text{ emission factor for each domestic wastewater treatment/discharge pathway or system } (\text{EF}_j) = \text{B}_0 \bullet \text{MCF}_j \quad (1)$$

where F_j emission factor, kg CH_4/kg BOD; j each treatment/discharge pathway or system; B_0 maximum CH_4 producing capacity, kg CH_4/kg BOD; MCF_j methane correction factor (fraction).

$$\text{Total organically degradable material in domestic wastewater (TOW)} = \text{P} \bullet \text{BOD} \bullet 0.001 \bullet \text{I} \bullet 365 \quad (2)$$

where TOW, total organics in wastewater in inventory year, kg BOD/year; P, country population in inventory year (person); BOD, country-specific per capita BOD in inventory year, g/person/day; 0.001, conversion from gram BOD to kg BOD, I, correction factor for additional industrial BOD discharged into sewers (for collected the default is 1.25; for uncollected the default is 1.00).

The general equation for estimating CH_4 emissions from domestic wastewater was calculated by using Eq. (3) (Eq. (3): Eq. 6.1 of IPCC, 2006: total CH_4 emissions from domestic wastewater).

$$\text{Total CH}_4 \text{ emissions from domestic wastewater } (\text{CH}_4 \text{ emissions}) = \left[\sum_{ij} (U_i \cdot T_{ij} \cdot EF_j) \bullet (\text{TOW} - \text{S}) - \text{R}\right. \quad (3)$$

where CH_4 emissions, CH_4 emissions in inventory year, kg CH_4/year; TOW, total organic wastewater in inventory year, kg BOD/year; EF_j, emission factor, kg CH_4/kg BOD; S, organic component removed as sludge in inventory year, kg BOD/year; U_i, fraction of population in income group *i* in inventory year; $T_{i,j}$, degree of utilization of treatment/discharge pathway or system, *j*, for each income group fraction *i* in inventory year; i, income group: rural, urban high income and urban low income; j, each treatment/discharge pathway or system; R, amount of CH_4 recovered in inventory year, kg CH_4/year.

2.2 Nitrous oxide (N_2O) emission calculating principles

Estimation of nitrogen in effluent, estimation of emission factor, and emissions of indirect N_2O emissions from wastewater are steps for calculating N_2O emissions. It is associated with the microbial conversion of nitrogen compound in the wastewater. It occurs as emissions from treatment plants or from wastewater after disposal of effluent into waterways, lakes, or the sea. The emission factor (0.005) is taken for domestic wastewater nitrogen effluent, referring to the default value recommended by IPCC [4]. The factor 44/28 is the conversion of kg N_2O-N into kg N_2O. A simplified equation is given in Eq. (5). Emission factors of N_2O were evaluated by incorporating N loads in influent of the SBR WWTP.

$$\text{Total nitrogen in the effluent} \; (N_{EFFLUENT}) = (P \bullet \text{Protein} \bullet F_{NPR} \bullet F_{NON-CON} \bullet F_{IND-COM}) - N_{SLUDGE} \tag{4}$$

where $N_{EFFLUENT}$, total annual amount of nitrogen in the wastewater effluent, kg N/year; P, human population; Protein, annual per capita protein consumption, kg/person/year; F_{NPR}, fraction of nitrogen in protein, default = 0.16, kg N/kg protein; $F_{NON\text{-}CON}$, factor for non-consumed protein added to the wastewater; $F_{IND\text{-}COM}$, factor for industrial and commercial co-discharged protein into the sewer system; N_{SLUDGE}, nitrogen removed with sludge (default = zero), kg N/year.

$$N_2O \text{ emissions from wastewater effluent} \; (N_2O \text{ emissions}) = N_{EFFLUENT} \bullet EF_{EFFLUENT} \bullet 44/28 \tag{5}$$

where N_2O emissions, N_2O emissions in inventory year, kg N_2O/year; $N_{EFFLUENT}$, nitrogen in the effluent discharged to aquatic environments, kg N/year, $EF_{EFFLUENT}$, emission factor for N_2O emissions from discharged to wastewater, kg N_2O-N/kg N; 44/28, the factor 44/28 is the conversion kg N_2O-N into kg N_2O.

2.3 Carbon dioxide (CO_2) emission calculating principles

The two main factors causing CO_2 production from wastewater treatment plants are the type of treatment process and electricity consumption. During anaerobic treatment, the BOD_5 in the wastewater is either converted to CO_2 or CH_4, or some of it enters the biomass and is also converted to CO_2 and CH_4 by endogenous respiration. Other sources of carbon dioxide emissions are caused by sludge digesters and digestion gas combustion. In aerobic process, CO_2 is produced by decomposition of organic substances. Since CO_2 emissions from the wastewater treatment plant are biogenic, they are not included in the national total emissions and are not considered in the IPCC Guidelines. Biogenic origin means that it is part of the natural carbon cycle and the food chain passing from plants to animals and humans, or natural atmospheric CO_2 source.

3. Emission calculating principles for constructed wetlands

3.1 Methane (CH_4) emission calculating principles

The direct methane emissions are the function of the amount of degradable carbon in the wastewater and sludge, and an emission factor. The emission factor is a function of the maximum CH_4 producing potential (B_o) and the methane correction factor (MCF) for the wastewater treatment and discharge system. The B_o value of 0.6 kg CH_4/kg BOD removal and the uncertainty range of ±30% is recommended by IPCC [16]. The MCF indicates that the extent to which the CH_4 producing capacity is realized in each type of treatment and discharge pathway and system. The CH_4 emissions from constructed wetlands and CH_4 emission factors for constructed wetlands are given in Eq. (6) (Eq. (6): Eq. 6.1 of IPCC, 2014: CH_4

emissions from constructed wetlands) and Eq. (7) (Eq. (7): Eq. 6.2 of IPCC, 2014: CH_4 emission factor for constructed wetlands), respectively.

$$\text{CH}_4 \text{ emissions from constructed wetlands}$$
$$(\text{CH}_4 \text{ emissions}) = \sum_j (TOW_j \cdot EF_j) + \sum_{ij} (TOW_{ij} \cdot EF_j) \tag{6}$$

where CH_4 emissions, CH_4 emissions in inventory year, kg CH_4/year; TOW_j, total organics in wastewater entering CW in inventory year, kg BOD/year or kg COD/year; EF_j, emission factor, kg CH_4/kg BOD (for domestic wastewater only) or kg CH_4/kg COD (for domestic and industrial wastewater). If more than one type of CW is used in an industrial sector, this factor would need to be a TOW_{ij} weighted average: i, industrial sector; j, type of CW.

$$\text{CH}_4 \text{ emission factor for constructed wetlands } (\text{EF}_\text{j}) = \text{B}_0 \bullet \text{MCF}_\text{j} \tag{7}$$

where EF_j, emission factor, kg CH_4/kg BOD or kg CH_4/kg COD; j, type of CWs; B_0, maximum CH_4 producing capacity, kg CH_4/kg BOD or kg CH_4/kg COD; MCF_j, methane correction factor (fraction).

3.2 Nitrous oxide (N_2O) emission calculating principles

Nitrogen oxides are associated with microbial conversion of nitrogen compound in wastewater and emerge as emissions from wastewater discharge to waterways, lakes, or seas and treatment plants or wastewater. The emission factor (0.005) is taken for domestic wastewater nitrogen effluent, referring to the default value recommended by IPCC [16]. The factor 44/28 is the conversion of kg N_2O-N into kg N_2O. A simplified equation is given in Eq. (8).

$$\text{N}_2\text{O emissions from constructed wetlands}$$
$$(\text{N}_2\text{O emissions}) = \sum_j (N_j \cdot EF_j \cdot 44/28) + \sum_{i,j} (N_{i,j} \cdot EF_j \cdot 44/28) \tag{8}$$

where N_2O emissions, N_2O emissions in inventory year, kg N_2O/year; N_j, total nitrogen in domestic wastewater entering CWs in the inventory year, kgN/year; $N_{i,j}$, total nitrogen in industrial wastewater entering CWs in the inventory year, kgN/year; EF_j, emission factor, kg N_2O-N/kg N. If more than one type of CW is used in an industrial sector, this factor would need to be a $N_{i,j}$ weighted average: i, industrial sector; j, type of CW; 44/28, the factor 44/28 is the conversion of kg N_2O-N into kg N_2O.

4. Results

4.1 SBR wastewater treatment plant's GHG emissions

Eqs. (1)–(3) and default maximum CH_4 producing capacity for domestic wastewater of 0.6 kg CH_4/kg BOD and 0.25 kg CH_4/kg COD and ($T_{i,j}$) values given in Table 6.5 of IPCC, "Suggested values for urbanization and degree of utilization of treatment discharge pathway or method ($T_{i,j}$) for each income group for selected countries," are used for the calculation of methane emissions from SBR WWTP [4].

Eqs. (4) and (5) and the values given in **Table 1** are used for N_2O emission calculations. Finally, indirect GHG emissions from the consumption of electricity are calculated by the use of emission factor of 0.91 tCO_2e Mwh^{-1} [4]. Total emissions of full-scale SBR WWTP operated from 2012 to 2015 are given in **Table 2** [17].

4.2 Horizontal subsurface constructed wetland's GHG emissions

Eight-month average (April–December) methane and nitrous oxide emissions at the outlet of full-scale constructed wetland were calculated by using Eqs. (6)–(8). **Table 3** gives the calculated GHG emissions in terms of CH_4 and N_2O [18].

	Definition	Default value	Range
$EF_{EFFLUENT}$	Emission factor (kg N_2O-N/kg-N)	0.005	0.0005–0.25
P	Number of people in country	Country specific	±10%
Protein	Annual per capita protein consumption	Country specific	±10%
F_{NPR}	Fraction of nitrogen in protein	0.16	0.15–0.17
$F_{NON\text{-}CON}$	Non-consumed protein adjustment factor	1.1 for countries with no garbage disposals1.4 for countries with garbage disposals	1.0–1.5
$F_{IND\text{-}COM}$	Co-discharge factor for industrial nitrogen into sewers. Higher for countries with significant fish processing plants	1.25	1.0–1.5

Table 1.
N_2O methodology default data [4].

Year	CH_4 (tCO_2e)	N_2O (tCO_2e)	Electricity usage (tCO_2e Mwh^{-1})	Total (tCO_2e)
2012	74.87	0.0143	69.34	144.22
2013	248.99	0.0143	69.34	318.34
2014	87.68	0.0143	387.1	474.79
2015	68.41	0.0143	928.2	996.62

Table 2.
Total emissions of the SBR wastewater treatment plant.

Parameter (kg/d)	2012	2013	2014	2015
BOD	312	406	292	224
TN	45	67	23	346
CH_4 emission	18.72	24.36	17.52	13.44
N_2O emission	0.56	0.83	0.29	4.30

Table 3.
Horizontal subsurface constructed wetland's GHG emissions.

5. Conclusions

This review chapter demonstrates the potential climate change impact of a sequential batch reactor and constructed wetland treating domestic wastewater by giving methods for calculation of their greenhouse gas emissions in terms of N_2O and CH_4.

If methane is to be recovered for energy use, the net emission of methane should be calculated by subtracting the recovered and flared amount of methane from the gross methane emission. In other words, methane that is not released is calculated as the amount used for biogas (and is thus included in CO_2 emissions from energy production).

Author details

Başak Kiliç Taşeli
Environmental Engineering Department, Faculty of Engineering, Giresun University, Giresun, Turkey

*Address all correspondence to: basak.taseli@giresun.edu.tr

References

[1] Available from: https://www.indiamart.com/hydro-drops/technologies-offered.html [Accessed: 14 May 2019]

[2] Chen Y, Wen Y, Zhou Q, Vymazal J. Effects of plant biomass on denitrifying genes in subsurface-flow constructed wetlands. Bioresource Technology. 2014;**157**:341-345

[3] Vymazal J. Plants used in constructed wetlands with horizontal subsurface flow: A review. Hydrobiologia. 2011; **674**:133-156

[4] IPCC. Guidelines for national greenhouse gas inventories. In: Eggleston HS, Buendia L, Miwa K, Ngara T, Tanabe K, editors. The National Greenhouse gas Inventories Programme, the Intergovernmental Panel on Climate Change. Switzerland: The Intergovernmental Panel on Climate Change; 2006. ISBN: 978-92-9169-139-5

[5] Ravishankara AR, Daniel JS, Portmann RW. Nitrous oxide (N_2O): The dominant ozone-depleting substance emitted in the 21st century. Science. 2009;**326**:123-125

[6] Itokawa H, Hanaki K, Matsuo T. Nitrous oxide production in high-loading biological nitrogen removal process under low COD/N ratio condition. Water Resources. 2001;**35**: 657-664

[7] Law YY, Ye L, Pan YT, Yuan ZG. Nitrous oxide emissions from wastewater treatment processes. Philosophical Transactions of the Royal Society B. 2012;**367**:1265-1277

[8] Michiel RJ Doorn, Sirintornthep Towprayoon, Sonia Maria Manso Vieira, William Irving, Craig Palmer, Riitta Pipatti et.al. IPCC. Guidelines for national greenhouse gas Inventories. In: Chapter 6: Wastewater Treatment and Discharge. Switzerland: The Intergovernmental Panel on Climate Change; 2006. pp. 6.1-6.28

[9] Wang J, Zhang J, Xie H, Qi P, Ren Y, Hua Z. Methane emissions from a full-scale A/A/O wastewater treatment plant. Bioresource Technology. 2011; **102**:5479-5485

[10] IPCC. Metz B, Davidson OR, Bosch PR, Dave R, Meyer LA. Summary for policymakers. In: Climate change 2007: Mitigation; Contribution of Working Group III to the Fourth Assessment Report of the Intergovernmental Panel on Climate Change. Cambridge University Press; 2007

[11] Dixon A, Simon M, Burkitt T. Assessing the environmental impact of two options for small-scale wastewater treatment: Comparing a reedbed and an aerated biological filter using a life cycle approach. Ecological Engineering. 2003; **20**:297-308

[12] Machado AP, Urbano L, Brito AG, Janknecht P, Salas JJ, Nogueira R. Life cycle assessment of wastewater treatment options for small and decentralized communities: Energy-saving systems versus activated sludge. Water Science and Technology. 2007; **56**(3):15-22

[13] Maltais-Landry G, Maranger R, Brisson J. Effect of artificial aeration and macrophyte species on nitrogen cycling and gas flux in constructed wetlands. Ecological Engineering. 2009; **35**:221-229

[14] Sovik AK, Augustin J, Heikkinen K, Huttunen JK, Necki JM, Karjalainen SM, et al. Nitrous oxide and methane from constructed wetlands in Europe. Journal of Environmental Quality. 2006;**35**: 2360-2373

[15] Teiter S, Mander Ü. Emission of N_2O, N_2, CH_4, and CO_2 from constructed wetlands for wastewater treatment and from riparian buffer zones. Ecological Engineering. 2005; **25**(5):528-541

[16] IPCC: 2013 Supplement to the 2006 IPCC Guidelines for National Greenhouse Gas Inventories: Wetlands Hiraishi, T, Krug T, Tanabe K, Srivastava N, Baasansuren J, Fukuda M, Troxler TG editors. IPCC, Switzerland, 2014

[17] Taşeli BK. Point source pollution and climate change impact from sequential batch reactor wastewater treatment plant. Global NEST Journal. 2018;**20**(1):33-41

[18] Taşeli BK. Greenhouse gas emissions from a horizontal subsurface flow constructed wetland for wastewater treatment in small villages. Global NEST Journal, (under revision)

Chapter 6

The Use of Industrial Waste for the Bioremediation of Water Used in Industrial Processes

Rosa Hernández-Soto, José A. Hernández, Alba N. Ardila-Arias, Mercedes Salazar-Hernández and María del Carmen Salazar-Hernandeza

Abstract

Recently the interest in the remediation of liquid effluents from industries such as paint manufacturing, leather tanning, etc. has increased, because the quality of the water used in these processes is highly compromised and is generally discarded without any process of purification, causing an inadequate use of water and contributing to the hydric stress of the planet. Therefore, it is necessary to find alternatives for the remediation of water used in industrial processes; one of the methods that has been widely accepted given its high efficiency, low cost, and versatility compared to others is the bioadsorption using materials derived from various processes used for the elimination of metals such as Cr, Co, Cu, Ni, etc. from liquid effluents. Among the materials used for this purpose are rice husk, orange, and wheat as well as apatite (hydroxyapatite and brushite), derived from animal bones, which have shown good capacity (>90%) to adsorb metals from aqueous solutions. Through the characterization by DRX, FTIR, and SEM, of the brushite and studies in equilibrium and kinetics of adsorption, it has been demonstrated that this material has a good capacity to remove metals present in water.

Keywords: brushite, removal, isotherm, kinetics, metals

1. Introduction

The contamination of water bodies due to the presence of heavy metals is a serious problem, because every day this vital resource is scarce and because of the high toxicity of these compounds for the health of living beings. The metals present in water are a risk factor for the development of diseases such as cancer and dermatitis; in addition, they may be accumulated in the human body because they cannot be metabolized [1–7]. Solid-liquid removal processes such as chemical precipitation, filtration, and adsorption, among others, have been widely used for the removal of metals such as nickel (Ni), iron (Fe), copper (Cu), zinc (Zn), cobalt (Co), and chromium (Cr) of liquid effluents [8–11]; however, some of these methods have disadvantages such as high operating cost and low efficiency; however, methods such as coagulation and precipitation are already used in various industrial processes for the removal of metals from industrial effluents [12]. In recent years, adsorption has been one of the most used metal ion removal techniques, since it is a simple, effective, and inexpensive process compared to other methods [13–17]. Adsorption processes have been experimented with an extensive amount of materials such as adsorbents, among which activated carbon stands out due to its high capacity for capturing metal ions; however, this material has the disadvantage of generating large quantities of sludge, since the removal of metals trapped in activated carbon can only be done with processes that are often expensive such as leaching [5, 9, 18, 19]. For this reason, the use of different materials that are economical, are easy to obtain, and have high efficiency in the removal of metal ions has been investigated. In recent decades, these studies have focused on the waste derived from the agricultural industry that produces large amounts of waste such as biomass, wheat husks, rice, orange, etc. [2, 4, 8, 9, 16, 18–30]; the use of residues from other industries has also been investigated, such as the case of apatites derived from the bone tissue of animals, which have been used for removal of dyes and metal ions obtaining promising results. The use of apatites in particular hydroxyapatite and brushite for the adsorption of heavy metals such as Cd, Cu, Ni, Pb, Co, Mn, and Fe, to name a few, has already been reported [31–35]; however, in most of the studies carried out, only the process of adsorption of metallic solutions of a single component has been analyzed, so the objective of the present work is to evaluate the capacity of brushite (nDCPD), obtained from bovine bone to remove Ni (II), Co (II), and Cu (II) of aqueous solutions, analyzing the selectivity of removal of metal ions in aqueous solutions with two or three different metals, determining the kinetic models and in equilibrium in which the removal of metals takes place and the structural changes suffered by nDCPD during the development of the different tests.

2. Experimental

2.1 Reagents

All the reagents that were used were of analytical grade. The water used for the preparation of solutions in the experiment was deionized. The solutions to be evaluated were prepared by dissolution of salts of nickel ($Ni(NO_3)_2\ 6H_2O$), cobalt ($Co(NO_3)_2{\cdot}6H_2O$), and copper ($Cu(NO_3)_2$) in concentrations from 0 to 1,000 ppm.

2.2 Preparation of the adsorbent

Brushite natural (nDCPD) was obtained from bovine bone, which was washed with hot water several times to remove tissue debris, and then it was dried at 353 K for 24 h. Next, the bones were crushed and sieved to obtain a particle size of 150 mesh (104 μm). Then, the powder obtained was treated with HCl and NaOH solution, 10^{-2} M, respectively, using a ratio of 30% w/v. Finally nDCPD was stored until its use [1].

2.3 Characterization of natural brushite

The X-ray diffraction patterns (XRD) were obtained in a Rigaku diffractometer (Ultima IV). The Fourier transform infrared studies of the samples were performed in an IR 100 Analyzer spectrophotometer (PerkinElmer), in a range of 400–4000 cm^{-1}. The scanning electron microscopy (SEM) images were obtained in a JOEL equipment (6510-Plus).

2.4 Co, Cu, and Ni adsorption isotherm

The different isotherm models used to describe the adsorption of Ni, Co, and Cu are concentrated in **Table 1**, for each one of the regression coefficients was calculated to evaluate the adjustment of each nonlinear model and the separation factor, R_L, which allows to predict the affinity between the adsorbent and adsorbate, using Eq. 1 [5]:

$$R_L = \frac{1}{1 + K_L C_0} \tag{1}$$

where K_L is the constant of the Langmuir model and C_0 is the initial concentration of Ni, Co, and Cu ions. To know the thermodynamics of the adsorption process, the free-energy parameters of apparent Gibbs were determined. ΔG (kJ/mol), ΔH (kJ/mol), and ΔS (kJ/mol K), which are directly related to changes in temperature, and help to understand the mechanism of adsorption of metal ions, by using Eqs. 2, 3, and 4 [27, 36]:

$$\Delta G = -RTln(\mathrm{k}) \tag{2}$$

$$k = 55.5 K_L \tag{3}$$

$$\ln(k) = -\frac{\Delta H}{RT} + \frac{\Delta S}{R} \tag{4}$$

where K_L is the constant of the Langmuir model (L/mol), R is the constant of the ideal gases, and T is the absolute temperature (K). The values of ΔH and ΔS can be determined with the slope and ordered to the origin of the graph ln k as a function of $1/T$.

For the study of the adsorption isotherms of the different metals in nDCPD, 1 g of sorbent was put in contact with 50 ml of the aqueous solution of the metal ions Co, Cu, and Ni, varying the concentration between 0 and 1000 ppm in a shaker (ZHWY-200D) with an agitation of 200 rpm at 25, 35, and 45°C for 24 h of contact time.

2.5 Batch removal kinetics

The experiments to establish the kinetics of metal ion removal in nDCPD and to know the evolution of the adsorption of Ni, Co, and Cu ions in the biomaterial were carried out in batches. They were carried out varying the concentration of nDCPD (C_{ads}), from 0 to 40 g/L, during 24 h at a speed of 200 rpm and at 25, 35, and 45°C. At specific times aliquots of aqueous solution were taken to separate the adsorbent material and the liquid supernatant by centrifugation at 10,000 rpm. The supernatant was analyzed with the help of a spectrophotometer (JENWAY 6705) to know the concentration of the different ions present in the solution. The amount of ions removed (q_e) by nDCPD was obtained by applying Eq. 5 [4]:

$$q_e = \frac{V(C_0 - C)}{m} \tag{5}$$

where C_0 and C represent the initial concentration and the concentration at time t or in equilibrium (mg/L), V is the volume of solution (L), and m is the mass of nDCPD (g).

The percentage of removal (%R) was calculated as Eq. 6 [5]:

$$\%R = \frac{(C_0 - C_e)}{C_0} \times 100 \tag{6}$$

Model	Equation
SIPS	$q_e = \frac{q_m (K_s C_e)^{n_s}}{1 + (K_s C_e)^{n_s}}$
Redlich-Peterson (R-P)	$q_e = \frac{K_R C_e}{1 + a_R C_e^{\beta}}$
Langmuir	$q_e = \frac{q_m K_L C_e}{1 + K_L C_e}$
Temkin	$q_e = A + Bln(C_e)$
Freundlich	$q_e = K_F C_e^{1/n}$

SIPS: K_S (L/mg), q_m (mg/g), n_S (dimensionless); Redlich-Peterson: K_R (L/g), a_R $(L/mg)^{\beta}$, β (dimensionless); Langmuir: K_L (L/mg), q_m (mg/g), R_L (dimensionless); Freundlich: K_F $[(mg/g)(L/mg)]^{1/n}$, n (dimensionless); Temkin: A (L/mg), B (kJ/mol).

Table 1.
Nonlinear adsorption isotherm models [40, 41].

3. Results and discussion

3.1 Adsorption isotherms

Adsorption data obtained at different temperatures help to understand the interaction between nDCPD and Ni, Co, and Cu ions; analyzing the data with different adsorption models, the model with better fit is obtained. In **Table 1**, the most used isotherm models are described according to the literature, and in **Figures 1–3** for Ni, Co, and Cu, respectively, the behavior of the removal of metal ions at different temperatures is shown. The parameters derived from the adjustments together with the coefficient of determination, R^2, are shown in **Tables 2–4**, for Ni, Co, and Cu, respectively, using the SigmaPlot software (®Version 11) at a temperature of 15, 30, and 45°C. Based on the results obtained, it is observed that the most appropriate models to adjust the adsorption in nDCPD of Ni metal ions, in the three temperatures analyzed (**Figure 1** and **Table 2**), are SIPS, Redlich-Peterson, and Langmuir and Freundlich, suggesting that Ni ions are adsorbed on the surface of nDCPD causing the formation of a monolayer and that the adsorption process is physical. On the other hand, the results obtained in the removal of Co ions (**Figure 2** and **Table 3**), show that the models with the best fit of the data are SIPS, Redlich-Peterson and Langmuir in the three temperatures analyzed, which implies that the cobalt ions are adsorbed on the surface of nDCPD, in a physical process independent of temperature. Finally, in the removal of Cu ions (**Figure 3** and **Table 4**), only an appropriate fit was obtained with the SIPS model, which suggests that the removal of copper ions in equilibrium is carried out by a physisorption process on the surface of the adsorbent. The obtained results also show that the removal of all the ions analyzed using nDCPD is favored given that the R_L values are between 0 and 1, which indicates that the adsorption is favorable; this tendency can be seen independently of the temperature used. The results of metal ion removal obtained in the present study using nDCPD are comparable with those obtained using other agroindustrial residues, since for Ni, removals between 6.88 and 120 mg/g have been obtained, while the value obtained in this work was of 22.48 mg/g. In the case of Co, a removal range of 2.55–45.44 mg/g has been reported, and in our case a value of 41.81 mg/g was obtained, finally, for Cu, removal values have been reported in the range of 0.98–163.01 mg/g, while in the present study a value of 41.88 mg/g of adsorbent was obtained [3–5, 13, 20–23, 26]. On the other hand, when comparing the results obtained with those reported by other authors who also used hydroxyapatites and brushite as adsorbent material, it was observed that the

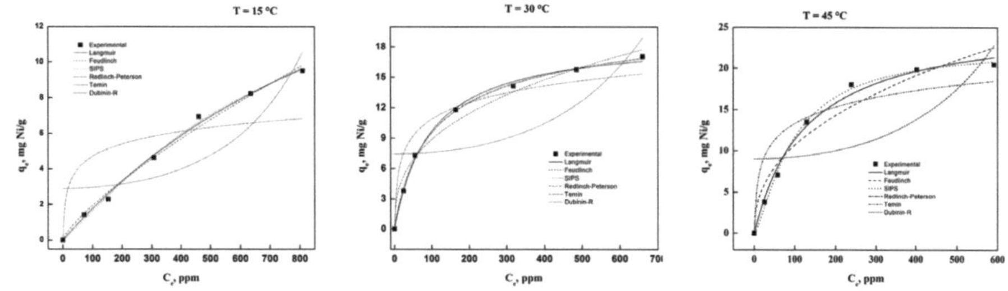

Figure 1.
Adjustment of the experimental data of Ni adsorption using the different models of isotherms at different temperatures.

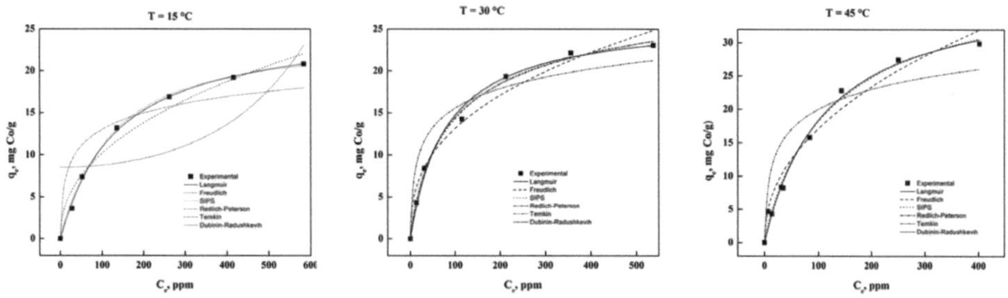

Figure 2.
Adjustment of the experimental data of Co adsorption using the different models of isotherms at different temperatures.

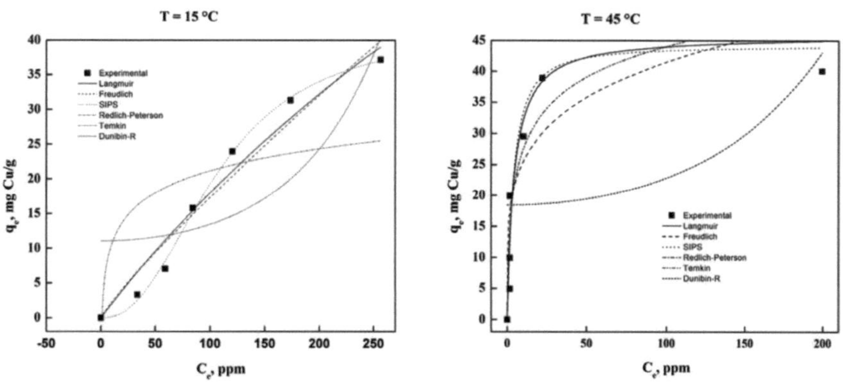

Figure 3.
Adjustment of the experimental data of Cu adsorption using the different models of isotherms at different temperatures.

model that was best adjusted was that of Langmuir, with a capacity of Ni adsorption reported between 9.9 and 40.0 mg/g, these data are comparable with those obtained in the present work. In the case of Co, an adsorption with apatites of 8.8–20 mg/g has been reported, a value lower than that obtained with natural nDCPD obtained in the present study, which is 41.8 mg/g. Finally, in the case of Cu ions, a range of adsorption has been documented in this type of materials of 26.6–343.64 mg/g, a value comparable to that obtained in the present work with nDCPD [7, 15, 32–35]. From the results obtained, it can be inferred that brushite is a material with good characteristics for the removal of metals from wastewater; it could also be observed that nDCPD favors the adsorption of Cu over Co and Ni.

Models	Parameters		
	15°C	30°C	45°C
SIPS			
K_S	0.005	0.016	0.0020
q_m	18.2426	20.6152	22.4789
n_S	1.1611	0.8644	1.3671
R^2	0.9969	0.9985	0.9974
Redlich-Peterson			
K_R	0.0191	0.2423	0.2024
a_R	0.007	0.0240	0.0078
β	1.0	0.9046	1.0
R^2	0.9943	0.9990	0.9913
Langmuir			
K_L	0.0007	0.0078	0.0105
q_m	27.7198	18.9551	25.9445
R^2	0.9962	0.9975	0.9913
R_L	0.588–0.935	0.114–0.562	0.087–0.488
Freundlich			
K_F	0.0552	1.6384	1.5515
n	1.2921	2.7263	2.3854
R^2	0.9923	0.9869	0.9980
Temkin			
A	1.873×10^{-8}	2.683×10^{-8}	2.3779×10^{-10}
B	1.0213	2.3617	2.8940
R^2	0.6402	0.9078	0.8337

Table 2.
Equilibrium parameters of adsorption models at different temperatures for Ni ions.

3.2 Effect of temperature on the removal of metal ions

In the literature it has been widely reported that one of the most important variables in the process of solid-liquid removal is temperature, since it directly influences some thermodynamic parameters (**Table 5**) of great interest in the removal process. The negative values of ΔG indicate that the removal processes of the metal ions of the aqueous solutions are carried out spontaneously; the increase of the value in the negative scale parallel to the increase of the temperature shows the direct dependence with this variable. The positive value of the ΔS indicates that the sites of the solid-liquid interface during the uptake of metal ions increase due to the randomness in the adsorbent [4, 5, 13, 34, 37], while the positive value of ΔH reveals that the process has an endothermic nature, which will influence the increase in ion removal as the temperature increases, which makes the process feasible and spontaneous at temperatures above room temperature [27, 36, 37]. Several reports indicate that the metal adsorption process in brushite is spontaneous; however, some authors indicate that the heat of adsorption in this process is negative, therefore, as the temperature increases, the removal capacity decreases.

Models	Parameters		
	15°C	30°C	45°C
SIPS			
K_S	0.0470	0.0208	0.0117
q_m	23.8607	30.1446	41.8104
n_S	1.1294	0.8178	0.9105
R^2	0.9895	0.9958	0.9895
Redlich-Peterson			
K_R	0.1884	0.4259	0.3462
a_R	0.0073	0.0326	0.0089
β	1.0	0.8886	1.0
R^2	0.9980	0.9953	0.9891
Langmuir			
K_L	0.0073	0.0089	0.0124
q_m	25.6863	26.5263	39.0000
R^2	0.9980	0.9933	0.9891
R_L	0.121–0.578	0.101–0.529	0.075–0.447
Freundlich			
K_F	1.3457	2.2983	2.1685
n	2.2770	2.6411	2.2291
R^2	0.9723	0.9773	0.9753
R_L			
Temkin			
A	2.0398×10^{-8}	1.8935×10^{-8}	4.696×10^{-8}
B	2.8230	3.3781	-6.3526×10^{-6}
R^2	0.8390	0.9084	0.4984

Table 3.
Equilibrium parameters of adsorption models at different temperatures for Co ions.

[4, 5, 13], However, other authors agree with that found in this work, where the heat of adsorption is positive and, as the temperature of the process increases, the adsorption of the ions in the material increases [32].

3.3 Effect of contact time and amount of adsorbent

In the process of solid-liquid adsorption, it is convenient to know the necessary contact time between the adsorbent and the adsorbate to achieve a maximum interaction between both and get removed as much metal ions as possible. **Figures 4–6** show the adsorption curves of each metal ion, where it is observed that about 10 h after the start of the process, the adsorption speed is high and that after this time the rate of removal of the ions. It decays, probably, because the sites where the capture of the ions is carried out are saturated, and therefore the balance is reached regardless of the concentration of absorber that has been used. This type of behavior is common among various adsorbent materials [37]. Also, it is observed that the adsorption capacity of the adsorbent material decreases significantly with the

Models	Parameters		
	15°C	30°C	45°C
SIPS			
K_S	1.396×10^{-5}	ND	0.2080
q_m	41.4493		41.3266
n_S	2.3971		1.1676
R^2	0.9980		0.9200
Redlich-Peterson			
K_R	0.2036	ND	9.9421
a_R	0.0013		0.2322
β	1.0		1.0
R^2	0.9632		0.9170
Langmuir			
K_L	0.0013	ND	0.2322
q_m	154.8825		42.8164
R^2	0.9632		0.9170
R_L	0.435–0.885		0.004–0.041
Freundlich			
K_F	0.2914	ND	14.5249
n	1.1274		4.7001
R^2	0.9548		0.7741
Temkin			
A	2.9527×10^{-8}	ND	10.1004
B	4.5983		6.7337
R^2	0.5789		0.8472

Table 4.
Equilibrium parameters of adsorption models at different temperatures for Cu ions.

Metal	T (K)	-ΔG (kJ/mol)	ΔH (kJ/mol)	ΔS (kJ/mol K)
Ni	288.15	18.52	69.63	308.53
	303.15	25.56		
	318.15	27.62		
Co	288.15	24.15	13.39	130.08
	303.15	25.91		
	318.15	28.07		
Cu	288.15	20.20	131.74	572.27
	303.15	ND		
	318.15	36.02		

Table 5.
Thermodynamic parameters for the equilibrium adsorption process of Ni, Co, and Cu ions.

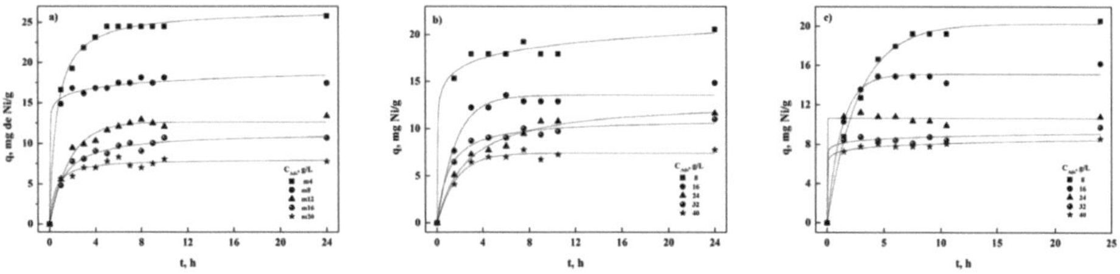

Figure 4.
Behavior of the removal of Ni ions at different temperatures and concentrations of adsorbent: (a) 15, (b) 30 and (c) 45°C.

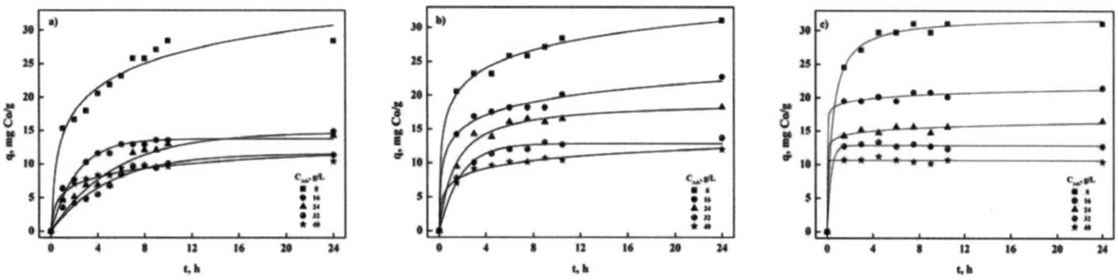

Figure 5.
Behavior of the removal of Co ions at different temperatures and concentrations of adsorbent: (a) 15, (b) 30 and (c) 45°C.

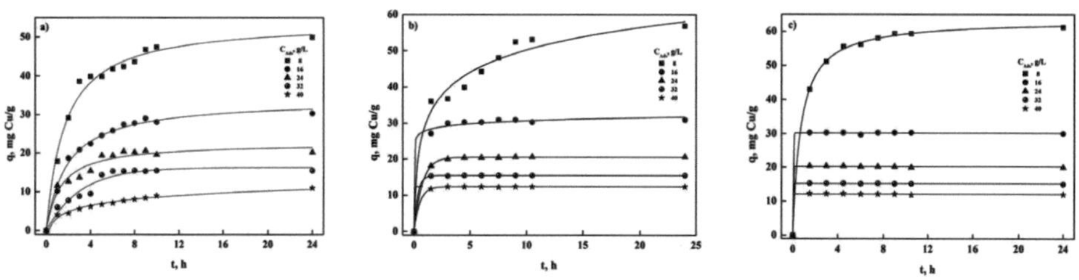

Figure 6.
Behavior of the removal of Cu ions at different temperatures and concentrations of adsorbent: (a) 15, (b) 30 and (c) 45°C.

increase in the concentration of absorbent independent of the temperature at which the removal process is carried out, so that the adsorption potential of the Ni ions decreases from 25.8 to 7.78 mg/g at 15°C, while at 30 and 45°C, it decreases from 20.6 to 7.78 and 20.6 to 8.6 mg/g, respectively. In the case of Co ions, the adsorption capacity decreases from 28.4 to 10.4 mg/g, from 31.04 to 11.97 mg/g, and from 31.04 to 10.4 mg/g at 15, 30, and 45°C, respectively. Finally, in the case of Cu ions, the decrease in the adsorption capacity at 15, 30, and 45°C was 49.9–11.0, 56.8–12.4, and 61.2–12 mg/g, respectively. This type of behavior is observed in the adsorption of metals [1, 26, 28, 29].

In the aqueous solutions with the three metal ions present, a behavior similar to that observed in the solutions prepared with a single metal compound was observed; however, the time necessary to reach equilibrium was around 5 h, and the adsorption capacity of the ions in solution was also significantly decreased by increasing the concentration of sorbent in the solution. The adsorption capacity achieved (**Figure 7**), for Ni ions, was from 32.4 to 6.2 mg/g, for Co ions from 41.9 to 10 mg/g, and for Cu ions from 59.9 to 12.1 mg/g at 45°C, which indicates that

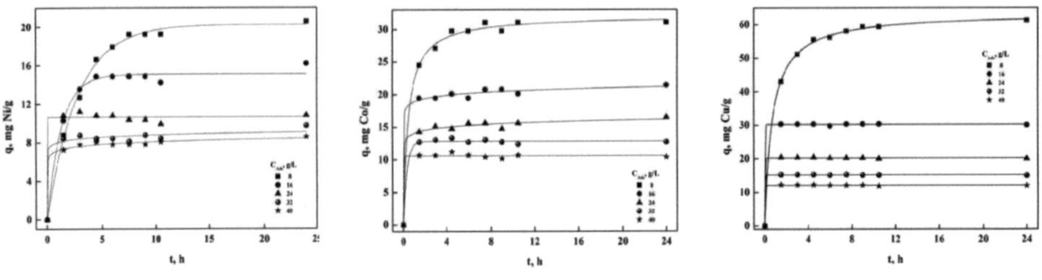

Figure 7.
Behavior of the different metallic ions of Ni, Co, and Cu found in the solution at 45°C.

it is not necessary to saturate the solution with nDCPD to achieve greater metal uptake [32–34]. On the other hand, in the analysis of the Co-Cu binary solution, a behavior similar to the previous cases was observed, achieving a greater adsorption capacity with less amount of absorbent, so that at a temperature of 45°C (**Figure 8**), for Co ions, it was possible to adsorb 32–9.9 mg/g and for Cu ions from 57 to 11.8 mg/g.

In the removal of ions in solutions, it was observed that the percentage of removal increases with the increase in the amount of nDCPD used, achieving removal percentages close to 100% for Cu ions and 95% for Co ions; however, for Ni about 70% removal was only achieved (**Figure 9**), suggesting that sites available for nickel capture are quickly saturated due to the large amount of chromium in the effluent. The changes caused by the variation of the temperature depend on the metal ion, since the maximum removal of the Ni ions is between 15 and 30°C, while for the Cu and Co ions, it is given at 30°C. Similar observations have been reported in adsorption studies conducted with other biomaterials [29, 38]. The selectivity shown in the present study was the following: Cu > Co > Ni, with percentages of removal in solutions composed of the three ions of 96% (Cu), 83% (Co), and 59% (Ni). Additionally, in **Figure 10**, it is observed that in the solutions composed only by Co or Cu, the time necessary to reach the equilibrium decreases considerably; in addition, the selectivity is maintained toward the removal of Cu compared to the Co, having a percentage of removal of 95–79%, respectively. Finally, the changes caused by the variation of the temperature depend on the metal ion, since the maximum removal of the Ni ions is between 15 and 30°C, while for the Cu and Co ions, it was at 30°C (**Figure 11**).

3.4 Kinetic study

To understand the process of removal of the different metal ions in aqueous solution, as well as the relationship that occurs between the metals and the interaction of the metal ions with the adsorbent material, the data obtained in the experimentation was evaluated at different temperatures and with different concentrations of adsorbents with the adsorption models that are concentrated in **Table 6** [6]. The kinetic parameters calculated with the different models (**Table 1**), were used to adjust the experimental adsorption values obtained for the three metal ions with the different concentrations of adsorbent and temperatures utilized, by using SigmaPlot software (Version 11®). Based on the obtained values, it is distinguished that the models that present the best adjustments to the obtained data are pseudo first order, Elovich and pseudo-second order, so it can be inferred that the removal mechanism is due to the ions being adsorbed in one or two sites on the surface that is heterogeneous since each site has different adsorption energy

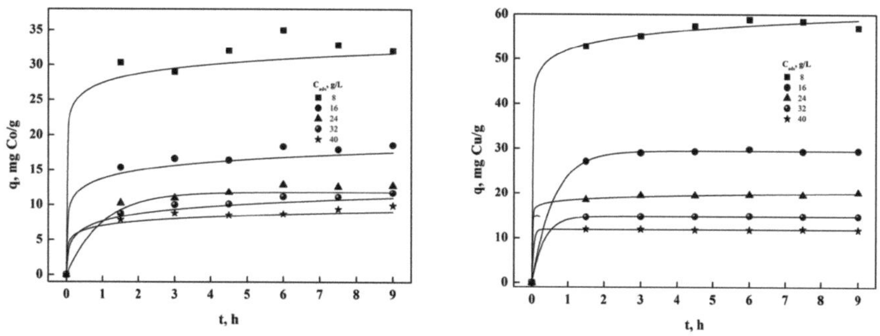

Figure 8.
Behavior of the different metal ions of Co and Cu found in the solution at 45°C.

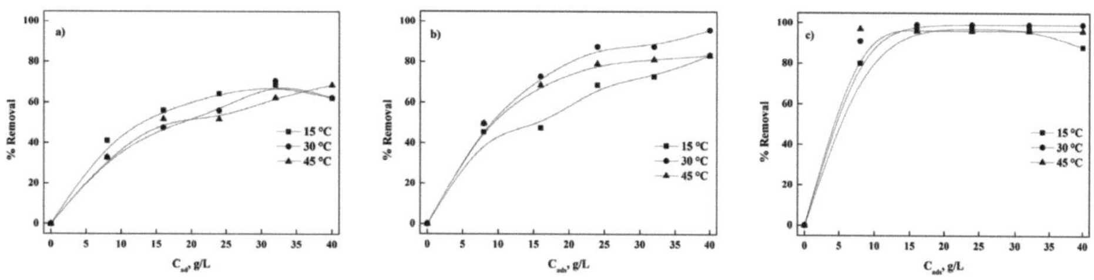

Figure 9.
Removal of the different metals in aqueous solution at 15, 30, and 45°C.

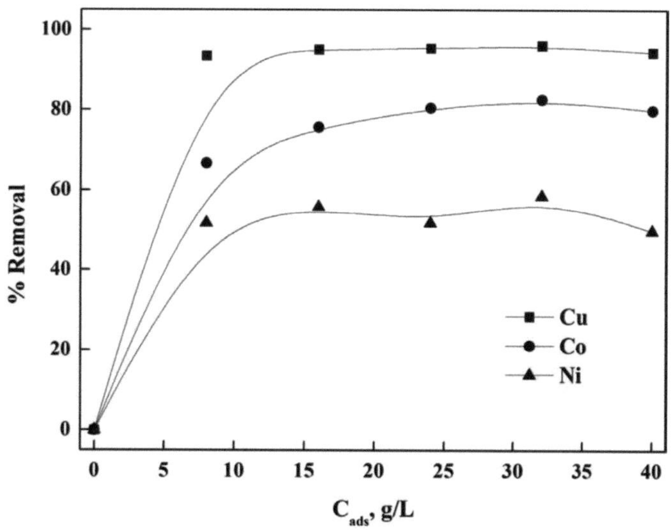

Figure 10.
Removal of Cu, Co, and Ni ions that are in the same solution at 45°C.

[1, 4, 5]. Similarly, it can be noticed that the models that do not adequately describe the process of metal removal in nDCPD are those that are related to the mass transfer both external and internal, independent of the temperature, the concentration of the adsorbent material, and the presence of another metal ion in the same solution; so it can be inferred that the removal process of these metals does not present problems of mass transfer.

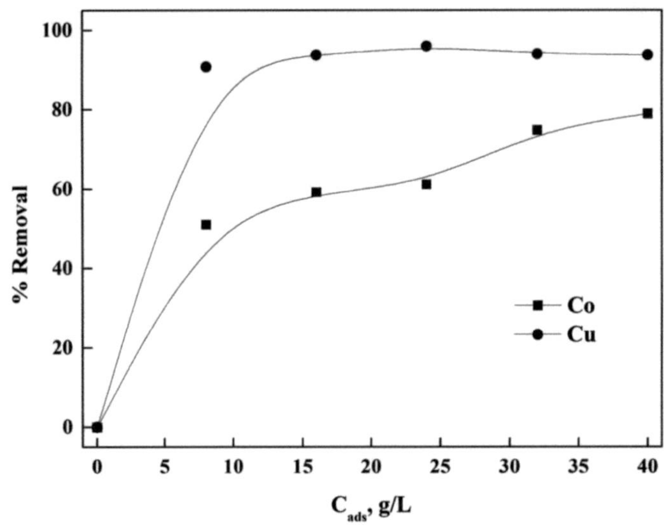

Figure 11.
Removal of Cu and Co ions in the aqueous solution at 45°C.

Model	Equation	Ref.
Pseudo-first order	$\frac{dq}{dt} = k_1(q_e - q)$	[35]
Pseudo-second order	$\frac{dq}{dt} = k_2\,(q_e - q)^2$	
Elovich	$\frac{dq}{dt} = \alpha e^{-\beta q}$	
Intraparticle diffusion (DI)	$q = k_{id}t^{0.5}$	
External diffusion (DE)	$ln\left(\frac{C}{C_0}\right) = -k_{ext}t$	[37]

Table 6.
Adsorption isotherm models used for the analysis of experimental data.

3.5 Characterization of nDCPD

Figure 12 shows the SEM image of the nDCPD doped with the Ni, Co, and Cu ions, where the main characteristics of nDCPD (**Figure 12a**), monoclinic disk morphology [1, 15, 29, 39], which does not present significant changes in the presence of Ni (**Figure 12b**), Co (**Figure 12c**), and Cu (**Figure 12d**) ions can be observed, probably because the vast majority of metal ions are adsorbed on the surface of nDCPD. The XRD patterns of nDCPD doped with Ni, Co, and Cu are shown in **Figure 13**, and according to the International Center of Diffraction Data Chart (JCPDS) database, some changes are observed in the samples with metal ions compared to that of nDCPD without doping [1]. The peaks at ~32° and 40° increase in intensity, while at ~34°, the intensity of the peaks directly related to the presence of Ni, Co, and Cu ions decreases. While at 26, 29, and 53°, new peaks appear in the doped brushite samples, which also caused a decrease in the network parameter of 4.99 nm, a value lower than the 5.77 nm obtained for nDCPD, without doping, together with the particle size from 9.26 to 4.45, 4.08, and 4.35 Å, for Ni, Co, and Cu, respectively, which was determined with the Scherrer Equation [39, 40]. This may be due to the fact that the Ca ions of nDCPD without doping (1.12 Å) are replaced during the adsorption process by the ions of Ni (0.78 Å), Co (0.63 Å), and Cu (0.69 Å), which implies that part of the ions are retained inside the structure of nDCPD [40].

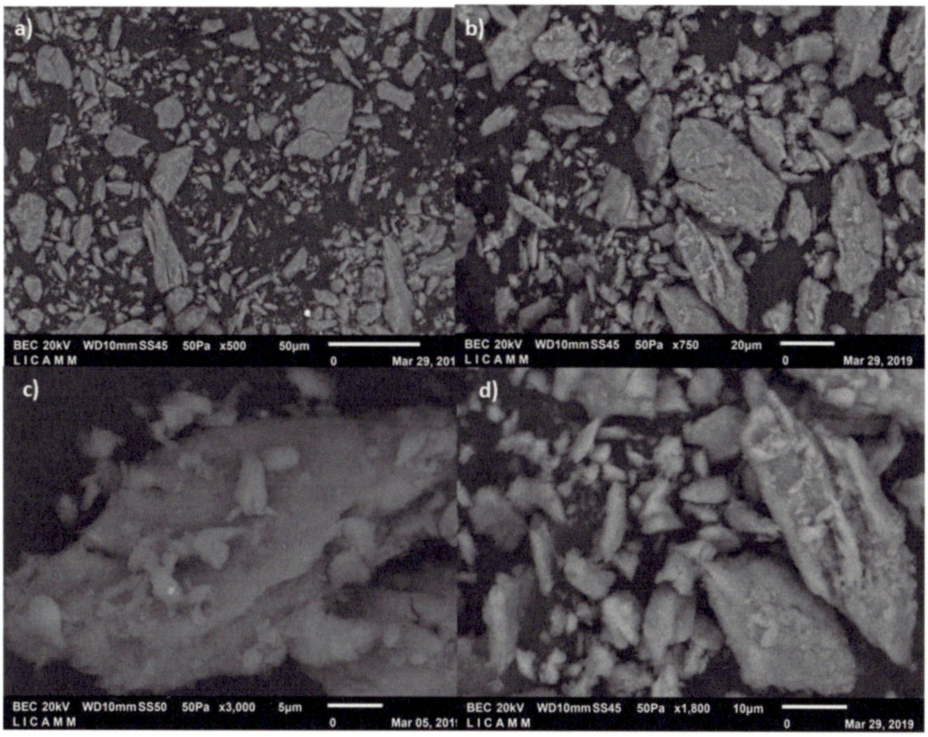

Figure 12.
Micrographs obtained in SEM: (a) brushite, (b) Ni, (c) Co, and (d) Cu.

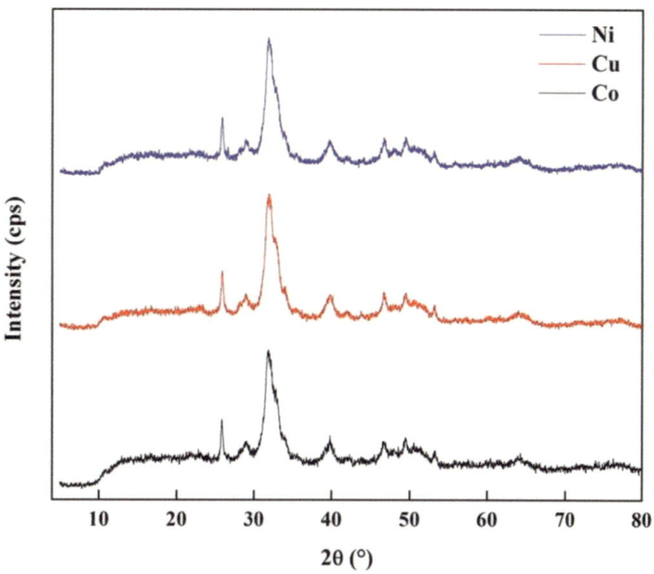

Figure 13.
X-ray diffraction patterns of nCDPD doped with the different metal ions.

In **Figure 14**, the infrared spectra of nDCPD are presented, where the characteristic peaks of the groups that make up apatite are observed, among which the presence of a doublet stands out at 577 and 526 $cm-^{1}$, characteristic of the vibrations of flexing of the phosphate groups; while the signal at 790 $cm-^{1}$ is due to bending out of the P-O-H plane; at 987 and 873 $cm-^{1}$, there are peaks related to

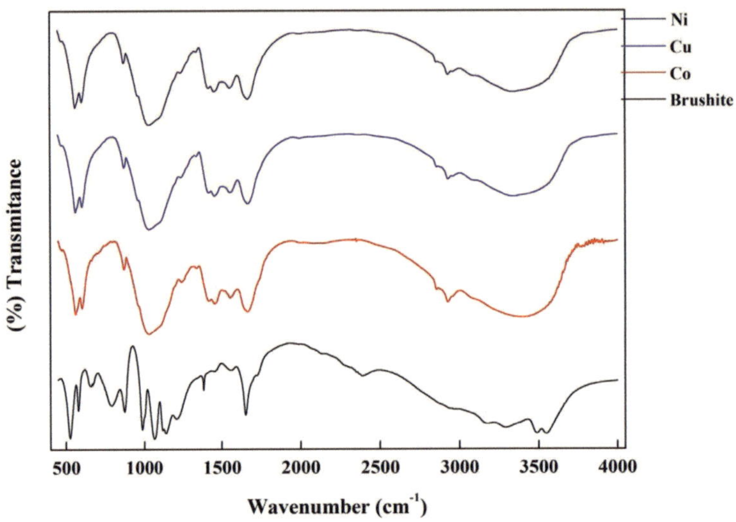

Figure 14.
Infrared spectrum of pure nCDPD doped with Ni, Co, and Cu.

the stretching of P-O (H) in the HPO$-^2{}_4$. On the other hand, the signals found at 1138, 1120, 1065, and 1004 cm$-^1$ are attributed to the stretching of the P-O link. At 1215 cm$-^1$ the signal generated by the flexion of the O-H group plane is presented, and the peaks observed at 665 and 1651 cm$-^1$ are related to the binding vibrations and physical vibrations of the water molecule. Additionally, a shoulder at 1725 cm$-^1$ was observed, which is related to the flexural stress of the water molecule, and the signals obtained at 3548, 3484, 3278, and 3166 cm$-^1$ are related to the extension of the water molecule in the apatite. Finally, a band at 2944 cm$-^1$ was also observed, directly related to the stretching of PO-H that occurs in HPO [1, 41, 42]. In the case of the nDCPD spectra with the different metal ions, it was observed that several peaks decrease in intensity or disappear as a result of the presence of Ni, Co, and Cu; these signals are related to the PO$^3-_4$ and OH groups, with which it can be inferred that the groups present on the surface of this apatite participate in the adsorption of metals directly.

4. Conclusion

Based on the thermodynamic parameters obtained at different temperatures and concentrations of brushite (nDCPD), used as an adsorbent material in the removal of different heavy metals (Ni, Co, and Cu), from aqueous solutions, it can be concluded that the adsorption of heavy metals with nDCPD is feasible and that the adsorption capacity of nDCPD increases at a higher temperature. The percentages of removal achieved show that nDCPD has a higher affinity for adsorbing Cu and Co compared to Ni. From the data obtained from the adsorption isotherms, it can be inferred that the process of removing the metal ions in solution on the surface of nDCPD is carried out, forming a heterogeneous monolayer since the model that best fits to the data obtained in the SIPS is a combination of the Langmuir and Freundlich models. The experimental data obtained also show that there are different models to describe the kinetic process of adsorption, such as the pseudo-first-order model, which suggests that only one surface site is needed for adsorption to take place, as well as the pseudo-second-order model which indicates that two sites are needed for each ion molecule and that each site requires different adsorption energies as described by the Elovich model, which confirms the heterogeneity of the nDCPD surface determined from of the adsorption isotherms generated. The surface changes of nDCPD, evidenced by the XRD and FTIR analyses, suggest that the HPO^- and $PO^{3-}{}_4$ are directly involved in the Ni, Co, and Cu adsorption process, which allows us to conclude that nDCPD is an appropriate candidate to be used in the removal of heavy metals present in wastewater from different industries.

Acknowledgements

The authors thank the Interdisciplinary Professional Unit of Engineering Campus Guanajuato of the IPN, the Center for Nanoscience and Nanotechnology (CNyN) of the UNAM, and the Colombian Polytechnic "Jaime Isaza Cadavid" for the infrastructure provided for the realization of this project and the Research and Postgraduate Secretary for the financial support for this work (SIP: 20170577) and to the BIOCATEM Network.

Author details

Rosa Hernández-Soto[1]*, José A. Hernández[1], Alba N. Ardila-Arias[2], Mercedes Salazar-Hernández[3] and María del Carmen Salazar-Hernandeza[4]

1 Instituto Politécnico Nacional, UPIIG, Guanajuato, Mexico

2 Grupo de Investigación en Química Básica y Aplicada a Procesos Bioquímicos, Biotecnológicos y Ambientales, Politécnico Colombiano Jaime Isaza Cadavid, Medellín, Colombia

3 Departamento de Ingeniería en Minas, Metalurgia y Geología, División de Ingenierías, Universidad de Guanajuato, Guanajuato, Mexico

4 Departments of Materials Science, Chemistry, Environmental Science, Earth and Planetary Sciences, UPIIG - Instituto Politécnico Nacional Unidad Profesional Interdisciplinaria De Ingeniería Campus Guanajuato, Silao de la Victoria, Guanajuato, Mexico

*Address all correspondence to: rosyhdezs@yahoo.com

References

[1] Hernández-Maldonado JA, Torres-Garcia FA, Salazar-Hernández MM, Hernández-Soto R. Removal of chromium from contaminated liquid effluents using natural brushite obtained from bovine bone. Desalination and Water Treatment. 2017;**95**:262-273. DOI: 10.5004/dwd.2017.21480

[2] Yildiz S. Kinetic and isotherm analysis of Cu(II) adsorption onto almond shell (*Prunus dulcis*). Ecological Chemistry and Engineering S. 2017;**24**(1):87-106. DOI: 10.1515/eces-2017-0007

[3] Moreira VR, Lebron YAR, Freire SJ, Santos LVS, Palladino F, Jacob RS. Biosorption of copper ions from aqueous solution using *Chlorella pyrenoidosa*: Optimization, equilibrium and kinetics studies. Microchemical Journal. 2019;**145**:119-129. DOI: 10.1016/j.microc.2018.10.027

[4] Giza S. Biosorption of heavy metal from aqueous waste solution using cellulosic orange peel. Ecological Engineering. 2017;**99**:134-140. DOI: 10.1016/j.ecoleng.2016.11.043

[5] Sudha R, Srinivasan K, Premkumar P. Removal of nickel(II) from aqueous solution using Citrus Limettioides peel and seed carbon. Ecotoxicology and Environmental Safety. 2015;**117**:115-123. DOI: 10.1016/j.ecoenv.2015.03.025

[6] Qiu Y, Yang L, Huang S, Ji Z, Li Y. The separation and recovery of copper(II), nickel(II), cobalt(II), zinc(II), and cadmium(II) in a sulfate-based solution using a mixture of Versatic 10 acid and Mextral 984H. Chinese Journal of Chemical Engineering. 2017;**25**:760-767. DOI: 10.1016/j.cjche.2016.10.013

[7] Guo Y, Zhao H, Han Y, Liu X, Guan S, Zhang Q, et al. Simultaneous spectrophotometric determination of trace copper, nickel, and cobalt ions in water samples using solid phase extraction coupled with partial least squares approaches. Spectrochimica Acta Part A: Molecular and Biomolecular Spectroscopy. 2017;**173**:532-536. DOI: 10.1016/j.saa.2016.10.003

[8] Huan-Ping C, Chung-Cheng C, Aileen N. Biosorption of heavy metals on *Citrus maxima* peel, passion fruit shell, and sugarcane bagasse in a fixed-bed column. Journal of Industrial and Engineering Chemistry. 2014;**20**:3408-3414. DOI: 10.1016/j.jiec.2013.12.027

[9] Pedrosa Xavier AL, Herrera Adarme OF, Furtado LM, Dias Ferreira GM, Mendes da Silva LH, Alves Gurgel LV, et al. Modeling adsorption of copper(II), cobalt(II) and nickel(II) metal ions from aqueous solution onto a new carboxylated sugarcane bagasse. Part II: Optimization of monocomponent fixed-bed column adsorption. Journal of Colloid and Interface Science. 2018;**516**:431-445. DOI: 10.1016/j.jcis.2018.01.068

[10] Qiu X, Hu H, Yang J, Wang C, Cheng Z. Removal of trace copper from simulated nickel electrolytes using a new chelating resin. Hydrometallurgy. 2018;**180**:121-131. DOI: 10.1016/j.hydromet.2018.07.015

[11] Liu X, He Y, Zhao Z, Chen X. Study on removal of copper from nickel-copper mixed solution by membrane electrolysis. Hydrometallurgy. 2018;**180**:153-157. DOI: 10.1016/j.hydromet.2018.07.019

[12] Sikder T, Rahman M, Jakariya M, Hosokawa T, Kurasaki M, Saito T. Remediation of water pollution with native cyclodextrins and modified cyclodextrins: A comparative overview and perspectives. Chemical Engineering Journal. 2019;**355**:920-941. DOI: 10.1016/j.cej.2018.08.218

[13] Ivanets AI, Srivastava V, Kitikova NV, Shashkova IL, Sillanpää M. Kinetic and thermodynamic studies of the Co(II) and Ni(II) ions removal from aqueous solutions by Ca-Mg phosphates. Chemosphere. 2017;**171**:348-354. DOI: 10.1016/j.chemosphere.2016.12.062

[14] Vollprecht D, Mischitz R, Krois LM, Sedlazeck KP, Müller P, Olbrich T, et al. Removal of critical metals from waste water by zero-valent iron. Journal of Cleaner Production. 2019;**208**:1409-1420. DOI: 10.1016/j.jclepro.2018.10.180

[15] Martins JJ, Órfão JJM, Soares OGSP. Sorption of copper, nickel and cadmium on bone char. Protection of Metals and Physical Chemistry of Surfaces. 2017;**53**(4):618-627. DOI: 10.1134/S2070205117040153

[16] Moscatello NJ, Pfeifer BA. Yersiniabactin metal binding characterization and removal of nickel from industrial wastewater. Biotechnology Progress. 2017;**33**(6):1548-1554. DOI: 10.1002/btpr.2542

[17] Osińska M. Removal of lead(II), copper(II), cobalt(II) and nickel(II) ions from aqueous solutions using carbon gels. Journal of Sol-Gel Science and Technology. 2017;**81**:678-692. DOI: 10.1007/s10971-016-4256-0

[18] Bhatnagar A, Sillanpää M. Utilization of agro-industrial and municipal waste materials as potential adsorbents for water treatment: A review. Chemical Engineering Journal. 2010;**157**:277-296. DOI: 10.1016/j.cej.2010.01.007

[19] Bhatnagar A, Sillanpää M, Witek-Krowiak A. Agricultural waste peels as versatile biomass for water purification: A review. Chemical Engineering Journal. 2015;**270**:244-271. DOI: 10.1016/j.cej.2015.01.135

[20] Romero-Cano LA, García-Rosero H, Gonzalez-Gutierrez LV, Baldenegro-Pérez LA, Carrasco-Marín F. Functionalized adsorbents prepared from fruit peels: Equilibrium, kinetic and thermodynamic studies for copper adsorption in aqueous solution. Journal of Cleaner Production. 2017;**162**:195-204. DOI: 10.1016/j.jclepro.2017.06.032

[21] Femina Carolin C, Senthil Kumar P, Saravanan A, Janet Joshiba G, Naushad M. Efficient techniques for the removal of toxic heavy metals from aquatic environment: A review. Journal of Environmental Chemical Engineering. 2017;**5**:2782-2799. DOI: 10.1016/j.jece.2017.05.029

[22] Liang S, Guo X, Feng N, Tian Q. Adsorption of Cu^{2+} and Cd^{2+} from aqueous solution by mercapto-acetic acid modified orange peel. Colloids and Surfaces B: Biointerfaces. 2009;**73**:10-14. DOI: 10.1016/j.colsurfb.2009.04.021

[23] Annadurai G, Juang RS, Lee DJ. Adsorption of heavy metals from water using banana and orange peels. Water Science and Technology. 2002;**47**(1):185-190. DOI: 10.2166/wst.2003.0049

[24] De Gisi S, Lofrano G, Grassi M, Notarnicola M. Characteristics and adsorption capacities of low-cost sorbents for wastewater treatment: A review. Sustainable Materials and Technologies. 2016;**9**:10-40. DOI: 10.1016/j.susmat.2016.06.002

[25] Romero-Cano LA, Gonzalez-Gutierrez LV, Baldenegro-Perez LA. Biosorbents prepared from orange pressure peels using instant controlled drop for Cu(II) and phenol removal. Industrial Crops and Products. 2016;**84**:344-349. DOI: 10.1016/j.indcrop.2016.02.027

[26] Moreno-Piraján JC, Giraldo L. Heavy metal ions adsorption from wastewater using activated carbon from orange peel. E-Journal of Chemistry. 2012;**9**(2):926-937. DOI: 10.1155/2012/928073

[27] Farroq U, Kozinski JA, Ain Khan M, Athar M. Biosorption of heavy metal ions wheat based biosorbents—A review of the recent literature. Bioresource Technology. 2010;**101**:5043-5053. DOI: 10.1016/j.biortech.2010.02.030

[28] Habib A, Islam N, Islam A, Shafiqul Alam AM. Removal of copper from aqueous solution using orange peel, sawdust and bagasse. Pakistan Journal of Analytical & Environmental Chemistry. 2007;**8**(1-2):21-25. DOI: 10.21743/pjaec/2016.06.005

[29] Khalfaoui A, Meniai AH. Application of chemically modified orange peels for removal of copper(II) from aqueous solutions. Theoretical Foundations of Chemical Engineering. 2012;**46**(6):732-739. DOI: 10.1134/S0040579512060103

[30] Surovka D, Pertile E. Sorption of iron, manganese, and copper from aqueous solution using orange peel: Optimization, isothermic, kinetic, and thermodynamic studies. Polish Journal of Environmental Studies. 2017;**26**(2):795-800. DOI: 10.15244/pjoes/60499

[31] Hamidi E, Arsalane S, Halim M. Kinetics and isotherm studies of copper removal by brushite calcium phosphate: Linear and non-linear regression comparison. E-Journal of Chemistry. 2012;**9**(3):1532-1542. DOI: 10.1155/2012/928073

[32] Mobasherpour I, Salahi E, Pazouki M. Comparative of the removal of Pb^{2+}, Cd^{2+} and Ni^{2+} by nano crystallite hydroxyapatite from aqueous solutions: Adsorption isotherm study. Arabian Journal of Chemistry. 2012;**5**:439-446. DOI: 10.1016/j.arabjc.2010.12.022

[33] El Hamidi A, Mulongo Masamba R, Khachani M, Halim M, Arsalane S. Kinetics modeling in liquid phase, sorption of copper ions on brushite di-calcium phosphate dihydrate $CaHPO_4{\cdot}2H_2O$ (DCPD). Desalination and Water Treatment. 2016;**56**:779-791. DOI: 10.1080/19443994.2014.940391

[34] Huang Y, Chen L, Wang H. Removal of Co(II) from aqueous solution by using hydroxyapatite. Journal of Radioanalytical and Nuclear Chemistry. 2012;**291**:777-785. DOI: 10.1007/s10967-011-1351-0

[35] Fernane F, Boudia S, Aiouache F. Removal Cu(II) and Ni(II) by natural and synthetic hydroxyapatites: A comparative study. Desalination and Water Treatment. 2014;**52**:2856-2862. DOI: 10.1080/19443994.2013.807084

[36] Singh KK, Hasan SH, Talat M, Singh VK, Gangwar SK. Removal of Cr(VI) from aqueous solutions using wheat bran. Chemical Engineering Journal. 2009;**151**:113-121. DOI: 10.1016/j.cej.2009.02.003

[37] Elabbas S, Mandi L, Berrekhis F, Pons MN, Leclerc JP, Ouazzani N. Removal of Cr(III) from chrome tanning wastewater by adsorption using two natural carbonaceous materials: Eggshell and powdered marble. Journal of Environmental Management. 2016;**166**:589-595. DOI: 10.1016/j.jenvman.2015.11.012

[38] Pehlivan E, Pehlivan E, Tutor-Kahraman H. Hexavalent chromium removal by Osage Orange. Food Chemistry. 2012;**133**:1478-1484. DOI: 10.1016/j.foodchem.2012.02.037

[39] Mirković MM, Lazarević-Pašti TD, Došen AM, Ćebela MŽ, Rosić AA, Matović BZ, et al. Adsorption of malathion on mesoporous monetite obtained by mechanochemical treatment of brushite. RSC Advances. 2016;**6**:12219-12225. DOI: 10.1039/c5ra27554g

[40] Thangadurai V, Kopp P. Chemical synthesis of Ca-doped CeO_3-intermediate temperature oxide ion

electrolytes. Journal of Power Sources. 2007;**168**:178-183. DOI: 10.1016/j.jpowsour.2007.03.030

[41] Sopcak T, Medvecky L, Giretova M, Stulajterona R, Durisin J, Girman V, et al. Effect of phase composition of calcium silicate phosphate component on properties of brushite based composite cements. Materials Characterization. 2016;**117**:17-29. DOI: 10.1016/j.matchar.2016.04.011

[42] Arifuzzaman SM, Rohani S. Experimental study of brushite precipitation. Journal of Crystal Growth. 2004;**267**:624-634. DOI: 10.1016/j.crysgro.2004.04.024

Chapter 7

Contamination of Water Resources by Food Dyes and Its Removal Technologies

Sajjad Hussain, Nadeem Khan, Saima Gul, Sabir Khan and Hammad Khan

Abstract

Food dyes comprise different groups which impart color to a wide range of food products. Food products are mainly purchased and consumed by people because they are nutritive and flavorsome and have an attractive color. Food color stimulates appetite and enhances its esthetic appeal of food on table for customer. With sky rocketing industrialization and modernization, the worldwide production of dyes in 2010 was forecasted to be 2.1 metric tons. It has been estimated that 15% of total dyes produced worldwide are discharged to water bodies which adversely affect aquatic ecosystem. Dyes in water reduces its transparency, thereby declining light penetration in the water, hence influencing photosynthesis which consequently reduces dissolved oxygen which is an alarming situation for both aquatic flora and fauna. Dyes wastewater discharged from huge number of industries like textile, leathers, paint, food, pharmaceutical etc. and deteriorating the aquatic environment and pose threat to living organism. The presence of dye molecules in water channels is an emerging alarm to an environmental scientist. An environmental friendly and self-sustainable treatment method should be explored to address this problem. Therefore, this work elaborates the various methods used for removal and degradation of dyes in water, although some processes have a common shortcoming like production of secondary pollution to the environment. This chapter have tried to highlight the important application of food dyes, their contamination and their toxic effect. Herein we also focus on remediation techniques like separation (adsorption, filtration, etc.) and degradation (chemical, biological and electrochemical oxidation) of dyes in aqueous solution. The mechanism and pros and cons of different methods are explored and discussed briefly.

Keywords: food dyes, degradation, wastewater, contamination

1. Introduction

1.1 Pollution caused by dyes and their harmful effects on aquatic environment

Food dyes are promising colorants globally used to color a broad spectrum of food products. Food dyes not only impart color to food but also boost up appetite and promote attractiveness and esthetic appeal of food for consumer. According to recent financial reports, 7×10^5 tons of 10,000 different dyes and pigments are produced worldwide, and its market is skyrocketing each year and has exceeded over US $11 billion by 2008 [1]. Approximately 1-2% of dyes stuffs are lost during production and about 10-15% of dyes are discharge as effluent during dye applications [2]. From an environmental perspective, the discharge of dye effluents from dye manufacturing or consuming units into the water bodies poses potential threats to the quality of water and induces serious health problems to human, plants and animal life in particular and aquatic biota in general. The presence of dyes in wastewater even in trace amounts (less than 1 mg/dm^3 for few dyes) is remarkably objectionable and undesirable [3]. Most of the organic dyes used to color food products, for example, azo dyes, have aromatic centres in their molecular structures, and their metabolic and degradation products like aromatic amines (anilines), benzidines and benzene sulphonic acids are well-established carcinogens, mutagens and DNA adducts and hence induces subsequent deadly effects on cells [4]. Moreover dyes can make water colored, thereby reducing the transparency (sunlight penetration) and aeration of water body, which badly affects the efficacy of crucially important photosynthesis, consequently reducing significantly the dissolved oxygen (DO) levels in water. The discharge of dye effluents in water bodies poses direct and indirect consequences to aquatic ecosystem. The direct effects includes depletion of dissolved oxygen levels, decreased reoxygenation potential, leaching of dyestuff from soil into groundwater, reduced light penetration into water which hinders photosynthesis (which gives red signal to aquatic flora and fauna) and esthetic issue of water downstream [5, 6]. The bird's eye view of indirect effects caused by dye effluents includes death of aquatic organisms, genotoxicity and microtoxicity imposed by colored allergens, depression of human immune system, allergic reactions, hyperactivity in kids (ADHD), bladder cancer in humans and deadly process of water eutrophication [7].

1.2 Application of dyes

Both natural and synthetic dyes have indeed revolutionized the modern world both at domestic and industrial sectors. More recent and independent reviews have reported that over 100,000 different dyes and pigments are in practice, and their annual production has exceeded over 7×10^7 tons. Nowadays the use of dyes is indispensable because dyes are widely used in electronics; textile; rubber; food; leather; cosmetics; paper and pulp; photography; solar cells/solar penal, i.e. dye-sensitized solar cells; pharmaceuticals; pigment; agriculture research; paints; printing inks; cosmetics; coloring of plastics products; and many other fields of domestic and industrial interest [8, 9]. Similarly dyes are also used to assess the efficacy of sewage and wastewater treatment plants as well as agriculture research. Textile industries consume more than 70% of total dyes produced worldwide [10]. Dyes in textile, food or in any other industries are used in minute quantities; for instance, it has been reported that one ton of textile dye is more than enough to color 42,000 clothes/suits [11].

2. Techniques incorporated for dye removal from wastewater

Purification of water from dyes tuff is of tremendous importance from both perspectives, i.e. water purification and its reusability. Pollution caused by dyes has attracted the attention of environmental chemists worldwide more specifically in developed countries like the USA, United Kingdom, Germany, China, Japan, Scandinavian countries, etc., which have incorporated rapid industrializations and are more vulnerable to dye pollution. Nevertheless dyes cause severe water pollution; hence these dyes need to be eventually removed to minimize or eliminate their dreadful impacts. Consequently the worldwide environmental standards so far have turned more stringent, seeking more precise technological tools to cope with this global environmental issue, and that is why it has attracted the attention of environmental scientists worldwide from past two decades. Various techniques like physicochemical methods, enzymatic degradation, microbiological treatment, chemical methods and advanced oxidation techniques have been developed so far [12–14]. But unfortunately most of the organic dyes are non-biodegradable, thermally stable and refractory with respect to biochemical oxidation due to their large/complex sizes and inert nature. These characteristics have rendered organic dyes reluctant with respect to decolourization by using conventional methods of wastewater treatment like filtration, coagulation, biochemical and physicochemical techniques. Thus from the last two decades, research for more efficient and sophisticated techniques for removal of these dyes from wastewater to minimize or eliminate the water pollution has been turned the core interest to environmental scientists worldwide [15]. The dye abatement techniques in water are broadly classified into two principal types.

2.1 Segregation techniques

Segregation techniques involve separation of dyes pollutants via multiple physical techniques like adsorption, filtration/nanofiltration or coagulation processes. These traditional water purification methods like filtration, ion exchange, coagulation by lime or by using salts of iron/aluminium, adsorption over activated carbon and other low-cost adsorbents have been proven to be very effective decolourization methods. Among techniques decontamination of dyes by adsorption offer the most promising, simple and efficient technique. Similarly the use of low-cost adsorbents such as vegetable matter, saw dust, fruits peels, hen feathers, other carbonaceous materials, etc. make the dye removal process more economical and feasible [16, 17], but their utility is limited by production of associated hazardous sludge which poses an another issue of their disposal at suitable safe site. Similarly the use of adsorption and coagulations techniques for decolourization of wastewater is limited by regular regeneration of adsorbent and coagulant materials after dye adsorption and coagulation, respectively, which is much needed (to make the process economical and efficient), so this in turn gives rise to other issues [18, 19].

2.2 Degradation techniques

2.2.1 Chemical techniques

Over the past two decades, chemical techniques are also withdrawing the attention of environmental chemists; these techniques include chemical oxidation by using hypochlorite (OCl^{-1}), chlorine dioxide or ozone which has always shown high efficiency and reproducibility. But this technique has a major disadvantage with respect to relatively high prices and production of secondary pollutants like chlorinated hydrocarbons which are known carcinogens; similarly the difficulty in storage and transportation of reactants causes a substantial inconvenience for safe operation [20]; thus environmental scientists are looking for more, convenient, advanced, safe, efficient and sophisticated techniques which must be incorporated for removal of dyes from wastewater [12] .

2.2.2 Advanced oxidation processes (AOPs)

The term advanced oxidation processes is actually coined to address the process of oxidation of organic pollutants primarily via in situ generations of highly reactive hydroxyl free radicals ($^{*}OH$) from H_2O_2. These hydroxyl free radicals are strong oxidizing species by virtue of powerful oxidation potential of (2.80 V) vs. SHE. Different advanced oxidation processes are ozonation; photocatalytic methods using O_3/UV, TiO_2/UV and H_2O_2/UV; and Fenton's reagent (H_2O_2/Fe^{+2}). Fenton's reagent offers a novel dye degradation technique [21]. The phenomenal Fenton's reagent was named after Fenton who introduced it for the first time almost 100 years ago in 1884. Fenton's reagent is a precursor of hydroxyl free radicals; these madly reactive hydroxyl free radicals attack dye substrate molecules at the sites of multiple bonds and carry out either excessive hydroxylation of dye contaminants or causes dehydrogenation of dye molecules, hence producing stable inorganic materials directly or indirectly converting the dye pollutant into biodegradable and safe materials. Fenton's processes offer a remarkable efficiency and can be extended to much broader spectrum of dyes due to its fabulous non-selectivity [22]. Recently photo-Fenton's process, i.e. Fenton's coupled with light (UV or visible), electro-Fenton, sono-Fenton and sono-electro-Fenton are emerging water treatment techniques.

Ozonation is also reliable, safe and effective wastewater treatment tool which utilizes ozone gas as a strong oxidant usually used to disinfect water in swimming pools. In aqueous medium the mode of action of ozone is quite complicated. Molecular ozone can also oxidize the dyestuff in water by virtue of selective, direct or by indirect decomposition through a chain reaction mechanism by generating in situ free hydroxyl radicals (OH^{*}) [23]. AOPs show a tremendously high degree of decolourization efficiency with associated photocatalytic degradation of dyestuff [24]. But their sky kissing prices and operational difficulties put a question mark and limit the use of these photocatalytic techniques for decolourization of wastewater.

2.2.3 Microbiological and enzymatic degradation techniques

On the contrary to AOPs microbiological methods like activated sludge process, aerobic and anaerobic decomposition of pure and mixed cultures using fungi and bacteria and enzymatic degradation techniques have so far shown excellent decolourization and degradation efficiencies of dyes in wastewater; furthermore these methods are gaining popularity by virtue of their simplicity, ease of operation and applicability [25]. Biodegradation processes may be aerobic and anaerobic, and sometimes the combinations of both aerobic and anaerobic biological treatments are used for dye removal from water [26]. The main mechanism of dye removal in biological treatment is the adsorption of dyestuff on to biomass, but at saturation point, the adsorption potential of dyes by biomass drops [27]. Unfortunately the utility of these microbiological methods is quite inadequate because most of the dyes are stable and resistant towards biodegradation due to their huge sizes, complicated and conjugated benzenoid structures with extensive electron delocalization and high degree of stability. Similarly due to high degree of specificity and sensitivity of the enzymes of the microorganisms, these techniques cannot be extended over a broader spectrum of dyes and needs an extensive study and homework of enzymes, their nature, selection and applicability. Thus biological abatement of dyes in wastewater has become difficult owing to the bio-refractory nature and stability of dyes. That is why environmental chemists were chewing their nails to explore more advanced, effective and non-selective techniques to achieve dye abatement goals. Thanks to electrochemical techniques which have taken the bull of organic pollutants by horns and have controlled the aquatic pollution to a great extent [22].

2.2.4 Catalytic degradations

Catalytic degradations assisted by suitable promoter and coupled techniques like photocatalytic, sonocatalytic and photoelectrocatalytic degradations of dyes have so far shown excellent efficiencies. But here again catalytic poisoning and recovery of catalyst materials causes issues.

2.2.5 Electrochemical techniques

The term electrochemical techniques refers to application of DC current from a suitable source and carrying out an entire degradation of pollutants by electro-oxidation or reduction processes to inorganic materials. The main pollutants for electrochemical methods are not only dyes, but also other pollutants like pharmaceuticals, pesticides, herbicides, herbicides, detergents and many other harmful contaminants are the subjects of investigation. Electrochemical techniques usually incorporated for wastewater abatements are electrochemical reduction, electrochemical oxidation, electrocoagulation methods, photoelectrocatalytic and photo-assisted Fenton's oxidation techniques. Electrochemical oxidation techniques are further subdivided into two types which include direct and indirect oxidation techniques [28]. Over the past 10 years, the electrochemical methods have received a remarkable attention. This is because these methods have promising water decontamination potentials, have shown great novelty due to their versatility and potential cost-effectiveness and offer the most promising, clean, safe, efficient and green technologies for the decolourization of wastewater. Wastewater abatement using electrochemical methods has a fantastic advantage of environmental compatibility because its sole reagent, i.e. the electron, is a safe, clean and green reagent produced in situ and works most efficiently. Furthermore these methods are gaining attention of environmental chemists due to their excellent efficiencies, flexibility of automation and safe applicability over a broad spectrum of organic dyes [28]. These techniques need mild conditions for their operation, no heating of the samples are required and work under ambient conditions of both parameters, i.e. temperature and pressure. Nowadays a vast variety of electrochemical techniques are in practice such as electrochemical oxidation (EO), electrocoagulation (EC) using a variety of anodes, active chlorine indirect oxidation, etc. Recently emerging techniques which utilize twin technologies of both electrochemical cells and suitable light like UV light or sono-electrochemical degradations of dyes in wastewater are gaining much attention and appreciation [29, 30]. These photo- and sono-assisted electrochemical setups have been categorized as electrochemical advanced oxidation processes (EAOPs) [21]. The use of electrochemical methods for abatement of contaminants in wastewater was pioneered by Nilsson et al. 1973 [31] by electrochemical oxidation of phenolic-based wastes; later in the early 1980s, these studies were proceeded in collaboration with Chettier and Watkinson [32].

Over the past two decades, much of the research regarding decolourization of wastewater by electrochemical oxidation has been focused on the use of different anodic materials, their relative efficiencies, exploration of various factors affecting process efficiency (like PH, temperature, nature and concentration of electrolytes, etc.), kinetics and mechanism of oxidation of a variety of pollutants in water. The electrochemical oxidation is of two types, the first type is direct oxidation also called anodic oxidation by using suitable anode material. Direct oxidation is carried out by physically adsorbed hydroxyl free radical ($^{*}OH$) or chemisorbed active oxygen in oxide lattice. The second type of electrochemical oxidation is indirect oxidation using appropriate oxidant material such as hypochlorite (OCl^{-1}) which is formed anodically [33, 34]. The main degradation products of electrochemical oxidation of organic pollutants are CO_2 and water; thus during electrochemical oxidation of dyes, neither sludge formation occurs nor further treatment of degradation product is required; moreover no harmful by-products are formed; this is one of the greatest superiorities of electrochemical oxidation over other methods used for decolourization of wastewater [35]. A flow sheet representation of various techniques utilized for dyes abatement is shown below (**Figure 1**).

2.2.6 Electrochemical oxidation and its types

Environmental chemists took a long comfortable breath when they incorporated electrochemistry for removal of organics from wastewater. Electrochemical oxidation effectively degrades organics in wastewater. Electrochemical oxidation may be either indirect or direct oxidation processes. Indirect electro-oxidation also called mediated oxidation involves electro-generations of oxidizing species (EOS) in water at electrodes (anode) like active chlorine species, physisorbed hydroxyl free radicals or chemisorbed active oxygen atoms (in metal oxide anode lattice) which are very efficient, and oxidation mediated by these EOS gives rise to partial or complete decontamination of organics. Indirect oxidation processes can be elaborated by twin approaches:

1. Electrochemical transformation

2. Electrochemical combustion

During electrochemical transformation, the stable and refractory organic contaminants are converted into biodegradable materials (most often in to carboxylic acids) by chemisorbed active oxygen atoms. Then during electrochemical combustion which is also called electrochemical incineration, the organic contaminants are entirely mineralized to CO_2, water and inorganic ions by virtue of physisorbed hydroxyl free radicals ($\dot{OH}$). These hydroxyl free radicals are potentially strong oxidants with a powerful oxidation potential of (E_o = 2.80 V) vs. SHE; that is why this radical has been ranked as second most powerful oxidizing agent after fluorine in electrochemical series. The ($\dot{OH}$) radicals induce excessive dehydrogenation and hydroxylation of organic contaminants which ultimately oxidizes to yield CO_2. The mechanism of indirect oxidation of a dye pollutant at a metal oxide (MO_x) anode can be summarized as follows [36, 37]:

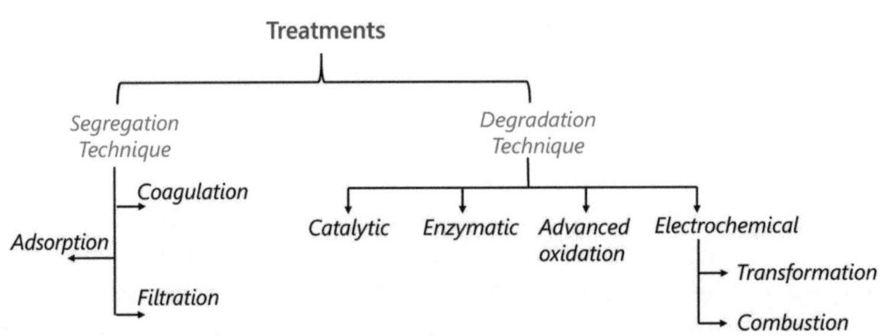

Figure 1.
Different methods used to remove dyes from wastewater.

First of all water is electrolyzed at metal oxide interface generating physically adsorbed (physisorbed) hydroxyl free radicals ($O^{\cdot}H$) as shown below;

$$MO_x + H_2O \rightarrow MO_x(O^{\cdot}H) + H^{+1} + 1e^{-1} \quad (1)$$

Now the next step is a crucial one which entirely depends on selection of anode material; moreover this step exclusively differentiates between two limiting classes of electrode (anode) materials, i.e. active and non-active electrodes.

3. Active anodes like DSA types of anodes which carry out indirect oxidation of organics through formation of strong oxidants like hydroxyl free radicals, aqueous chlorine, i.e., $Cl_2(aq)$, hypochlorous acid (HOCl) or hypochlorite ion (OCl^-), ClO_2^{-1}, ClO_3^{-1}, ClO_4^{-1}, etc. Now at active electrodes like DSA where higher oxidation states of metal electrode atoms (M) are available, the physically adsorbed hydroxyl free radicals undergoes transformation to chemically adsorbed (Chemisorbed) active oxygen with increase in oxidation state of metal atom (M) to form higher oxide (MO_{x+1}) accompanied by liberation of a proton and an electron as shown below:

$$MO_x(O^{\cdot}H) \rightarrow MO_{x+1}O^{\cdot} + H^{+1} + 1e^{-1} \quad (2)$$

This surface redox couple $MO_x/MO_{x+1}\dot{O}$ also called chemisorbed active oxygen atoms, i.e. $(MO_{x+1}\dot{O})$, selectively attack dye molecule (R) and partially oxidize it to intermediate compounds (RO), while the metal oxide anode *(MO_x)* (MO_x) is regenerated as shown below:

$$MO_{x+1}O^{\cdot} + R \rightarrow MO_x + RO \quad (3)$$

As in active electrodes like DSA, the chemisorbed active oxygen atoms are chemically bonded to electrode surface; they do not have full freedom to attack dye molecules; that is why they carry out very poor oxidation of dyes in the presence of supporting electrolytes like Na_2SO_4 in the absence of chloride medium, but in the presence of NaCl as supporting electrolytes, the situation is entirely different. The second reason for decreased efficiency of DSA and likewise active electrodes have low oxygen evolution overpotentials (they start the unlikely Oxygen Evolution Reaction easily at low voltages).

4. At non-active anodes such as BBD and PbO_2 where the possibility of higher oxidation states for anode are entirely excluded (e.g. in PbO_2 electrode, the lead atom has oxidation state equal to +4, and it does not show oxidation state greater than +4), the physisorbed hydroxyl free radicals, i.e. $MO\,x(O^{\cdot}H)$, attack dye pollutant "R" non-selectively and completely mineralizes dye contaminants into organic species like CO_2 and H_2O, while the metal oxide anode *(MO_x)* is regenerated as shown below.

$$\alpha MO_x(O^{\cdot}H) + R \rightarrow \alpha MO_x + mCO_2 + nH_2O + xH^{+1} + ye^{-1} \quad (4)$$

As these $^*OH$ radicals are physisorbed (i.e. not chemically bonded but are attached to electrode surface through weak Van der Waals forces which breaks easily) and $^*OH$ free radicals enjoy full freedom of attack on dye contaminants, these electrodes show greater degradation of dye contaminants. The second reason is that the non-active anodes have higher oxygen evolution over potentials (do not start OER easily).

Both physisorbed and chemisorbed active oxygen atoms undergo a simultaneous undesirable and unlikely competitive side reaction which is called oxygen evolution reaction (OER). This reaction substantially reduces the efficiency of electrodes for electrochemical oxidation of organics. Now as a general rule, the anode materials like active electrodes (DSA and others) have low oxygen evolution over potentials (i.e. they are excellent electro-catalysts for undesirable process of oxygen evolution reaction, i.e. OER; that is why they are active anodes) and hence carryout partial oxidation of organics. On the contrary the non-active anodes such as BDD and PbO_2 anodes have high oxygen evolution over potentials (i.e. they are poor electro-catalysts for OER; that is why they are termed as non-active anodes), and they carry out complete mineralization of organics to inorganic species like CO_2 and H_2O. The oxygen evolution potential of DSA (RuO_2) is 1.47 volts (Vs SHE in 0.5 M H_2SO_4), while that of BDD is 2.3 volts (Vs SHE in 0.5 M H_2SO_4). That is why non-active electrodes are most widely used anodes for elimination of organic contaminants during wastewater treatment in the absence of chloride medium [38–40].

The well-known $^*OH$ free radical is a powerful oxidant (E^o = 2.80 V vs. SHE), but it is unstable and has much shorter life than HClO (E^o = 1.49 V vs. SHE); therefore in the dye decontamination from wastewater which is heavily loaded with inorganic salts like chlorides, the indirect electrochemical oxidation mediated by active chlorine species in situ provides an excellent tool for such wastewater abatement. The non-active electrodes like BDD, PbO_2 and SnO_2 are more active with respect to generation of physisorbed $^*OH$ free radicals for wastewater abatement; however they cannot be utilized for generation of active chlorine species because they preferably generate $^*OH$ free radical along with other mild oxidants like *peroxo-diphosphate*, *peroxodicarbonate*, *peroxodisulphate*, etc. On the contrary the active anode materials like platinum and DSA have remarkably high efficiency for oxidation of chloride ion, thereby generating active chlorine oxidants rather than generating physisorbed $^*OH$ free radicals at their surfaces [41].

When NaCl is used as supporting electrolyte with active electrodes like Ti-based DSA (which has been selected as model anode in this study), then chloride ions (Cl^-) undergo oxidation at anodes to produce a very strong dye killing oxidants in situ, i.e. aqueous chlorine (i.e. Cl_2 (aq)), HOCl, OCl^-, ClO_2^{-1}, ClO_3^{-1} and ClO_4^{-1} [30, 42, 43]. These species mediates the oxidation of organics as shown below.

Reaction at anode:

$$2Cl^{-1} \rightarrow Cl_2(aq) + 2e^{-1} \tag{5}$$

Reaction at cathode:

$$2H_2O + 2e^- \rightarrow H_2 + 2OH^{-1} \tag{6}$$

Reaction at bulk:

$$Cl_2(aq) + H_2O \rightarrow HOCl + H^{+1} + Cl^{-1} \tag{7}$$

$$HOCl \rightarrow OCl^{-1} + H^{+1} \quad (pK_a = 7.54) \tag{8}$$

HOCl has greater oxidation potential (1.49 V) than OCl^{-1} (0.94 V). The type of active chlorine species in water depends upon pH solution, for example, at pH less than 03, dissolved chlorine (Cl_2) gas exists in water in the form of aqueous chlorine, i.e. Cl_2 (aq). At pH range of 03–7.5, Cl_2 undergoes self-oxidation/self-reduction reaction (disproportionation reaction) and forms HOCl, while at PH greater than 7.5, hypochlorite ions (OCl^-) exist in solution which is a much weaker oxidizing agent than HOCl [44].

Bonfatti has presented an alternative approach to electrochemical oxidation of organics; the schematic sketch of electrochemical oxidation at non-active oxide electrode *(MO_x)* and with associated CER is shown in **Figure 2**. Here "R" represents a dye pollutant molecule [45].

Figure 2 shows that in the first step, the hydroxide from solution adsorbs on metal oxide to form MO_x ($^{\cdot}OH$) (see reaction 05 given below), and then it reacts with chloride ions (Cl^{-1}) to yield adsorbed hypochlorous acid radical, i.e. $MO_X(HO^{\cdot}Cl)$ (see reaction 10); this adsorbed hypochlorous radical attacks organic pollutant (R) and degrades it to CO_2 and H_2O, and chloride ions are regenerated (see reaction 11). This degradation reaction proceeds simultaneously with chlorine evolution reaction (CER) to produce Cl_2, some of which Cl_2 gas evolves (if applied current density is very high), while some chlorine gas dissolve in water in the form of aqueous chlorine Cl_2 (aq). This aqueous chlorine reacts with hydroxide ions to generate HOCl, OCl^{-1}, ClO_3^{-1}, ClO_3^{-1}, etc. (depending on pH of solution), which are active chlorine species and oxidize organic pollutants in situ.

$$MO_x + O^{\cdot}H \rightarrow MO_x(O^{\cdot}H) \tag{9}$$

$$MO_x(O^{\cdot}H) + Cl^{-1} \rightarrow MO_x(O^{\cdot}Cl)_{ads} + 1e^{-1} \tag{10}$$

$$MO_x(O^{\cdot}Cl) + R \rightarrow MO_x + CO_2 + H_2O + Cl^{-1} \tag{11}$$

The oxidation of organic compounds via active chlorine species had a serious concern about the formation of chlorinated organic compounds in water which are known as carcinogens, but Panizza and Bonfatti successfully demonstrated that by selecting optimal experimental conditions, the generations of toxic chlorinated compounds can be totally avoided in aqueous medium [46, 47]. For example,

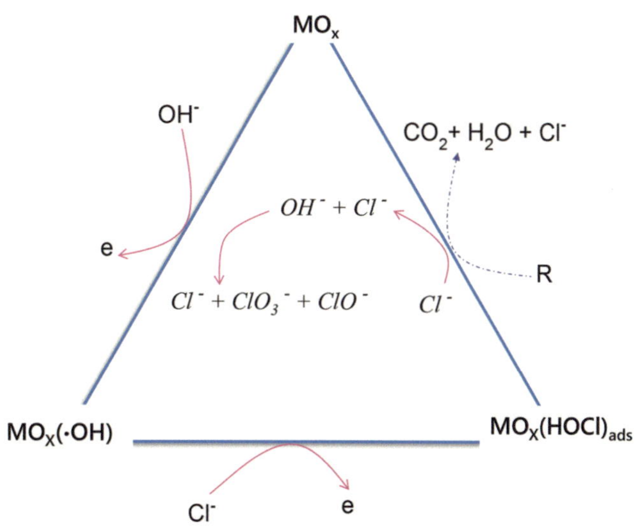

Figure 2.
Mechanism of electrochemical degradation of pollutants in chloride medium.

Panizza investigated electrochemical oxidation of 2-naphthol which was associated with the formation of few chlorinated organic compounds initially, but later as these reactions proceeded, chlorinated compounds were either mineralized to CO_2 or converted to volatile $CHCl_3$ which escaped off [48].

3. Conclusion

Advance oxidation process has a tremendous power to degrade dyes in aqueous solution. Some of these technologies like electro-Fenton, bio-electro-Fenton, photoelectrocatalytic and photocatalytic techniques and process based on peroxymonosulfate obtain good focus in the last decades. Moreover, the performance of some of these processes has been improved by coupling ultrasound, UV and biological process. The recent development of electrode materials and membranes enhances the efficiency of electrochemical process to remove dyes from wastewater. During the electrochemical process, the cathode also helps to generate active species ($S_2O_8^{2-}$, Cl^-, HO•), depending on experimental conditions. Additionally, it has been shown that the integration of different technologies is a good alternative for the degradation of dyes. From our point of view, a lot research has been done, and now it is time to enlarge and apply at pilot and industrial scale. Similarly, reactor should be design to work batch and continues mode and industrial water reusable for different purposes.

Author details

Sajjad Hussain[1,2]*, Nadeem Khan[1], Saima Gul[3], Sabir Khan[4] and Hammad Khan[1]

1 Faculty of Materials and Chemical Engineering, GIK Institute of Engineering Sciences and Technology, KPK, Pakistan

2 Faculty of Engineering, Architecture, Urban Planning and Geography, Federal University of Mato Grosso do Sul, Campo Grande, MS, Brazil

3 Department of Chemistry, Islamia College Peshawar, Pakistan

4 Institute of Chemistry Araraquara, State University of Sao Paulo (UNESP), Araraquara, SP, Brazil

*Address all correspondence to: sajjad.hussain@giki.edu.pk

References

[1] Pearce CI, Lloyd JR, Guthrie JT. The removal of colour from textile wastewater using whole bacterial cells: A review. Dyes and Pigments. 2003;**58**:179-196. DOI: 10.1016/S0143-7208(03)00064-0

[2] Husain Q. Potential applications of the oxidoreductive enzymes in the decolorization and detoxification of textile and other synthetic dyes from polluted water: A review. Critical Reviews in Biotechnology. 2006;**26**:201-221. DOI: 10.1080/07388550600969936

[3] Robinson T, McMullan G, Marchant R, Nigam P. Remediation of dyes in textile effluent: A critical review on current treatment technologies with a proposed alternative. Bioresource Technology. 2001;**77**:247-255. DOI: 10.1016/S0960-8524(00)00080-8

[4] Ratna, Padhi B. Pollution due to synthetic dyes toxicity & carcinogenicity studies and remediation. International Journal of Environmental Sciences. 2012;**3**:940-955. DOI: 10.6088/ijes.2012030133002

[5] Lachheb H, Puzenat E, Houas A, Ksibi M, Elaloui E, Guillard C, et al. Photocatalytic degradation of various types of dyes (alizarin S, Crocein Orange G, methyl red, Congo red, methylene blue) in water by UV-irradiated titania. Applied Catalysis B: Environmental. 2002;**39**:75-90. DOI: 10.1016/S0926-3373(02)00078-4

[6] de Oliveira GAR, Leme DM, de Lapuente J, Brito LB, Porredón C, de Rodrigues LB, et al. A test battery for assessing the ecotoxic effects of textile dyes. Chemico-Biological Interactions. 2018;**291**:171-179. DOI: 10.1016/j.cbi.2018.06.026

[7] Kobylewski S, Jacobson MF. Toxicology of food dyes. International Journal of Occupational and Environmental Health. 2012;**18**:220-246. DOI: 10.1179/1077352512Z.00000000034

[8] Hunger K. Industrial dyes: Chemistry, properties. Journal of the American Chemical Society. 2003a;**125**:10144. DOI: 10.1021/ja0335418

[9] Hunger K. Industrial dyes: Chemistry, properties, applications edited by Klaus Hunger (Kelkheim, Germany). Wiley-VCH: Weinheim. 2003. Xxiv + 600 pp. $185.00. ISBN 3-527-30426-6. Journal of the American Chemical Society. 2003b;**125**:10144-10144. DOI: 10.1021/ja0335418

[10] Chudgar RJ, Oakes J. Dyes, azo. In: Kirk-Othmer Encycl. Chem. Technol. Hoboken, NJ, USA: John Wiley & Sons, Inc.; 2014

[11] Drumond Chequer FM, de Oliveira GAR, Anastacio Ferraz ER, Carvalho J, Boldrin Zanoni MV, de Oliveir DP. Textile dyes: Dyeing process and environmental impact. In: Eco-Friendly Textile Dyeing and Finishing. InTech: IntechOpen; 2013. DOI: 10.5772/53659

[12] Azbar N, Yonar T, Kestioglu K. Comparison of various advanced oxidation processes and chemical treatment methods for COD and color removal from a polyester and acetate fiber dyeing effluent. Chemosphere. 2004;**55**:35-43. DOI: 10.1016/j.chemosphere.2003.10.046

[13] Rai HS, Bhattacharyya MS, Singh J, Bansal TK, Vats P, Banerjee UC. Removal of dyes from the effluent of textile and dyestuff manufacturing industry: A review of emerging techniques with reference to biological treatment. Critical Reviews in Environmental Science and Technology. 2005;**35**:219-238. DOI: 10.1080/10643380590917932

[14] Yagub MT, Sen TK, Afroze S, Ang HM. Dye and its removal from aqueous solution by adsorption: A review. Advances in Colloid and Interface Science. 2014;**209**:172-184. DOI: 10.1016/j.cis.2014.04.002

[15] Nidheesh PV, Gandhimathi R, Velmathi S, Sanjini NS. Magnetite as a heterogeneous electro Fenton catalyst for the removal of Rhodamine B from aqueous solution. RSC Advances. 2014;**4**:5698. DOI: 10.1039/c3ra46969g

[16] Gupta VK, Jain R, Nayak A, Agarwal S, Shrivastava M. Removal of the hazardous dye-Tartrazine by photodegradation on titanium dioxide surface. Materials Science and Engineering: C. 2011;**31**:1062-1067. DOI: 10.1016/j.msec.2011.03.006

[17] Hussain S, Gul S, Khan S, Rehman H, Ishaq M, Khan A, et al. Removal of Cr(VI) from aqueous solution using brick kiln chimney waste as adsorbent. Desalination and Water Treatment. 2015a;**53**:373-381. DOI: 10.1080/19443994.2013.837001

[18] Mittal A, Kurup L. Column operations for the removal and recovery of a hazardous dye 'acid red –27' from aqueous solutions, using waste materials - bottom ash and de-oiled soya. Ecol. Environmental Conservation. 2006;**12**:181-186

[19] Thiam A, Sirés I, Brillas E. Treatment of a mixture of food color additives (E122, E124 and E129) in different water matrices by UVA and solar photoelectro-Fenton. Water Research. 2015;**81**:178-187. DOI: 10.1016/j.watres.2015.05.057

[20] Moreira FC, Boaventura RAR, Brillas E, Vilar VJP. Electrochemical advanced oxidation processes: A review on their application to synthetic and real wastewaters. Applied Catalysis B: Environmental. 2017;**202**:217-261

[21] Nidheesh PV, Gandhimathi R, Ramesh ST. Degradation of dyes from aqueous solution by Fenton processes: A review. Environmental Science and Pollution Research. 2013;**20**:2099-2132. DOI: 10.1007/s11356-012-1385-z

[22] Li X, Jin X, Zhao N, Angelidaki I, Zhang Y. Novel bio-electro-Fenton technology for azo dye wastewater treatment using microbial reverse-electrodialysis electrolysis cell. Bioresource Technology. 2017;**228**:322-329. DOI: 10.1016/j.biortech.2016.12.114

[23] Ikhlaq A, Kasprzyk-Hordern B. Catalytic ozonation of chlorinated VOCs on ZSM-5 zeolites and alumina: Formation of chlorides. Applied Catalysis B: Environmental. 2017;**200**:274-282. DOI: 10.1016/j.apcatb.2016.07.019

[24] Pereira GF, El-Ghenymy A, Thiam A, Carlesi C, Eguiluz KIB, Salazar-Banda GR, et al. Effective removal of Orange-G azo dye from water by electro-Fenton and photoelectro-Fenton processes using a boron-doped diamond anode. Separation and Purification Technology. 2016;**160**:145-151. DOI: 10.1016/j.seppur.2016.01.029

[25] Erkurt EA, Erkurt HA, Unyayar A. Decolorization of Azo Dyes by White Rot Fungi. Berlin, Heidelberg: Springer; 2010. pp. 157-167

[26] van der Zee FP, Villaverde S. Combined anaerobic–aerobic treatment of azo dyes—A short review of bioreactor studies. Water Research. 2005;**39**:1425-1440. DOI: 10.1016/j.watres.2005.03.007

[27] Aksu Z. Application of biosorption for the removal of organic pollutants: A review. Process Biochemistry. 2005;**40**:997-1026. DOI: 10.1016/j.procbio.2004.04.008

[28] Martínez-Huitle C a, Brillas E. Decontamination of wastewaters containing synthetic organic dyes by electrochemical methods: A general review. Applied Catalysis B: Environmental. 2009;**87**:105-145. DOI: 10.1016/j.apcatb.2008.09.017

[29] Modirshahla N, Behnajady MA. Photooxidative degradation of malachite green (MG) by UV/H2O2: Influence of operational parameters and kinetic modeling. Dyes and Pigments. 2006;**70**:54-59. DOI: 10.1016/j.dyepig.2005.04.012

[30] Hussain S, Steter JR, Gul S, Motheo AJ. Photo-assisted electrochemical degradation of sulfamethoxazole using a Ti/Ru 0.3 Ti 0.7 O 2 anode: Mechanistic and kinetic features of the process. Journal of Environmental Management. 2017;**201**:153-162. DOI: 10.1016/j.jenvman.2017.06.043

[31] Nilsson A, Ronlán A, Parker VD. Anodic oxidation of phenolic compounds. Part III. Anodic hydroxylation of phenols. A simple general synthesis of 4-alkyl-4-hydroxycyclo-hexa-2,5-dienones from 4-alkylphenols. Journal of the Chemical Society, Perkin Transactions 1. 1973:2337-2345. DOI: 10.1039/P19730002337

[32] Sirés I, Brillas E, Oturan MA, Rodrigo MA, Panizza M. Electrochemical advanced oxidation processes: Today and tomorrow. A review. Environmental Science and Pollution Research International. 2014;**21**:8336-8367. DOI: 10.1007/s11356-014-2783-1

[33] Zhang F, Feng C, Li W, Cui J. Indirect electrochemical oxidation of dye wastewater containing acid Orange 7 using Ti/RuO2-Pt electrode. International Journal of Electrochemical Science. 2014;**9**:943-954

[34] Hussain S, Gul S, Steter JR, Miwa DW, Motheo AJ. Route of electrochemical oxidation of the antibiotic sulfamethoxazole on a mixed oxide anode. Environmental Science and Pollution Research. 2015b;**22**:15004-15015. DOI: 10.1007/s11356-015-4699-9

[35] Rocha JHB, Gomes MMS, Fernandes NS, Da Silva DR, Martínez-Huitle CA. Application of electrochemical oxidation as alternative treatment of produced water generated by Brazilian petrochemical industry. Fuel Processing Technology. 2012;**96**:80-87. DOI: 10.1016/j.fuproc.2011.12.011

[36] Comninellis C, Nerini A. Anodic oxidation of phenol in the presence of NaCl for wastewater treatment. Journal of Applied Electrochemistry. 1995;**25**:23-28. DOI: 10.1007/BF00251260

[37] Panizza M, Cerisola G. Direct and mediated anodic oxidation of organic pollutants. Chemical Reviews. 2009;**109**:6541-6569. DOI: 10.1021/cr9001319

[38] Garcia-Segura S, Vieira dos Santos E, Martínez-Huitle CA. Role of sp3/sp2 ratio on the electrocatalytic properties of boron-doped diamond electrodes: A mini review. Electrochemistry Communications. 2015;**59**:52-55. DOI: 10.1016/j.elecom.2015.07.002

[39] Tavares MG, da Silva LVA, Sales Solano AM, Tonholo J, Martínez-Huitle CA, Zanta CLPS. Electrochemical oxidation of methyl red using Ti/Ru 0.3Ti 0.7O 2 and Ti/Pt anodes. Chemical Engineering Journal. 2012;**204-206**:141-150. DOI: 10.1016/j.cej.2012.07.056

[40] Radjenovic J, Sedlak DL. Challenges and opportunities for electrochemical processes as next-generation

Technologies for the Treatment of contaminated water. Environmental Science & Technology. 2015;**49**:11292-11302. DOI: 10.1021/acs.est.5b02414

[41] Abdessalem AK, Oturan MA, Oturan N, Bellakhal N, Dachraoui M. Treatment of an aqueous pesticides mixture solution by direct and indirect electrochemical advanced oxidation processes. International Journal of Environmental Analytical Chemistry. 2010;**90**:468-477. DOI: 10.1080/03067310902999132

[42] Martínez-Huitle C, Ferro S, De Battisti A. Electrochemical incineration of oxalic acid. Electrochimica Acta. 2004;**49**:4027-4034. DOI: 10.1016/j.electacta.2004.01.083

[43] Rajkumar D, Kim JG. Oxidation of various reactive dyes with in situ electro-generated active chlorine for textile dyeing industry wastewater treatment. Journal of Hazardous Materials. 2006;**136**:203-212. DOI: 10.1016/j.jhazmat.2005.11.096

[44] da Silva RG, Aquino Neto S, de Andrade AR. Electrochemical degradation of reactive dyes at different DSA® compositions. Journal of the Brazilian Chemical Society. 2011;**22**:126-133. DOI: 10.1590/S0103-50532011000100017

[45] Bonfatti F, Ferro S, Lavezzo F, Malacarne M, Lodi G, De Battisti A. Electrochemical incineration of glucose as a model organic substrate. II. Role of active chlorine mediation. Journal of the Electrochemical Society. 2000b;**147**:592-596

[46] Bonfatti F, De Battisti A, Ferro S, Lodi G, Osti S. Anodic mineralization of organic substrates in chloride-containing aqueous media. Electrochimica Acta. 2000a;**46**:305-314. DOI: 10.1016/S0013-4686(00)00586-7

[47] Panizza M. Importance of electrode material in the electrochemical treatment of wastewater containing organic pollutants. In: Electrochemistry for the Environment. New York: Springer; 2010. pp. 25-54

[48] Panizza M, Cerisola G. Electrochemical oxidation of 2-naphthol with in situ electro- generated active chlorine. Electrochimica Acta. 2003;**48**:1515-1519

Chapter 8

Bi-Functionalized Hybrid Materials as Novel Adsorbents for Heavy Metal Removal from Aqueous Solution: Batch and Fixed-Bed Techniques

Yasser Hannachi, Afifa Hafidh and Salwa Ayed

Abstract

In this study, two new mesoporous hybrid gels were synthesized. The structural order, morphology, and textural properties of the prepared hybrid materials have been studied by ^{13}C CP MAS NMR, SEM, FTIR, and nitrogen adsorption–desorption analysis. The application for the heavy metal uptake from aqueous solution using the as-synthesized hybrid materials as an adsorbent is explored. Operating parameters influencing the adsorption procedure, for instance, solution pH, contact time, and temperature are contemplated. In order to gain an insight into the adsorption mechanism and reveal the rate-controlling steps, three models pseudo-first-order, pseudo-second-order, and intra-particle diffusion have been studied to fit. Langmuir, Freundlich, and Dubinin-Radushkevich (D-R) models are assigned to portray the adsorption isotherms. Besides, the feasibility of the synthesized adsorbents for a continuous process in fixed-bed column was investigated. Prior tests produced on electroplating effluents reveal that the as-prepared xerogel could be strongly used for the heavy metal uptake from real wastewater.

Keywords: xerogels, adsorption, kinetic, fixed-bed studies, selectivity, reusability

1. Introduction

The prompt industrial development and ongoing urbanization have induced growing problems relevant to waste output. Wastes are the generator of hazardous substances which migrate unimpeded through the ecosystem posing severe jeopardy to living organisms. Heavy metals are deemed as a standout among the most toxic groups of inorganic pollutants. The principal origin sources of these heavy metals are usually from electroplating, plastic manufacturing, mining, fertilizers, etc. [1]. The excessive magnitude of these contaminants in industrial wastewaters will severely devastate the ecosystem and human health via accumulation and spreading in the environment and food chain [2]. Faced with the strict environmental enactment, the treatment of wastewater before being discharged into the environment becomes paramount topics worldwide. Diverse treatment processes, for instance, coagulation, precipitation, ion-exchange, reverse osmosis, and electrolysis, are applied for the remediation of heavy metal ions from aqueous solutions [3–5]. Notwithstanding, by far, the majority of these procedures are restrained by both ecological and economic constraints. In contrast, adsorption has been qualified to be the most relevant and promising method, on account of its low cost, flexible operation, and reversibility. A wide variety of low-cost adsorbents, for example, agricultural by-products, industrial wastes, and natural mineral materials, were found to have good adsorption capacity [6]. However, these materials possess weak resistance to abrasive forces in column apply and leaching of few organics throughout retention process.

It is well known that the substantial factors affecting the adsorption process are pore size distribution, specific surface areas, and pore surface chemistry. In this endeavor, it is imperative to search materials with enormous porous structures, high specific surface areas, and low density. In the last few decades, ordered mesoporous silica has triggered a growing interest in the field of water treatment, owing to its diverse outstanding properties. These enclose tunable pore-size, high specific surface area, large pore volume, chemical inertness, thermal stability, and the ability to attach a plethora of different functional groups [7–9].

One of the basic techniques applied for fabricating organic-inorganic hybrid materials is the sol-gel method. Tetraalkoxysilanes (Si(OR)4) like tetraethyl orthosilicate (TEOS) or tetramethyl orthosilicate (TMOS) are widely employed as a precursor for preparing monolithic silica, owing to their hydrophobicity and strong covalent Si-O bonding [10]. It is worthy to state that the as-prepared hybrid materials by the sol-gel route are divided into two classes on account of the interaction type between organic and inorganic components. In class I, organic and inorganic components are strongly attached by covalent or dative covalent bonds, while in the second, these two components are weakly bonded through hydrogen or Van der Waals bonds. Indeed, these materials have generated considerable interest for their potential application in multiple fields such as: adsorption [11], drug delivery systems [12, 13], biosensors [14], nanotechnology, and nanomedicine applications [15, 16] and catalysis [17]. Organic-inorganic hybrid materials displayed high efficiency and outstanding selectivity towards target pollutant than the other silica gels. Nitrogen/thiol and magnetic functionalized mesoporous silica have been at the foreground of these composites. In the same vein, Benhamou et al. have shown that amine functionalized MCM-41 and MCM-48 exhibited a higher adsorption capacity than pristine. They also evinced that both hybrid materials have a higher affinity for Cu(II) and Pb(II) than for Cd(II) and Co(II) in single and mixed cation solutions [18]. Shahbazi and co-workers grafted aminopropyl (NH_2) and melamine-based dendrimer amines (MDA) to SBA-15 mesoporous silica. They observed that NH_2–SBA-15 and MDA–SBA-15 were over ten-fold better than the pristine SBA-15 in the adsorption of Pb(II), Cu (II), and Cd(II) [19]. Interestingly, they also showed that in column studies, the adsorption yield was swayed by the flow rate. Apart from the fact that magnetic silica-based materials exhibited excellent adsorption affinity towards heavy metal ions, such adsorbents compared to the nitrogen and thiol designer silicates can be easily removed from aqueous solution after adsorption. In the same context, Wang et al. [20] synthesized an amino-functionalized core-shell magnetic mesoporous SBA-15 silica composite which displayed a high adsorption capacity for Pb(II) ions. This adsorbent can be readily removed and regenerated. Despite the high adsorption capacity and the extra-ordinary selectivity towards target metal, as well as the capacity for simultaneous removal of aqueous pollutants, organic-inorganic hybrid materials are still not applied for a continuous process in a fixed-bed column. Another challenge to overcome is the difficulty in their large-scale production because of the complexity of synthesis methods and the control of the stability of the developed adsorbents. Therefore, great endeavors are required for the synthesis of hybrid silicate adsorbents which meet the listed challenges.

The main aims of the present study were the following:

Firstly, to synthesize and characterize new functionalized hybrid materials and secondly, to assess their ability to remove Zn (II), Cd(II), and Cu(II) ions from aqueous solution. To achieve these goals, experimental parameters affecting the adsorption such as pH solution and contact time were determined. Kinetic data were expressed by pseudo-first-order, pseudo-second-order, and intra-particle diffusion models. Adsorption equilibrium isotherms were evaluated according to Langmuir, Freundlich, and Dubinin-Radushkevich (D-R) models. In addition, the thermodynamic parameters, selectivity, and the reusability have been determined and extensively discussed. To the best of our knowledge, there are scant studies dealing with the applicability of silica hybrid materials for a continuous process in a fixed-bed column. Herein, we have successfully proven the feasibility of the synthesized adsorbents for large volumes of discharges.

2. Materials and methods

2.1 Chemical reagents

All reagents were of analytical grade and used as received without further purification. Hydrophobic tetraethyl orthosilicate (TEOS, 99%) was utilized as a silica precursor, while ethanol was a bridging medium. $Cd(NO_3)2.4H2O$, Cu(NO3) 2.3H2O, and Zn(NO3)2.6H2O were employed as metal sources for batch and column adsorption experiments. These reagents were supplied by Sigma-Aldrich, USA. The organic precursors 1,3,4-thiadiazole-2,5-diamine and 1,3,4-thiadiazole-2,5-dithiol were prepared according to the literature [21, 22].

2.2 Methods

2.2.1 Adsorbent synthesis

Pursuant to our foregoing studies, xerogels were synthesized using the following process [Helali 15 et 16]: 10 ml of deionized water, 20ml of ethanol, and 22.8 mL of TEOS were mixed under vigorous magnetic stirring. To the as-prepared mixture was added a necessary amount of organic precursor (10^{-1} M, 11.6 g of 1,3,4-thiadiazole-2,5-diamineor or 15 g of 1,3,4-thiadiazole-2,5-dithiol). Thereafter, the reactant mixture was stirred for 6 h at 78°C and at the last ripened for 48 h at 100°C; the resulting xerogels were labeled $\mathbf{M_1}$ and $\mathbf{M_2}$, and the synthesis mechanism is represented in **Figure 1**.

2.2.2 Characterization

Xerogel morphology was carried out using a scanning electronic microscope (Cambridge Instruments Stereoscan 120) operating at 15 kV. The textural properties of hybrid materials were determined from the N_2 adsorption-desorption isotherms recorded at 77 K with a Micrometrics ASAP-2000 volumetric apparatus. The specific surface areas were computed by the multi-point analysis (BET) (Brunauer et al., 1938) in the relative pressure interval of $0.03 < P/P^{\circ} < 0.3$. Howbeit pore size distribution was acquired from the adsorption-desorption branches of the isotherm

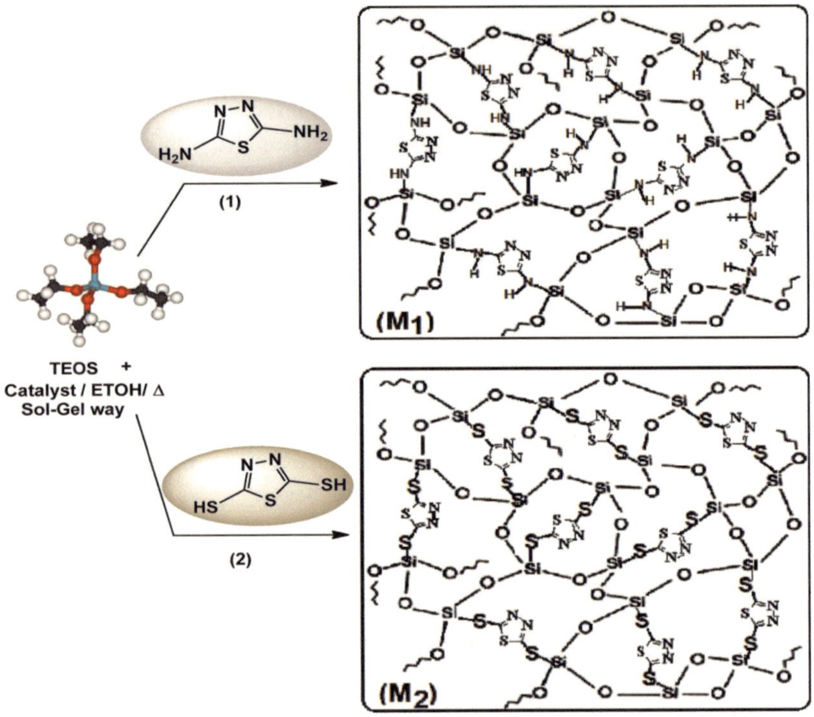

Figure 1.
Synthesis mechanism of hybrid materials $\mathbf{M_1}$ *and* $\mathbf{M_2}$.

through the BJH pattern. The total pore volume was evaluated at a relative pressure of $P/P^{\circ} = 0.99$.

The experimental parameters for the ^{13}C CP MAS NMR were 9 KHz SPIN rate, 5 s pulse delay. NMR spectroscopy was carried out on an MSL 500 Bruker Spectrometer. FT-IR spectra were collected on 550 Nicolet Magana Spectrometer in KBr pellets in the range of 4000–400 cm^{-1}. X-ray photoelectron spectroscopy (XPS) measurement was conducted on a VG ESCALAB MK II spectrometer in the pulse-count mode at a pass energy of 50 eV employing a Mg Kα (1253.6 eV) achromatic X-ray source. In order to evaluate the surface charges, the electro-kinetic potential was performed by Malvern instrument Zeta Nano ZS.

2.2.3 Adsorption experiments

Stock solutions were set up by dissolving the required metal mass in 1 L of double-distilled water. Aliquots were prepared by diluting standard stock solution to the desired concentrations (5–400 $mg.L^{-1}$). All experiments were done at room temperature in triplicate, and the average values were utilized for further estimation. For every essay, 0.01 g of xerogel was thoroughly blended in conical flasks containing 25 ml of test solution with various metal concentrations at a required pH adjusted prior to the experiment with 0.1 mol. L^{-1} of HNO_3 or 0.1 mol. L^{-1} of NaOH solution. The flasks were shaken for the coveted contact time in an electrically thermostatic reciprocating shaker (Selecta multimatic-55, Spain) at 150 rpm. The contact time for metal ions and the hybrid materials were ranged from 10 to 100 min.

For the adsorption isotherm contemplates, the initial metal concentration was run from 10 to 400 mg. L^{-1}. After each adsorption procedure, the gathered examples were centrifuged at 12000 rpm for 20min with the end goal to separate the solid from the liquid phase. The supernatant containing metal ions was evaluated employing an atomic absorption spectrophotometer (SHIMADZU AA-680, Japan). The percent adsorption of metal ion delineated is as follows (Eq. (1)):

$$adsorption\ (\%) = \frac{(C_i - C_f)}{C_i} \times 100 \tag{1}$$

where C_i and C_f are the initial and final (or equilibrium) adsorbate concentrations, respectively.

2.2.4 Fixed-bed column studies

Dynamic adsorption trial run in the fixed-bed column was conducted in a glass column (3.5 cm length and 1.2 cm diameter), stuffed with 1 g of the xerogel. The metal ion concentration (150 mg. L^{-1}) was pumped in a down flow mode at a flow rate of 20 mL min^{-1} by using a peristaltic pump (Flowtech India, model NFP01). Samples were gathered at determined time interims, and the residual adsorbate concentration (Ce) was measured spectrophotometrically (Eq. (2)). Column exploitation was ceased when the adsorbate concentration attained 95% of its initial concentration. The maximum column capacity (q_c) can be reckoned as follows Eq. (3):

$$q_t = A \times Q \times C \times \frac{1}{1000} \tag{2}$$

$$q_e = \frac{q_{total}}{M} \tag{3}$$

where q_{total} (mg) is the total quantity of adsorbed metal ions, A is the area under breakthrough curve ($C_e/(C_0)$ versus time, Q (mL.min^{-1}) is the flow rate, C_0 (mg.L^{-1}) is the influent concentration, and M (g) is the mass of adsorbent.

2.2.5 Desorption studies

As a means to examine the regeneration of the synthesized hybrid materials, the metal cation-loaded adsorbents are desorbed with 10 mL of HCl solution (0.5 M) for 60 minutes. Afterwards, the recovered xerogel is flushed with deionized water

Parameters	Values	WHO Standard
pH	2.8	5.5–6.5
Temperature	40	20–30
TSS (mg.L^{-1})	720	20
COD (mg.L^{-1})	371	280
BOD (mg.L^{-1})	90	40
Zn (mg.L^{-1})	100	5
Pb (mg.L^{-1})	10	0.2
Cd (mg.L^{-1})	2	0.1

Table 1.
Physicochemical characterization of electroplating wastewaters.

and parched in the air for the forthcoming experiment. Successive sorption-desorption cycles are rehashed 10 times to build up the genuine application and the high stability of the adsorbent.

2.2.6 Adsorption test with the electroplating wastewaters

The effluent specimens are gathered from the discharge exits of electroplating plant, Yaroslavl, Russia. The physicochemical parameters of the electroplating effluent are enlisted in **Table 1**.

Samples are stored in plastic bottles and cooled to 20°C; afterward, they are diluted ten times and alkalized with 0.5 mol.L^{-1} of NaOH solution, and finally, are filtered through a 0.35 μm membrane filter. All physicochemical features of the effluent are determined by a conventional procedure [23].

3. Results and discussion

3.1 Characterization of adsorbents

The morphology of xerogel adsorbents was portrayed by SEM, and the images are illustrated in **Figure 2**.

It very well may be obviously observed that all hybrid materials evinced wrinkled surface and irregular shaped particles with pore diameter over 1.5 nm, indicating the mesoporous structure. The textural properties of the synthesized materials, including specific surface area, total mesoporous volume, and average pore diameter were assessed by N2 adsorption-desorption isotherms and BJH method. Interestingly, all samples recorded typical type IV adsorption-desorption isotherms which is a characteristic pattern of mesoporous composites as stated in IUPAC classification. Moreover, the hysteresis loop is of type H_2 which is usually tied to the ink-bottle pores with bulky orifice of the border inner parts (**Figure 3**).

A clearly defined step is eventuated roughly at P/P° = 0.4 characteristic of the mesopore filling owing to capillary condensation [24]. Evenly, the pore size curve obtained from the isotherm branch showed a tight distribution centered at about 3.5 nm.

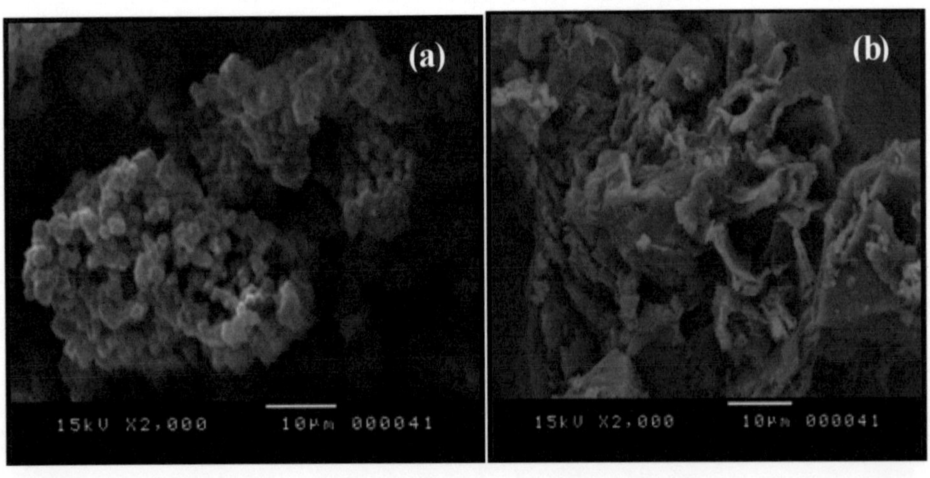

Figure 2.
SEM micrographs of pristine adsorbents: (a) $\mathbf{M_1}$ *and (b)* $\mathbf{M_2}$*.*

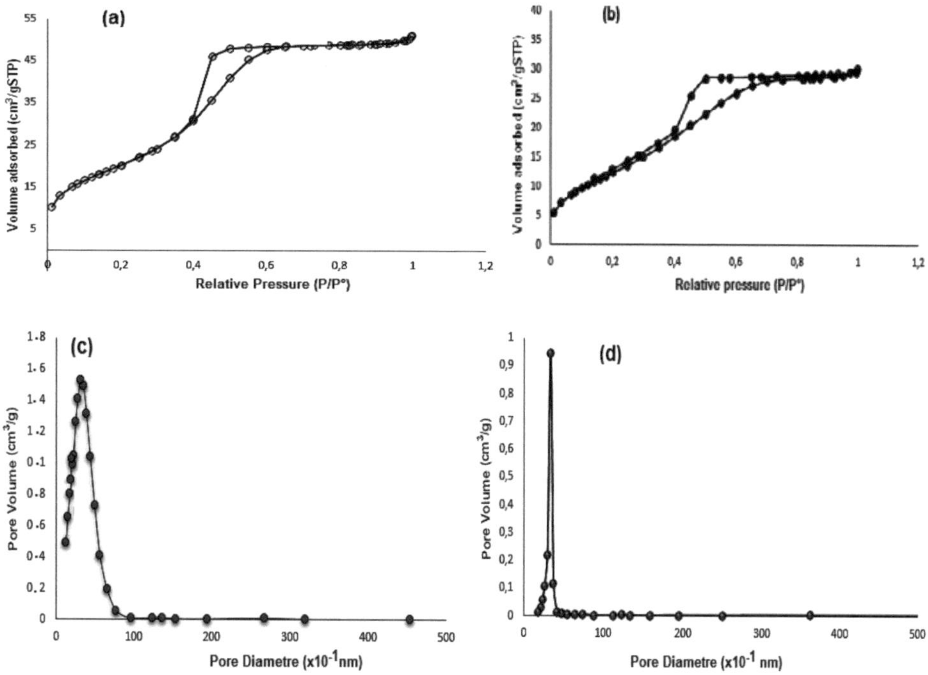

Figure 3.
N_2 absorption-desorption isotherms of (a) $\mathbf{M_1}$ *and (b)* $\mathbf{M_2}$ *and BJH pore size distribution of (c)* $\mathbf{M_1}$ *and (d)* $\mathbf{M_2}$*.*

Sample	S_{BET} ($m^2.g^{-1}$)	V_{por} ($cm^3.g^{-1}$)	d_{moy} (Å)
M_1	290	0.4	34.80
M_2	310	0.2	20

Table 2.
BET surface area, pore volume, and average pore size of xerogels $\mathbf{M_1}$ *and* $\mathbf{M_2}$*.*

Based on data in **Table 2**, all xerogels showed relatively high specific surfaces and pore volumes. More convincing underpins for the successful anchor of organic moieties into the siliceous network emanate from the ^{13}C CP MAS NMR analysis.

The ^{13}C NMR spectra of the synthesized materials are shown in **Figure 4**. These spectra recorded resonance peaks at 156.3–166.4 ppm (C=N) typical of sp^2 carbon atoms which characterized the cross links of the organic functional groups in the inorganic network, providing the possibility of creating Si-N and Si-S covalent bonds. The peaks at 13–58 ppm were attributed to CH_3CH_2O-Si- groups (TEOS). This outcome portrayed the incorporation of the organic bi-functional compounds in the inorganic network, providing the formation of Si-N and Si-S covalent bonds.

So as to confirm the attachment of the organic precursors onto the skeleton of silica species, FT-IR spectroscopy was carried out.

The FT-IR spectra of all samples are depicted in **Figure 5**.

All samples showed a typical band related to Si-O bonds, at 434–440 cm^{-1}, bending of O-Si-O at 754–760cm^{-1}, and strong bands at 1061–1134 cm^{-1} assigned to the stretching vibration of Si-O-Si groups. Besides, the broad and strong bands detected at 3383–3325 cm^{-1} were attributed to the stretching vibration of OH groups which are associated to Si-OH groups ensuing from the TEOS hydrolysis. The signals revealed at 1557–1576 cm^{-1} were ascribed to the stretching vibration of C=N groups (heterocyclic part). Another indicative band of the covalent Si-N bond

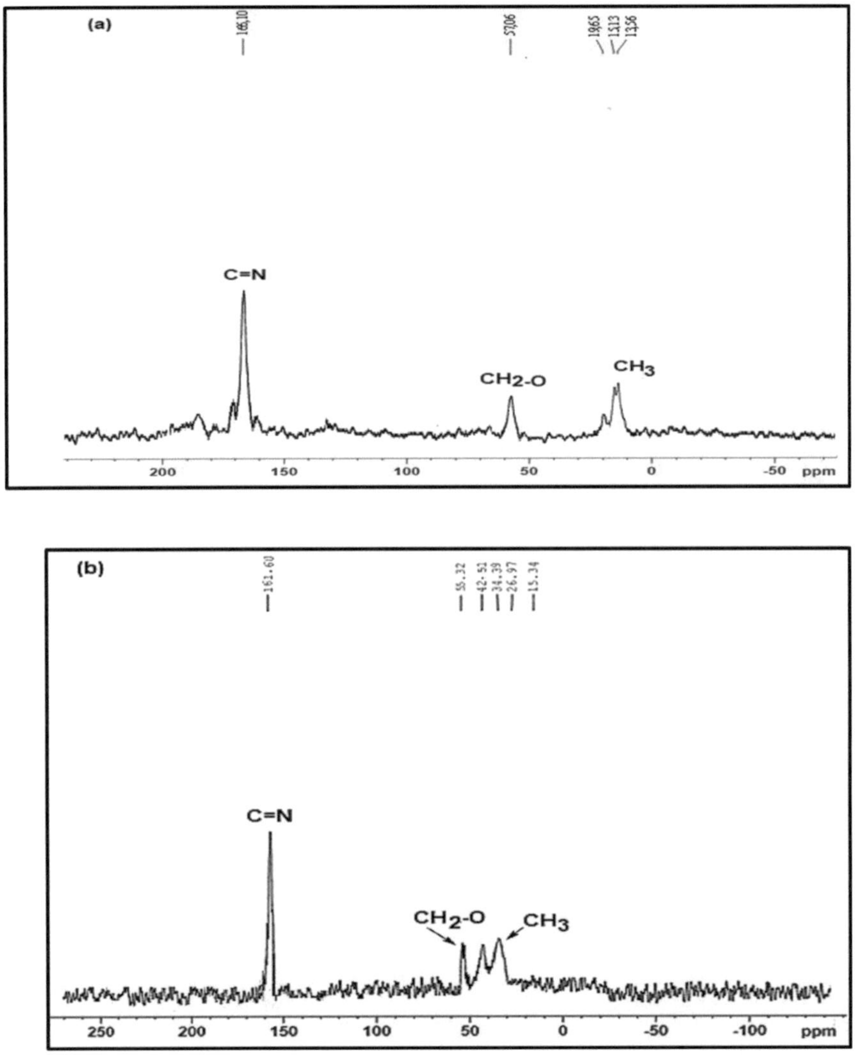

Figure 4.
^{13}C NMR CP MAS spectra of $\mathbf{M_1}$ (a) and $\mathbf{M_2}$ (b).

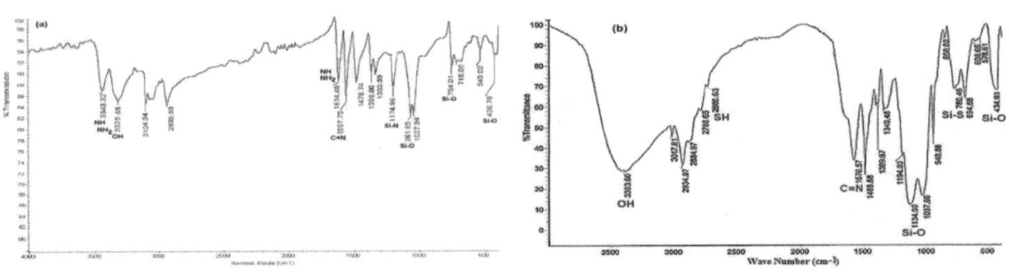

Figure 5.
IR spectra of: (a) $\mathbf{M_1}$ and (b) $\mathbf{M_2}$.

created between 1,2,4-thiadiazole heterocyclic molecules and the polysiloxane backbone emerged at 1175 cm^{-1}. The bands located at 780 and 2680 cm^{-1}, attributed to Si-S links and S-H groups, were observed in the $\mathbf{M_2}$ spectrum. In contrast, the signals at 1614 and 3343 cm^{-1} recognize the presence of Si-N links, NH and NH_2 groups, respectively, in $\mathbf{M_1}$ [7].

3.2 Effect of pH solution

The pH of the solution is regarded as one of the foremost adsorption factors, since it has a prominent impact on the metal ion solubility, as well as on the adsorbent surface metal binding sites.

The effect of pH on the removal efficiency of heavy metal ions by hybrid materials was assessed inside a scope of 2–10. Zeta potential is the best and reliable method to determine the adsorbent surface charge, which was characterized by point of zero charge (pHpzc). The pH of zero charge (pHZps) of the as-prepared xerogels $\mathbf{M_1}$ and $\mathbf{M_2}$ is recognized to be 4.6 ± 0.2 and 4.2 ± 0.2, respectively (**Figure 6**).

As illustrated in **Figure 6**, when pH ˂ pHZps, the adsorbent surface charge is positive by virtue of the amino and thiol group protonation. Therefore, electrostatic repulsive force emerges between heavy metal ions and the adsorbent surface, inciting abatement in the adsorption capacity. Be that as it may, at pH > pHZps, all samples earn a negative surface charge, promoting electrostatic attraction between metal cations and adsorbent and therefore, the absorption efficiency enhancement. It ought to be stressed that beyond pH 5 [25], the uptake yield of heavy metal ions diminishes through the metallic hydroxide precipitation which thwarts the diffusion of the metal ions into the adsorbent active site. Therethrough, the pH at 7 was selected as the ideal incentive in the resulting experiments (**Figure 7**).

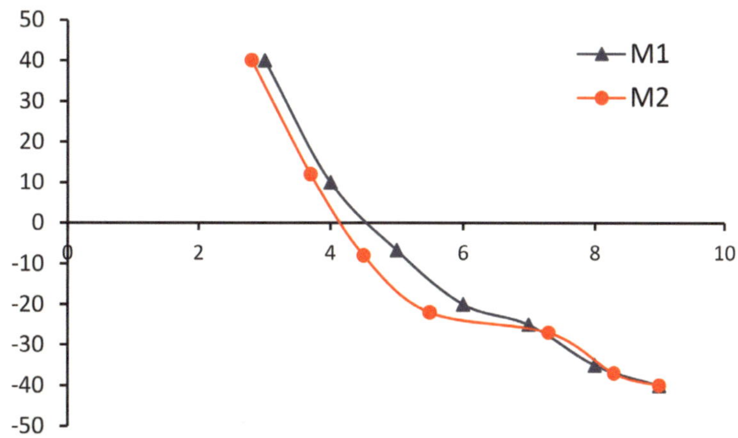

Figure 6.
Zeta potential of xerogels at different pH values.

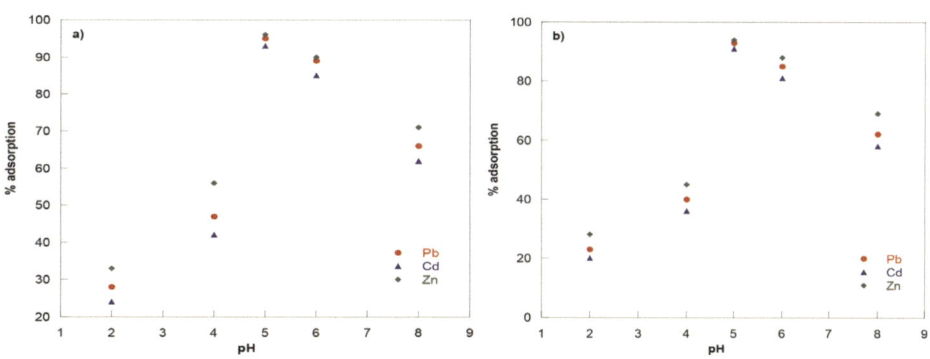

Figure 7.
Effect of pH on adsorption of metal ions onto both xerogels ((a)$\mathbf{M_1}$ and (b)$\mathbf{M_2}$) (metal concentration 20 mg.L^{-1}; adsorbent dosage 0.4 g.L^{-1}, contact time 1 h).

3.3 Adsorption kinetics

The contact time is well recognized as the dwelling time of sorbate uptake at the superficial adsorbent surface. To study the effect of contact time, 0.015 g of hybrid materials was thoroughly mixed in 25 mL of initial metal concentration 20 mg.L^{-1} and was shaken at a rotational speed of 150 rpm.

As portrayed in **Figure 8**, the heavy metal adsorption onto the three xerogels rapidly increased in the first 40 min; thereafter, it becomes slower and in the later stage reaches to saturation (equilibrium). Further increase in contact time did not ameliorate the uptake efficiency; this trend may be attributed to the fact that padding of void active sites becomes impossible owing to the electrostatic repulsion between the solute ions of the adsorbent and bulk phases [26]. Thus, 60 min is seen fit to attain equilibrium in ulterior trials.

In order to acquire an insight into the adsorption mechanism and reveal the rate controlling steps, three kinetic models including pseudo-first-order, pseudo-second-order, and intra-particle diffusion were checked.

The adsorption rates were first examined by Lagergren's pseudo-first-order [27] and its linearized integral form is spoken to as follows in Eq. (4):

$$\ln\left(q_e - q_t\right) = \ln q_e - k_1 t \quad (4)$$

where q_t and q_e (mg.g^{-1}) are the adsorption capacities at equilibrium (mg.g^{-1}) and t (min), respectively, and k_1 is the rate constant of the equation (min^{-1}). The rate constant, k_1, equilibrium adsorption capacity, q_e, and the correlation coefficient R^2, are determined experimentally by plotting $\ln\left(q_e - q_t\right)$ versus t. The allied parameters are listed in **Table 1**.

The pseudo-second-equation provided by Ho [28] can be stated by the pursuing equation Eq. (5):

$$\frac{t}{q_t} = \frac{1}{k_2 q_2^2} + \frac{t}{q_2} \quad (5)$$

where k_2 (g.mg^{-1} min^{-1}) is the rate constant and q_2 is the amount of adsorption equilibrium capacity (mg.g^{-1}). The values of q_e and k_2 can be determined graphically from the slope and the intercept of the plot t/q_t versus t at different temperatures.

The kinetic parameters gained from pseudo-first-order and pseudo-second-order are introduced in **Table 3**. It is obvious that the theoretical adsorption

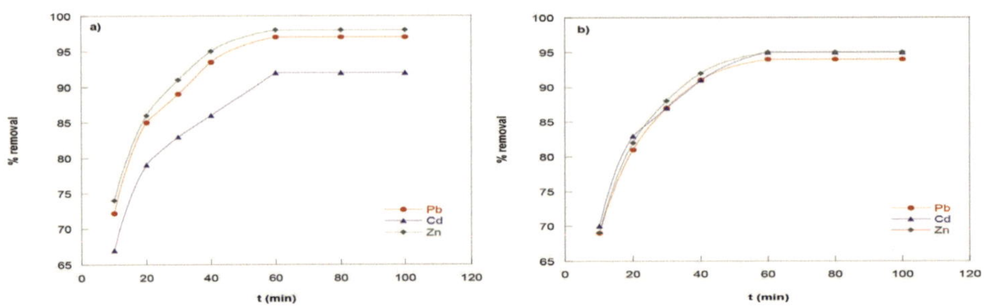

Figure 8.
Effect of contact time and temperature on adsorption of metal ions by the two adsorbents (metal concentration: 20 mg.L^{-1}; adsorbent dosage: 0.4 g.L^{-1}; pH: 5): (a) $\mathbf{M_1}$ and (b) $\mathbf{M_2}$.

capacities ((q_{th})) evaluated from the pseudo-second-order equation were very close to experimental (q_e (ex)) values; besides, the correlation coefficients related to the above-mentioned model were found to be higher than those ascertained from the pseudo-first-order equation. These unequivocally propose that the pseudo-second-order model, as opposed to the pseudo-first-order model, is more suitable to depict the adsorption procedure.

By the by, the aforementioned kinetic models cannot identify the diffusion mechanism and also, the rate controlling step of the adsorption kinetic process. In this regard, it is important to apply the Weber and Morris intra-particle diffusion model. This later assumes that the adsorption process might be controlled either by one of the resulting steps, namely film diffusion, pore diffusion, and adsorption onto the inner sites of adsorbent or a mix of a few stages (multi-step process) [29]. The rate parameter for intra-particle diffusion is displayed as follows Eq. (6):

$$q_t = K_{id} t^{0.5} + C \tag{6}$$

where K_{id} is the intra-particle rate constant ($mg.g.\text{min}^{0.5}$) obtained from the slope of the straight line q_t versus $t^{0.5}$ and C is the intercept corresponding to the boundary layer thickness. If the rate of adsorption is controlled only by the intra-particle diffusion, the value of C should be zero (C = 0) and the plots of q_t against $t^{0.5}$ provide a straight line passing through the origin.

As depicted in **Figure 9**, all the plots show multi-linear uptake revealing three adsorption stages. The first steep-sloped portion corresponds to the transport of

Metal ions	$q_{e\ exp}$ (mg g^{-1})	Pseudo-first-order kinetic			Pseudo-second-order kinetic		
		k_1(min^{-1})	$q_{e\ cal}$ (mg.g^{-1})	r	k_2(g mg^{-1} min^{-1})	$q_{e\ cal}$ (mg g^{-1})	r
M_1							
Pb^{2+}	285	$5.7.10^{-4}$	178	0.958	$17.3.10^{-4}$	278	0.998
Cd^{2+}	234	$2.8.10^{-4}$	167	0.953	$8.02.10^{-4}$	38	0.999
$Zn^{2\circ}$	410	$7.5.10^{-4}$	249	0.936	$19.9.\ 10^{-4}$	380	0.998
M_2							
Pb^{2+}	261	$4.6.10^{-4}$	185	0,926	$14.8.10^{-4}$	265	0.998
Cd^{2+}	214	$2.4.10^{-4}$	150	0,942	$7.3.\ 10^{-4}$	199	0.997
Zn^{2+}	360	$6.6.10^{-4}$	210	0,912	$17.8.\ 10^{-4}$	320	0.999

Table 3.
Pseudo-first-order and pseudo-second-order parameters for the adsorption of Pb (II), Cd (II), and Zn (II) onto two adsorbents at different temperatures.

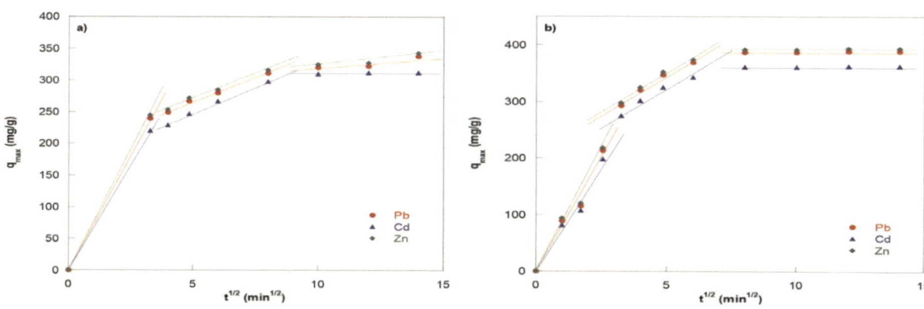

Figure 9.
Intra-particle diffusion plots model for metal ions adsorption onto (a) $\mathbf{M_1}$ *and (b)* $\mathbf{M_2}$.

Sample	K'_1 (mg.g^{-1}.min$^{-0.5}$)	K'_2 (mg.g^{-1}.min$^{-0.5}$)	K'_3 (mg.g^{-1}.min$^{-0.5}$)
M_1–Pb	10.252	4.792	2.103
M_2–Pb	12.348	6.453	3.501
M_1–Cd	11.457	5.395	3.424
M_2–Cd	12.567	6.637	3.667
M_1–Zn	10.684	5.138	2.735
M_2–Zn	12.729	6.841	3.829

Table 4.
Intra-particle diffusion adsorption rate constants of metal ions onto the two adsorbents.

metal ions from bulk solution to the adsorbent external surface via film diffusion. The second stage describes the progressive adsorption step, indicating the diffusion of adsorbate through the pores of xerogel (intra-particle diffusion). The third small-sloped section corresponds to the final equilibrium stage where the intra-particle diffusion commences progressively to slow down because of the quick abatement of metal cation concentrations. Intra-particle diffusion model parameters are enrolled in the **Table 4**.

The boundary layer parameter C was diverse to zero, showing that the intra-particle diffusion ought not to be the sole rate limiting step [30]. Be that as it may, it ought to be accentuated that the rate controlled step was governed by film-diffusion towards the start and afterward followed by intra-particle diffusion.

3.4 Adsorption isotherms

Adsorption isotherm is viewed as a standout amongst the most critical elements for determining the mechanism between adsorbent and adsorbate. In this research, Langmuir, Freundlich, and Dubinin-Radushkevich (D-R) isotherm models were utilized to assess the equilibrium data.

The Langmuir isotherm model expects the formation of monolayer coverage of adsorbate on the external surface of adsorbent and a finite number of equipotential sites [31]. The Langmuir model can take the following linear form Eq. (7):

$$\frac{C_e}{q_e} = \frac{1}{q_{max}K_L} + \frac{C_e}{q_{max}} \quad (7)$$

where q_e is the amount of metal cations adsorbed by unit weight of adsorbent (mg. g^{-1}), C_e is the equilibrium concentration of adsorbate in the solution (mg. L^{-1}), q_{max} is the maximum adsorption capacity at monolayer coverage (mg.g^{-1}), and K_L is the Langmuir adsorption constant (L.mg^{-1}). Linear plots of C_e/q_e against C_e were used to determine the value of q_{max} (mg·g^{-1}) and K_L (L·mg^{-1}). The data obtained with the correlation coefficients (R^2) are reported in **Table 5**.

The Freundlich isotherm is an experiential equation which presumes various affinities for the binding sites on the surface of the adsorbent accompanied by the interactions between adsorbed molecules. The linear form of the Freundlich adsorption isotherm [32] can be communicated as follows Eq. (8):

$$\log q_e = \log k_f + \frac{1}{n}\log C_e \quad (8)$$

where k_f *is* a constant related to the bonding energy and n is a measure of the deviation from linearity and the heterogeneity degree of adsorption sites. The

	Langmuir parameters			Freundlich parameters			D-R parameters		
	q_m (mg.g^{-1})	K_L(L mg^{-1})	R^2	K_F	1/n	R^2	q_m (mol.g^{-1})	E(KJ mol^{-1})	R^2
M_1									
Pb^{2+}	523	0.07	0.998	75.85	0.38	0.845	$3.73.10^{-4}$	13.1	0.993
Cd^{2+}	507	0.05	0.997	128	0.27	0.871	$2.64.10^{-4}$	13	0.991
Zn^{2+}	578	0.1	0.999	149	0.42	0.832	$4.69.10^{-4}$	13.63	0.994
M_2									
Pb^{2+}	509	0.09	0.997	135	0.27	0.836	$4.51.10^{-4}$	12.82	0.995
Cd^{2+}	493	0.04	0.998	67	0.41	0.883	$3.34.10^{-4}$	13.33	0.992
Zn^{2+}	549	0.09	0.998	140	0.37	0.842	$4.57.10^{-4}$	13.22	0.993

Table 5.
Langmuir, Freundlich, and D-R parameters for Pb (II), Cd (II), and Zn (II) adsorption onto mesoporous materials $\mathbf{M_1}$ *and* $\mathbf{M_2}$.

Freundlich equilibrium constants k_f and 1/n were determined from the slopes and intercepts of the linear plot of $\log q_e$ versus $\log C_e$. The values for Freundlich constants and correlation coefficients (R^2) for the adsorption process are also recorded in **Table 5**.

As can be seen, the adsorption capacity is higher for $\mathbf{M_1}$ than $\mathbf{M_2}$. This pattern could be related to the specific surface area as well as to the structure of the amino groups which displayed a high chelating ability to heavy metal ions. However, the isotherm parameters, together with the correlation coefficients showed that the Langmuir model gives a good fit to the adsorption isotherm. Additionally, the Freundlich adsorption capacity k_f is higher for $\mathbf{M_1}$ than $\mathbf{M_2}$. The n values which mirror the adsorption intensity also presented the same trend. The acquired values n for both adsorbates model that the adsorption process onto mesoporous material was favorable at considered conditions. However, compared to the R^2 values, with that obtained from the Langmuir model, it can be remarkably noted that the Langmuir isotherm model better fits the equilibrium data.

The Dubinin-Radushkevich (D-R) model is a semi-hypothetical equation which is by and large allied to a sorption induced by a pore padding mechanism. This model gives valuable information on the nature of the adsorption process (chemisorption or physisorption) [33]. The linear presentation of the D-R isotherm equation [34] is granted as Eq. (9):

$$\log q_e = \log q_m - \beta\,\varepsilon^2 \tag{9}$$

where q_e is the adsorption capacity (mol.g^{-1}), q_m is the maximum adsorption capacity (mol.g^{-1}), β is the activity coefficient which gives an idea about the mean free energy (mol^2.J^{-2}), and ε is the Polanyi potential [$\varepsilon = RTLn\left(1+\frac{1}{C_e}\right)$]. The values of q_m and β, can be generated from the slope and the intercept of the plot $\ln q_e$ versus ε^2. The adsorption mean free energy (E, kJ.mol^{-1}) can be calculated using Eq. (10):

$$E = \frac{1}{\sqrt{-2\beta}} \tag{10}$$

In addition, the magnitude of E (kJ.mol^{-1}) value provided data about the type of adsorption mechanism, either physical or chemical. The maximum adsorption capacity q_m, the adsorption free energy E, and the coefficients of linearity are computed and spoken to in **Table 5**. As observed from the table, the high correlation coefficients (≥ 0.99) propose that the adsorption equilibrium data fitted well the D-R isotherm model. Moreover, the mean adsorption energy values were in the range of 13–14 kJ·mol^{-1} for all samples. In perspective of the acquired outcomes, it tends to be reasoned that the adsorption processes of metal ions onto the as-prepared xerogels might be proceeded by chemisorption (binding surface functional groups) [35].

4. Column studies

It is well known that the column trials were conducted to predict the necessary residence time of effluent treatment with a specific flow rate and concentration (**Figure 10**).

It ought to be stressed that the breakthrough curves characterize the dynamic performance of saturated columns; delineated as the ratio of effluent concentration to influent concentration over time. The time where the pollutant concentration in effluent reached 5% is called breakthrough time (tb); however, the time taken for the effluent concentration to attain 95% of initial pollutant concentration is appealed exhaustion time. The Breakthrough curves for metal ion adsorption onto both adsorbents are portrayed in **Figure 6** and their column data are tabulated in **Table 6**.

As portrayed in **Figure 10**, all breakthrough curves exhibited an S-shaped profile; besides, earlier breakthrough and exhaustion times were observed for $\mathbf{M_2}$

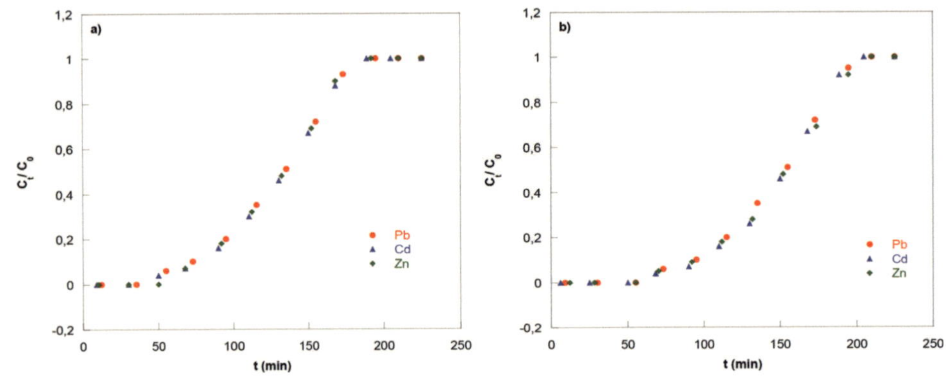

Figure 10.
Breakthrough curves for metal ion adsorption onto mesoporous materials $\mathbf{M_1}$ *(b) and* $\mathbf{M_2}$ *(a).*

	Metal	Breakthrough time	Exhaustion time	Breakthrough capacity (mg.g^{-1})
M1	Zn (II)	45	185	442
	Pb (II)	42	179	410
	Cd (II)	40	175	403
M2	Zn (II)	38	175	420
	Pb (II)	36	172	406
	Cd (II)	35	170	394

Table 6.
Column data for metal ion adsorption.

xerogel, while $\mathbf{M_1}$ displayed a longer breakthrough time. It is clear that breakthrough capacities calculated from column studies were lesser than those settled from the batch method. This pattern might be because of the impact of the prolonged residence time of the sorbate as well as the agitation speed which improve the adsorption in the batch technique. It is worthy to state that the grand breakthrough capacity of $\mathbf{M_1}$ is related to its longer breakthrough time.

4.1 Column regeneration

The regeneration ability is an essential factor for metal recovery and the applicability of adsorbents. The metal charged column was regenerated with 0.1 M HCl (40 mL) and then with 0.5 M HNO_3 (20 mL) at a flow rate of 7 $mL.min^{-1}$. Afterwards, each column was washed with 60 mL of hot deionized water and then dried in an oven at 60°C. The adsorption efficiency of the exhausted column was checked five times. The uptake yield decreased from 96%–94% to 90%–88% for $\mathbf{M_1}$ and 93%–91% to 87%–86% for $\mathbf{M_2}$ after five adsorption-desorption cycles (**Figure 11**). The acquired outcomes uncovered that the as-prepared xerogels could be effortlessly regenerated and continuously used in the metal cation removal process without an obvious decrease in the total adsorption performance.

4.2 Thermodynamic parameters

The mechanism of adsorption can be checked through determining thermodynamic parameters like Gibbs free energy ($\Delta G°$), enthalpy ($\Delta H°$), and entropy ($\Delta S°$). These parameters can be determined from the following equations: Eqs. (11) and (12):

$$LnK_L = \frac{\Delta S}{R} - \frac{\Delta H}{RT} \text{ (Van't Hoff equation)} \quad (11)$$

$$\Delta G° = -\mathrm{RT}\, LnK_L \quad (12)$$

where K_L is the Langmuir constant ($L.mol^{-1}$), T is the absolute temperature (K), and R is the gas constant. By plotting LnK_L against $1/T$, it is possible to determine graphically the value of $\Delta H°$ from the slope, and the value of $\Delta S°$ from the intercept (**Figure 12**). The calculated parameters are given in **Table 7**.

The values of Gibbs free energy change $\Delta G°$ were negative at various temperatures indicating that the adsorption of the two pollutants onto the as-synthesized adsorbent was feasible and spontaneous. Notwithstanding, the abatement of $\Delta G°$ values with temperature could be clarified by a diminishment in the mobility of the metal and the adsorption driving force [36]. The negative value of $\Delta G°$ affirmed the

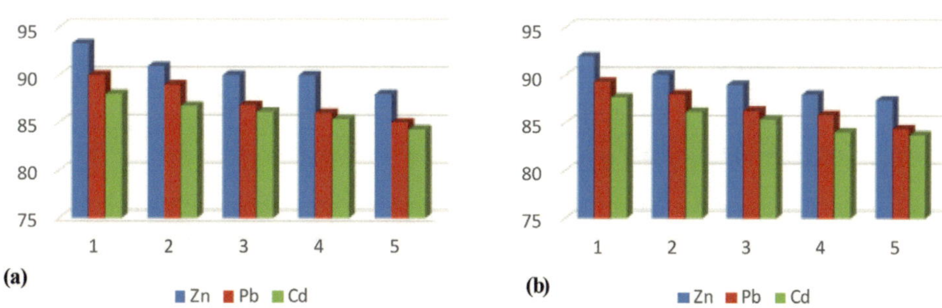

Figure 11.
Adsorption-desorption efficiency of xerogels after 5 cycles: (a) $\mathbf{M_1}$ and (b) $\mathbf{M_2}$.

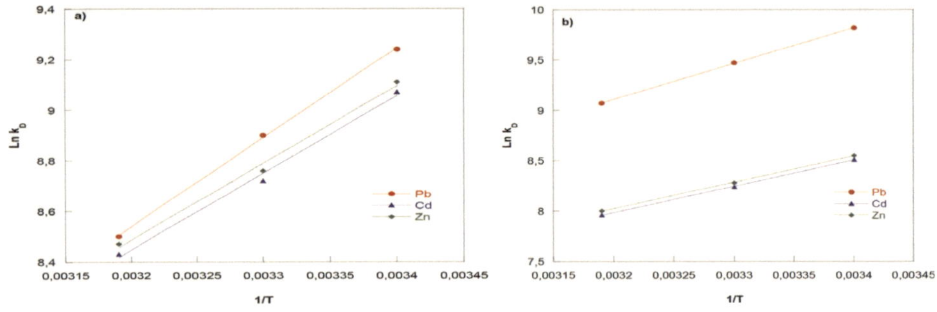

Figure 12.
Determination of thermodynamic parameters for the adsorption of metal cations onto the two adsorbents: (a) $\mathbf{M_1}$ *and (b)* $\mathbf{M_2}$.

Pb (II)	T(K)	$\Delta G^0 (kJ.mol^{-1})$	$\Delta H^0 (kJ.mol^{-1})$	$\Delta S^0 (J.mol^{-1}.K^{-1})$
M_1	293	−22.51	−74.67	−162 .3
	303	−21.82		
	313	−20.94		
M_2	293	−23.92	−62.23	−143.31
	303	−22 .89		
	313	−22.19		

Cd(II)	T(K)	$\Delta G^0 (kJ.mol^{-1})$	$\Delta H^0 (kJ.mol^{-1})$	$\Delta S^0 (J.mol^{-1}.K^{-1})$
M_1	293	−22.08	−69.08	−157.71
	303	−21.72		
	313	−21.55		
M_2	293	−20.72	−59.86	−131.34
	303	−20 .53		
	313	−20.39		

Zn(II)	T(K)	$\Delta G^0 (kJ.mol^{-1})$	$\Delta H^0 (kJ.mol^{-1})$	$\Delta S^0 (J.mol^{-1}.K^{-1})$
M_1	293	−23.16	−80.12	−172.35
	303	−22.93		
	313	−22.75		
M_2	293	−21.87	−66.74	−152.27
	303	−21 .64		
	313	−21.41		

Table 7.
Thermodynamic parameters for heavy metal adsorption onto the two adsorbents $\mathbf{M_1}$ *and* $\mathbf{M_2}$.

exothermic nature of the adsorption process; besides, its magnitude revealed the type of adsorption mechanism (physisorption or chemisorption). Since the $\Delta H°$ value was over 20 kJ.mol^{-1}, this indicates that the adsorption process of metal ions onto the xerogels occurred by means chemisorption [30]. The observed negative ΔS ° reflected a lessening in the randomness at the solid/solution interface during the adsorption process [37].

4.3 Adsorption mechanism

SEM, FTIR, and XPS analysis have been extensively used to identify the possible metal cation-adsorbent interactions. In order to examine the morphology structure of the adsorbents, SEM micrographs were taken after metal ion adsorption (**Figure 13**).

These micrographs indicated clearly the deformation and the presence of many shiny small particles over the surface of both supports $\mathbf{M_1}$ and $\mathbf{M_2}$ after the adsorption process. Moreover, there was also a decrease in the pore sizes after metal adsorption. This observation evidenced the surface coverage of adsorbents by metal ions.

To gain further insights into the mechanism involved in the metal ions uptake process, the FTIR spectra were analyzed, and the band positions for each adsorbent exposed to metal ions are listed in **Table 8**.

In the $\mathbf{M_1}$ xerogel IR spectrum, the strong band that occurred at around 3325 cm^{-1} attributed to NH and NH_2 stretching vibration was shifted and becomes weaker after metal adsorption. This is likely due to the chelation between amino groups and metal ions. Besides, the peak at about 1614 cm^{-1} ascribed to NH_2 and NH groups disappeared suggesting that the adsorption process is mainly dominated by the coordination of nitrogen with metal cations. However, the characteristic peak at 2680 cm^{-1} assigned to the stretching vibration of sulfhydryl group (S-H) was disappeared. This result revealed that metal ions reacted with (S-H) groups on the surface of $\mathbf{M_2}$ xerogel. No obvious shift of the Si-O group after lead adsorption onto the two supports was observed.

To deepen the understanding of the mechanism of metal uptake, XPS analysis before and after metal ion adsorption were performed.

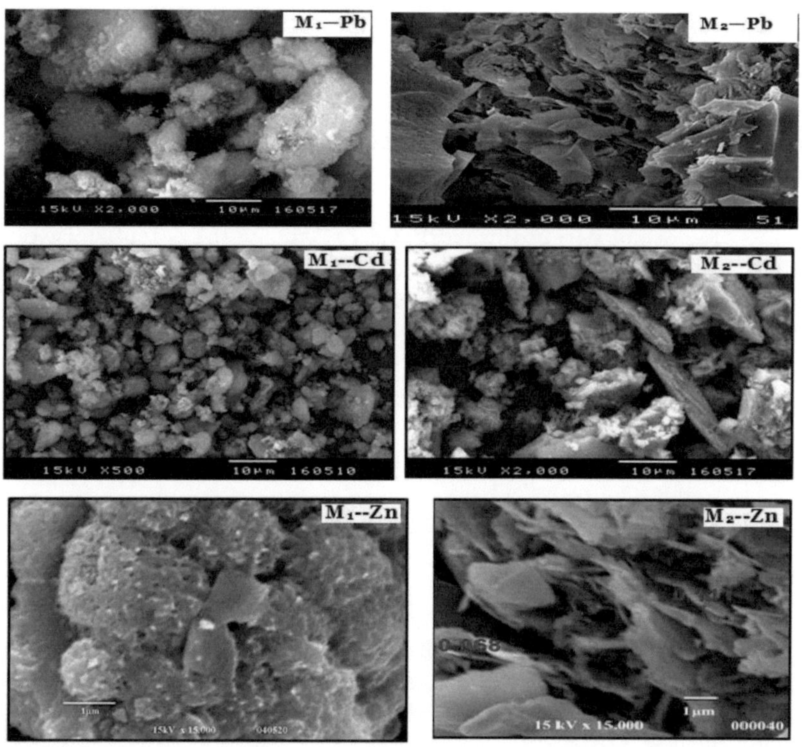

Figure 13.
SEM micrographs of the two adsorbents after metal-ion adsorption.

ν (cm^{-1})	OH and NH	C=N	Si—N	S—H	Si—O
$M_{1(pristine)}$	3343–3325	1557	1174	—	1058-951-460
Pb (loaded)	3303	1559	1170	—	1061-951-464
Cd (loaded)	3310	1561	—	—	1060-948-462
Zn (loaded)	3307	1558	—	—	1059-950-459
$M_{2(pristine)}$	3383	1576	—	2680	1134-946-434
Pb (loaded)	3380	1572	—	—	1131-947-430
Cd(loaded)	3383	1574	—	—	1132-949-433
Zn (loaded)	3380	1577	—	—	1135-944-436

Table 8.
Band positions before and after metal cation adsorption.

As displayed in **Figure 14** a single peak was clearly observed at 398.2 eV, corresponding to the presence of N atom in primary and secondary amine groups. After metal ion adsorption, a new peak with higher binding energy was appeared at about 400.2 eV which may be attributed to the complexation between NH_2 and metal ions (R-NH_2—M^{2+}). On the other hand, the S2p spectra exhibited a faint peak at 167.3 eV assigned to oxidized sulfur. Another peak was observed at 162.6 eV which corresponds to the unbounded S atom in thiol groups. After metal uptake, the peak ascribed to oxidized sulfur becomes much stronger as well as its ratio area, indicating the ion exchange reaction between (S-H) groups on the $\mathbf{M_2}$ xerogel surface and metal ions. Additionally, the XPS spectra of Pb4f, Cd 3d, and Zn2p were also obtained.

As portrayed in **Figure 15**, the binding energies for Pb $4f_{7/2}$, Cd $3d_{5/2}$, and $Zn2p_{3/2}$ were 137.9 eV, 404.7 eV, and 1021.1 eV, respectively. This result is in agreement with the FTIR analysis which suggests that metal ions form a bidentate complex on amino functionalized xerogel.

5. Performance of the synthesized adsorbents for treatment of electroplating wastewater

The most important part of this work was to evaluate the potential use of both synthesized adsorbents for the treatment of real effluent, via electroplating wastewater. The whole analysis was conducted under the same predetermined conditions (**Figure 16**). The treatment of paint wastewater utilizing hybrid material is represented in **Table 9**.

It is obvious that the adsorption rate decreased from 94%–92% to 78%–76% for $\mathbf{M_1}$ and 92%–92% to 75%–73% for $\mathbf{M_2}$. This pattern might be clarified by the way that in real discharge, organic matter and other pollutants may rival metal ions leading to lessening removal yield.

5.1 Selectivity of adsorbents

The selectivity of the adsorbent increases its interest for commercial use. In this context, the selectivity of the two xerogels was carried out by removing an aqueous solution containing a mixture of three metal ions (Pb (II), Cd (II), and Zn (II)) under the predetermined optimized conditions, that is, pH = 5, t = 60 min, adsorbent mass = 0.015 g, and T = 20°C.

It is apparent from **Figure 17** that the adsorption efficiency of the two adsorbents towards different ions exposed the following order ($Cd^{+2} < Pb^{+2} < Zn^{+2}$). The

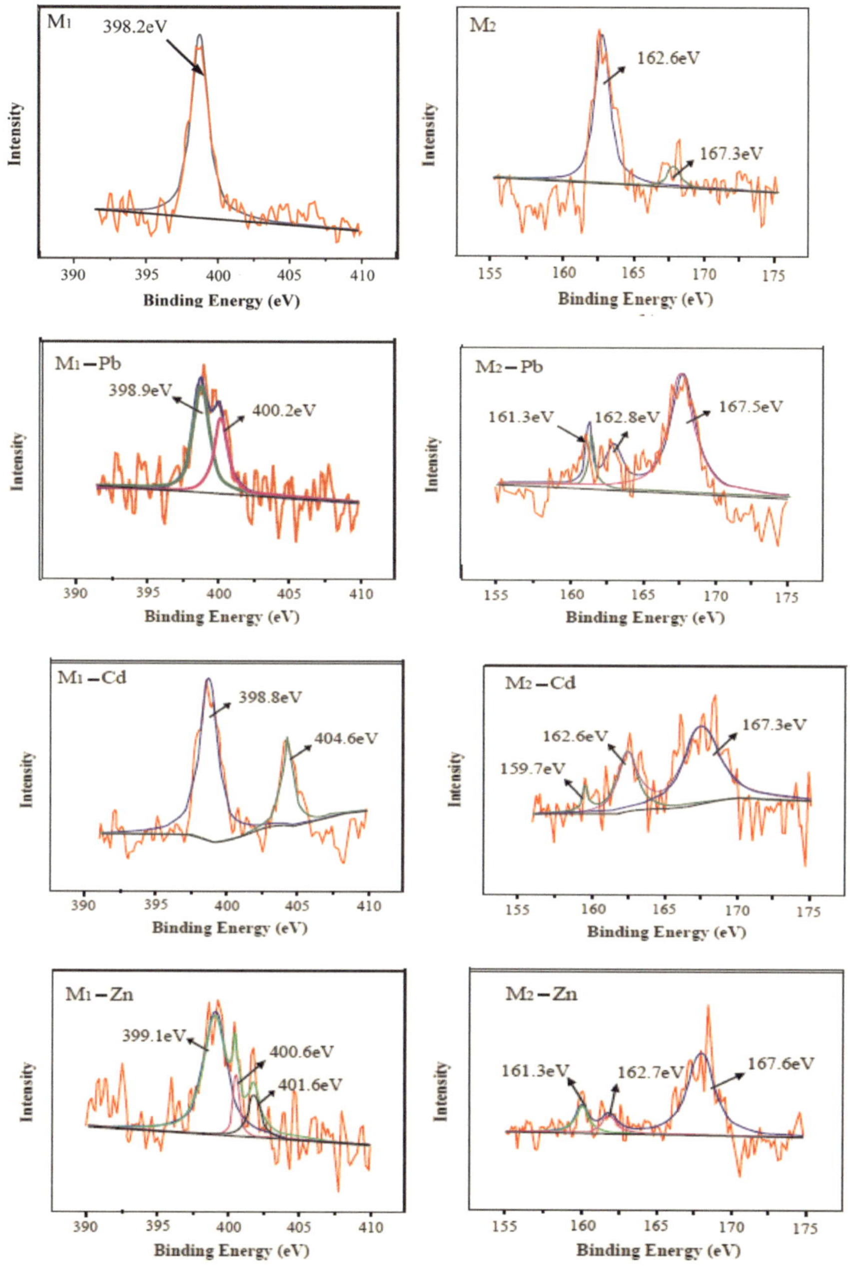

Figure 14.
XPS spectra of N1 s and S2p core level spectra before and after metal cation adsorption.

great selectivity of both xerogels for Zn (II) ions could be ascribed to their low hydrolysis constant and high covalent index. In this regard, the as-prepared adsorbents are relevant for practical application under industrial conditions.

5.2 Comparison with different adsorbents

The uptake efficiency of the as-prepared xerogels for the removal of three metal ions (Pb (II), Cd (II), and Zn (II)) was compared with other stated adsorbents (**Table 10**).

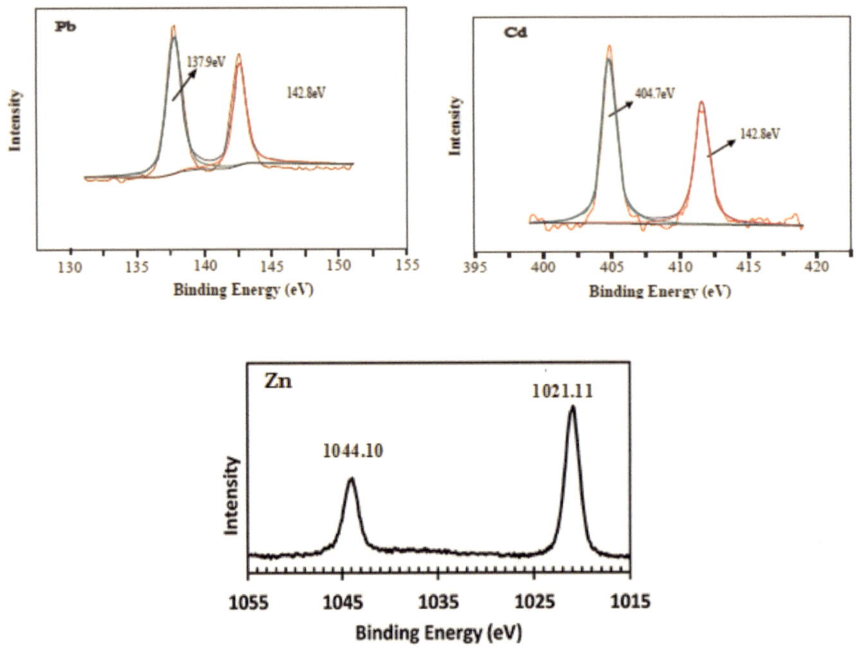

Figure 15.
XPS spectra of Pb4f, Cd 3d, and Zn2p after metal ion adsorption.

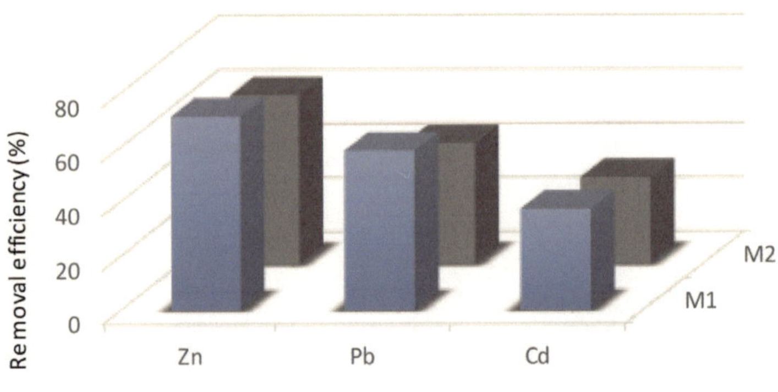

Figure 16.
Efficiency of the as-prepared xerogels for metal ion removal from electroplating wastewater.

Parameters	Values	WHO standard
pH	5.8	5.5–6.5
Temperature	27	20–30
TSS ($mg.L^{-1}$)	419	20
COD ($mg.L^{-1}$)	320	280
BOD ($mg.L^{-1}$)	78	40
Zn ($mg.L^{-1}$)	4.5	5
Pb ($mg.L^{-1}$)	0.15	0.2
Cd ($mg.L^{-1}$)	0.08	0.1

Table 9.
Physicochemical characterization of treated electroplating wastewaters.

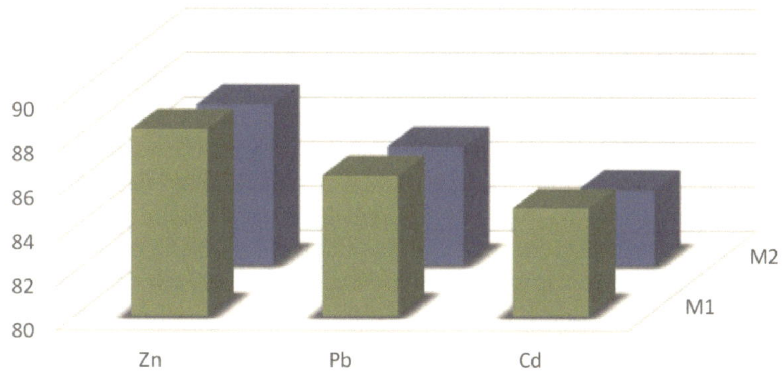

Figure 17.
*Selectivity of **M1** and **M2** xerogels (adsorbent dosage: 0.4 g. L^{-1}; contact time: 60 min; pH: 5; temperature: 20°C).*

Adsorbent	pH	Metal ions	q_m (mg/g)	References
SBA15-NH_2	5	Pb^{2+}	54.6	[38]
PEG-S	6	Pb^{2+}	241.36	[39]
Amino xerogel	5	Pb^{2+}	523	This work
Sulfhydryl xerogel	5	Pb^{2+}	509	This work
MIONPs-NH_2	6	Cd^{2+}	33.72	[40]
MC/Al_2O_3	3–6	Cd^{2+}	49.98	[41]
Amino xerogel	5	Cd^{2+}	507	This work
Sulfhydryl xerogel	5	Cd^{2+}	493	This work
SiNAL4	6	Zn^{2+}	86.51	[42]
SG-MCF	6	Zn^{2+}	39.96	[43]
Amino xerogel	5	Zn^{2+}	578	This work
Sulfhydryl xerogel	5	Zn^{2+}	549	This work

Table 10.
Comparison of adsorption capacity of both mesoporous materials M_1 and M_2 for Pb(II), Cd(II), and Zn (II) with that of other adsorbents.

It can be remarkably noted that the synthesized xerogels exhibited considerably higher adsorption capacity for metal cations than other sorbents specified previously.

This pattern might be attributed to high specific surfaces as well as the number of chelating fragments on the surface of the synthesized adsorbents. Besides, the facility of the synthesis method and the lower adsorption parameters such as contact time, pH solution, and adsorbent dosage made them more appropriate for industrial utilization.

6. Conclusion

The primary targets of this work were to synthesize novel functional organic-inorganic hybrid materials and to check their ability to remove metal ions from aqueous solution. The structural order, morphology, and texture of the prepared hybrid gels were studied by FTIR, ^{13}C CP MAS NMR spectroscopy, SEM, and nitrogen adsorption-desorption analysis.

The adsorption kinetic studies abide by the pseudo second-order model and exhibit a three-stage adsorption process. Moreover, the adsorption rate of metal cation was controlled by the diffusion rate inside the pore. The Langmuir model showed the best fit for the entire experimental data. The free energy values (E) of metal ion adsorption onto $\mathbf{M_1}$ and $\mathbf{M_2}$ xerogels generated by the D-R equation revealed that the adsorption proceeded principally by chemisorption. In column studies, the breakthrough efficiencies of both xerogels were comparable to those calculated from batch techniques and can be reused for at least 5 cycles with a slight decrease in the uptake capacity.

Thermodynamic parameters depicted the spontaneity and the exothermic nature of the adsorption process at 20–40°C. The FTIR and XPS analysis revealed that the chelation between the metal ions and the ligating nitrogen atoms of amino functionalized xerogel was the main mechanism involved in cadmium uptake. Otherwise, the proposed mechanism for lead adsorption onto sulfhydryl xerogel was probably through the ion exchange reaction between metal ions and (—SH) groups. Prior tests accomplished on electroplating wastewater evinced that the xerogel adsorbents possess an exceptional performance in heavy metal uptake from real wastewater. The findings reported in this work showed that the as-prepared xerogels could be widely applied for treatment of industrial wastewater owing to their cost-effectiveness, prominent reusability, good selectivity, and high adsorption efficiency.

Author details

Yasser Hannachi[1]*, Afifa Hafidh[2] and Salwa Ayed[1]

1 Physico-Chemistry Laboratory of Microstructures and Microsystems, IPEST, Tunis, Tunisia

2 Laboratory of Materials and Environment, Tunis Preparatory Institute for Engineering Studies, Tunis, Tunisia

*Address all correspondence to: hannachiyasser@gmail.com

References

[1] Manirethan V, Raval K, Rajan R, Thaira H, Balakrishnan MR. Kinetic and thermodynamic studies on the adsorption of heavy metals from aqueous solution by melanin nanopigment obtained from marine source: *Pseudomonas stutzeri*. Journal of Environmental Management. 2018;**214**:315-324. DOI: 10.1016/j. jenvman. .2018.02.084

[2] Abdel-Raouf MS, Abdul-Raheim ARM. Removal of heavy metals from industrial waste water by biomass-based materials: A review. Journal of Pollution Effects and Control. 2017;**5**: 180-193. DOI: 10.4172/2375-4397. 1000180

[3] Chatterjee S, Sivareddy I, De S. Adsorptive removal of potentially toxic metals (cadmium, copper, Nckel and zinc) by chemically treated laterite: Single and multicomponent batch and column study. Journal of Environmental Chemical Engineering. 2017;**5**: 3273-3289. DOI: 10.1016/j. jece.2017.06.029

[4] Elkhatib E, Mahdy A, Sherif F, Elshemy W. Competitive adsorption of cadmium (II) from aqueous solutions onto nanoparticles of water treatment residual. Journal of Nanomaterials. 2016;**2016**:1-10. DOI: 10.1155/2016/8496798

[5] Torab-Mostaedi M, Asadollahzadeh M, Hemmati A, Khosravi A. Equilibrium, kinetic, an thermodynamic studies for biosorption of cadmium and nickel on grapefruit peel. Journal of the Taiwan Institute of Chemical Engineers. 2013;**44**:295-302. DOI: 10.1016/j. jtice.2012.11.001

[6] Hegazi HA. Removal of heavy metals from wastewater using agricultural and industrial wastes as adsorbents. HBRC Journal. 2013;**9**:276-282. DOI: 10.1016/j. hbrcj.2013.08.004

[7] Hafidh A, Touati F, Hosni F, Hamzaoui AH, Somrani S. New silica hybrids elaborated by sol-gel process from bifunctional thiadiazole and 1,2,4-triazole precursors. Phosphorus, Sulfur and Silicon and the Related Elements. 2018;**193**:155-163. DOI: 10.1080/ 10426507.2017.1393422

[8] Hafidh A, Touati F, Hamzaoui AH. Synthesis of new silica xerogels based on bifunctional 1,3,4-thiadiazole and 1,2,4-triazole adducts. Journal of Sulfur Chemistry. 2019;**40**:18-30. DOI: https:// doi.org/10.1080/17415993.2018.1499742

[9] Hellali T, Hafidh A, Etteyeb N, Touati F, Hamzaoui AH, Somrani S. New silica-based hybrid materials. Phosphorus, Sulfur and Silicon and the Related Elements. 2017;**192**:1018-1026. DOI: 10.1080/10426507.2017.1315423

[10] Paul NE, Diagboya E, Dikio D. Silica-based mesoporous materials; emerging designer adsorbents for aqueous pollutants removal and water treatment. Microporous and Mesoporous Materials. 2018;**266**: 252-267. DOI: 10.1016/j. micromeso.2018.03.008

[11] Da'na E. Adsorption of heavy metals on functonlized – Mesoporous silica: A review. Microporous and Mesoporous Materials. 2017;**247**:145-157. DOI: 10.1016/j.micromeso.2017.03050

[12] Kao HM, Chang CH, Saikia D, Liao SH, Chao PY, Chen YH, et al. Highly carboxylic acid functionalized ethane-bridged periodic mesoporous organosilicas: Synthesis, characterization and adsorption properties. Chemistry, an Asian Journal. 2012;**7**:2111-2117. DOI: 10.1002/ asia.201200244

[13] Vityazev FV, Fedyuneva MI, Golovchenko VV, Patova OA, Ipatova EU, Durnev EA, et al. Pectin-silica gels

as matrices for controlled drug release in gastrointestinal tract. Carbohydrate Polymers. 2017;**157**:9-20. DOI: 10.1016/j.carbpol.2016.09.048

[14] Cristoph S, Fernandes FM, Coradin T. Immobilization of proteins in biopolymer-silica hybrid materials: Functional properties and applications. Current Organic Chemistry. 2015;**19**: 1669-1676. DOI: 10.2174/1385272819666150429231937

[15] Shi J, Tian Y, Liu H, Yang D, Zhang S, Wu Y, et al. Shielding of enzyme by a stable and protective organosilica layer on monolithic scaffolds for continuous bioconversion. Industrial and Engineering Chemistry Research. 2017; **56**:10615-10622. DOI: 10.1021/acs.iecr.7b03033

[16] Santos HA, Makila E, Airaksinen AJ, Bimbo LM, Hirvonen J. Porous silicon nanoparticles for nanomedicine: Preparation and biomedical applications. Nano. 2014;**9**:535-534. DOI: 10.2217/nnm.223

[17] Farghaly AA, Collinson MM. Mesoporous hybrid polypyrrole-silica nanocomposite films with a strata-like structure. Langmuir. 2016;**32**:5925-5936. DOI: 10.1021/acs.langmuir.6b00872

[18] Benhamou A, Baudu M, Derriche Z, Basly JP. Aqueous heavy metals removal on amine-functionalized Si-MCM-41 and Si-MCM-48. Journal of Hazardous Materials. 2009;**171**:1001-1008. DOI: 10.1016/j.jhazmat.2009.06.106

[19] Shahbazi A, Younesi H, Badiei A. Functionalized SBA-15 mesoporous silica by melamine-based dendrimer amines for adsorptive characteristics of Pb(II), Cu(II) and Cd(II) heavy metal ions in batch and fixed bed column. Chemical Engineering Journal. 2011;**168**: 505-518. DOI: 10.1016/j.cej.2010.11.053

[20] Wang W, Tian G, Zhang Z, Wang A. A simple hydrothermal approach to modify palygorskite for high-efficient adsorption of methylene blue and Cu (II) ions. Chemical Engineering Journal. 2015;**265**:228-238. DOI: 10.1016/j.cej.2014.11.135

[21] Adediji JF, Adebayo MA, Ajayi YO, Yusuf LA. Novel mixed ligand of 2,5-diamino-1,3,4-thiadiazole schiff base incorporating benzoic acid: Synthesis and antimicrobial activity. Journal of Chemical and Pharmaceutical Research. 2012;**4**:1501-1504

[22] Ahmed A, Ahmed R, Majed I. Synthesis and characterization of new Schiff bases and using them us tri dentate ligands in metal complexes. International Journal of Applied Science and Technology. 2014;**4**:108-112

[23] Rice EW, Baird RB, Eaton AD, Clesceri LS. Standard Methods for the Examination of Water and Wastewater. 23rd ed. American Public Health Association. American Water Works Association. Washington DC: Water Environment Federation; 2017

[24] Xue B, Zhang J, Tang X, Yang C, Chen Q, Man X, et al. Micro-pore structure and gas accumulation characteristics of shale in the Longmaxi formation, Northwest Guizhou. Petroleum Research. 2016;**1**:191-204. DOI: 10.1016/s2096-2495(17)30042-x

[25] Weng CH. Modeling Pb(II) adsorption onto sandy loam soil. Journal of Colloid and Interface Science. 2004; **272**:262-270. DOI: 10.1016/j.jcis.2003.11.051

[26] Rabiul Awual M, Alharthi NH, Okamoto Y, Karim MR, Halim ME, Hasan MM, et al. Ligand field effect for dysprosium(III) and lutetium (III) adsorption and EXAFS coordination with novel composite nanomaterials. Chemical Engineering Journal. 2017;**320**: 427-435. DOI: 10.1016/j.cej.2017.03.075

[27] Lagergren S. Zur theorie der sogenannten, adsorption geloster stoffe, Kungliga Sevenska Ventenskapasa kademiens. The Hand. 1898;**24**:1-39

[28] Ho YS. Second-order kinetic model for the sorption of cadmium onto tree fern: A comparison of linear and non-linear methods. Water Research. 2006; **40**:119-125. DOI: 10.1016/j.watres.2005.10.040

[29] Weber WJ, Morris JC. Kinetics of adsorption carbon from solutions. Journal of the Sanitary Engineering Division. 1963;**89**:31-60

[30] Das B, Mondal NK, Bhaumik R, Roy P. Insight into adsorption equilibrium, kinetics and thermodynamics of lead onto alluvial soil. International journal of Environmental Science and Technology. 2014;**11**:1101-1114. DOI: 10.1007/s13762-013-0279-z

[31] Langmuir I. The adsorption of gases on plane surface of glass, mica, and platinum. Journal of the American Chemical Society. 1918;**40**:1361-1403. DOI: 10.1021/ja02242a004

[32] Freundlich H, Helle WJ. Adsorption in solution. Ubber die adsorption in lusungen. Journal of the American Chemical Society. 1939;**61**: 2-28

[33] Hannachi Y, Rezgui A, Boubaker T. Biosorption potential of the Mediterranean plant (*Posidonia oceanica*) for the removal of Cu^{2+} ions from aqueous media: Equilibrium, kinetic, thermodynamic and mechanism analysis. Korean Journal of Chemical Engineering. 2014;**31**:1211-1218. DOI: 10.1007/s11814-014-0054-y

[34] Dubinin MM, Zaverina ED, Radushkevich LV. Sorption and structure of active carbons. Adsorption of organic vapors. Zhurnal Fizicheskoi Khimii. 1947;**21**:1351-1362

[35] Kundu S, Gupta AK. Arsenic adsorption onto Iron oxide-coated cement (IOCC): Regression analysis of equilibrium data with several isotherm models and their optimization. Chemical Engineering Journal. 2006; **122**:93-106. DOI: 10.1016/j.cej.2006.06.002

[36] Hannachi Y, Rezgui A, Dekhil AB, Boubaker T. Removal of cadmium (II) from aqueous solutions by biosorption onto the brown macroalga (*Dictyota dichotoma*). Desalination and Water Treatment. 2015;**54**:1663-1673. DOI: 10.1080/19443994.2014.891078

[37] Lawal OS, Sanni AR, Ajayi IA, Rabiu OO. Equilibrium, thermodynamic and kinetic studies for the biosorption of aqueous lead (II) ions onto the seed husk of *Calophyllum inophyllum*. Journal of Hazardous Materials. 2010;**177**: 829-835. DOI: 10.1016/j.hazmat.2009.12.108

[38] Hermàndez-Morales V, Navara R, Acosta-Silva YJ, Macias-Sànchez SA, Pérez-Bueno JJ, Pawelec B. Adsorption of lead (II) on SBA-15 mesoporous molecular sieve functionalized with NH_2 groups. Microporous and Mesoporous Materials. 2012;**160**:133-142. DOI: 10.1016/j.micromeso.2012.05.004

[39] Tian H, Feng Q, Chen Y, Yang H, Li X, Lu P. Synthesis of amino-functionalized mesoporous materials with environmentally friendly surfactants by evaporation-induced self-assembly and their application to the adsorption of lead(II) ions. Journal of Materials Science. 2015;**50**:2768-2778. DOI: 10.1007/s10853-015-8832-4

[40] Lin S, Liu L, Yang Y, Lin K. Study on preferential adsorption of cationic-style heavy metals using amine-functionalized magnetic iron oxide nanoparticles (MIONPs-NH_2) as efficient adsorbents. Applied Surface Science. 2017;**407**:29-35. DOI: 10.1016/j.apsusc.2017.02.173

[41] Yang W, Tang Q, Wei J, Ran Y, Chai L, Wang H. Enhanced removal of Cd (II) and Pb (II) by composites of mesoporous carbon stabilized alumina. Applied Surface Science. 2016;**369**: 215-223. DOI: 10.1016/j.apsusc.2016.01.151

[42] Radi S, El Massaoudi M, Bacquet M Degoutin S, Adarsh NN, Robeyns K, Garcia Y. A novel environment-friendly hybrid material based on a modified silica gel with a bispyrazole derivative for the removal of ZnII, PbII, Cd(II) and Cu(II) traces from aqueous solutions. Inorganic Chemistry Frontiers. 2017;**4**: 1821-1831. DOI: 10.1039/c7qi00322f

[43] Li M, Li M, Feng C, Zeng Q. Preparation and characterization of multi-carboxyl-functionalized silica gel for removal of Cu (II), Cd (II), Ni (II) and Zn (II) from aqueous solution. Applied Surface Science. 2014;**314**: 1063-1069. DOI: 10.1016/j.apsusc.2014.06.038

Chapter 9

Formulation of Corrosion Inhibitors

Yun Chen and Wenzhong Yang

Abstract

Corrosion inhibitors are widely used in the production process due to their significant effect. In this chapter, the formulation of typical corrosion inhibitors in acid solution, near-neutral solution, alkaline solution, and oil and gas systems will be discussed, respectively. Firstly, the importance of pickling corrosion inhibitors for thermal equipment in industrial production in different situations is discussed in the "Inhibitors for acid solution" section, and the types of pickling inhibitors are mainly applied in different kinds of acid media, such as sulfuric acid, hydrochloric acid, nitric acid, phosphoric acid, hydrofluoric acid, citric acid, and sulfamic acid. Secondly, in the "Inhibitors for near-neutral solution" section, the cooling water inhibitors principally include chromate, stabilized phosphate, alkaline zinc/organic, molybdate, all organic, soft and lean water, and environmental and closed-loop programs. The hot-water inhibitor is also mentioned here. Then in the "Inhibitors for alkaline solution" section, boiler water inhibitor, oxygen scavenger, and corrosion inhibitors for condensate line are talked over. Finally, in the "Inhibition for oil and gas systems" section, drilling fluid, fracturing acidizing, oil and gas well, and oil field which produced water treatment inhibitors are introduced here. The corrosion inhibition mechanism and development trend of inhibitors are also discussed.

Keywords: inhibitors for acid solution, inhibitors for near-neutral solution, inhibitors for alkaline solution, inhibition for oil and gas systems

1. Introduction

Pickling is widely used in various production processes, such as for pickling and rust removal of metal pipes, plates, and wires, and products in the metallurgical and mechanical industries, scale cleaning of heat exchange equipment, heat transfer equipment, and cooling equipment in various industrial sectors, especially for pickling of thermal equipment such as boilers in the electric power sector. It is meaningful using pickling, for example, from a socioeconomic point of view, fuel consumption due to fouling can be reduced; from the perspective of environmental protection, fuel exhaust and atmospheric pollution are reduced; from a safety point of view, thermal equipment such as boilers and heat exchangers can be used well. As we know, various types of dirt are gradually formed in the process, and the poor thermal conductivity of the dirt causes the local temperature of the furnace tube to rise, which reduces the strength of the steel and often causes a pipe burst accident and affects the operation of the boiler. Therefore, pickling plays a very important role in the boiler operation of power plants. In addition, the use of industrial circulating cooling water technology to increase the concentration of running water can save a lot of industrial water. In circulating cooling water, in order to avoid corrosion and scaling of equipment, it is often necessary to add water stabilizer. Also, in the process of crude oil and natural gas collection, the current drilling and oil recovery technology requires fracturing and acidification of limestone oil and gas fields with high concentration of hydrochloric acid solution. As the 15–28% hydrochloric acid solutions are often used in the oil wells, the corrosion problem of tubing and oil production equipment is serious in deep well acidification technology due to the high temperature in the well. Therefore, metal materials and products are subjected to various environments during production, storage, transportation, and use. Metal products are easy to vary from different degrees of corrosion. Corrosion problems in tubing and oil production equipment are serious, sometimes causing fracture accidents, resulting in construction failure.

Corrosion inhibitor is an anticorrosion chemical; it can be added to the corrosive environment to cause physical and chemical interaction with the metal surface, thereby significantly reducing the corrosion of the metal material. At the same time, the use of the corrosion inhibitor does not require special equipment, nor does it need to change the properties of the metal. Compared with other anticorrosion methods, corrosion inhibitors are easy to use, cost-effective, and widely used in industrial production and social life. Therefore, it is widely used in various industrial processes, such as pickling, cooling water systems, acidification of oil and gas wells, water injection in oil fields, storage and transportation of metal products, and so on. With the advancement of society and the enhancement of human environmental awareness, the development and application of corrosion inhibitors have received more and more attention. In recent years, countries around the world have paid great attention to the research and application of corrosion inhibitors, which can be used alone or in combination with other anticorrosion materials.

Herein, in view of the current requirements, the characteristic formulation of corrosion inhibitors in acid solution, near-neutral solution, alkaline solution, and oil and gas systems is discussed. The types of pickling inhibitors in different kinds of acid media, inhibitors for the cooling water, alkaline solution, and oil and gas systems are principally talked about in details, respectively.

2. Inhibitors for acid solution

In various corrosive media, acidic gases and liquids are of strong corrosive medium. The rate of metal corrosion in acidic medium is much faster than that in other media, especially in sulfuric acid, hydrochloric acid, nitric acid, phosphoric acid, hydrofluoric acid, citric acid, and sulfamic acid, which are commonly used in pickling and removing iron oxide, descaling and decontaminating of boiler equipment, and so on. In addition, high level of H_2S and CO_2 in oil and gas wells is also the corrosive acidic medium. Since acid has a corrosive effect on metal equipment, especially the corrosion effect of inorganic acid is more serious, and the released hydrogen will diffuse into the interior of the metal, causing hydrogen embrittlement of the equipment to be washed. In addition, as a large amount of acid gas will generate, it will deteriorate working conditions inevitably. Therefore, a suitable corrosion inhibitor should be added during pickling to inhibit the corrosion of the metal in the acidic medium, reduce the amount of acid used, improve the pickling effect, and prolong the service life of the thermal equipment. **Table 1** shows the commonly used corrosion inhibitors in corrosive acidic medium.

Inhibitor	Acid concentration and temperature	Inhibitor dosage	Inhibition efficiency	Metals
Advanced pyridine alkali	12% HCl + 5% HF, 40°C	0.2%	<0.1 mm/a	Carbon steel
4-Methylpyridine residual	10% HCl + 6% HF, 30°C	0.2%	<0.1 mm/a	Carbon steel
Nitroaniline	2–3 mol/L HNO_3	0.002 mol/L	High efficiency	Copper
1-Phenyl-3-thioformamide	20–35% HNO_3	0.0005%	High efficiency	Aluminum
Reactant of amine and alkynol	5–15% HCl, 93°C	0.01–0.25%	99.4%	Carbon steel
Methenamine + $CuCl_2$	2–25% HCl	0.6 + 0.02%	99%	Carbon steel
Methenamine + $SbCl_2$	10–15% HCl	0.8 + 0.001%	99%	Carbon steel
Furfural	0.2–6 mol/L HCl	5–10/230 mL	High efficiency	Copper and alloy
Methenamine + KI	20% H_2SO_4	0.6% (8:1)	99%	Carbon steel
Methenamine + thiourea + Cu^{2+}	10% H_2SO_4	0.14 + 0.097 + 0.003%	99%	Carbon steel
1,3-Dibutyl-2-thiourea + OP	10–20% H_2SO_4, 60–80°C	0.5 + 0.25%	High efficiency	Carbon steel

Table 1.
Operating conditions and properties of some common corrosion inhibitors.

2.1 Picking inhibitor in sulfuric acid

For the sulfuric acid solution, organic inhibitors are mainly organic amine, amide imidazoline quaternary ammonium salt, rosin amine, Mooney alkali, thiourea derivatives and acetylenic compounds, alkaloids, and so on.

In sulfuric acid solution, adding Cl^-, Br^-, or I^- to acetylenic alcohol and pyridine sulfate inhibitors can significantly improve their corrosion inhibition efficiency. Compounds containing halides, such as alkylbenzylpyridine chloride, cetylpyridinium chloride, urotropine and potassium iodide, and ethylquinoline iodide, are also the novel corrosion inhibitors. In addition, inorganic corrosion inhibitors such as arsenic acid and its salts, antimony trichloride, tin dichloride, and boron trifluoride are also the preferred sulfuric acid pickling inhibitors. Animal protein (KC) and quinoline base (CHM) are sulfuric acid pickling inhibitors as well. Satpati and Ravindran [1] found that 1,2,3-benzotriazole can inhibit the SS304 stainless steel in sulfuric acid solution up to 97%.

2.2 Picking inhibitor in hydrochloric acid

Most of the corrosion inhibitors effective for carbon steel hydrochloric acid pickling are organic heterocyclic compounds containing N, O, S, and P atoms, and most are used as nitrogen-containing compounds. There are three common hydrochloric acid pickling inhibitors:

1. Ammonia compound corrosion inhibitors, including alkyl amines and aromatic amines, saturated and unsaturated nitrogen ring compounds or niscine and quaternary ammonium, amide, polyamine, etc., prepared by ethylene nitride condensed polyamines, such as urotropine.

2. Sulfur-containing compound corrosion inhibitor thiourea and derivatives, in the acid wash, Fe^{3+} is a strong depolarizer, if accumulated more, it will aggravate the corrosion of the steel and cause pickling. The phenylthiourea and NH_4HF_3 complex can form a complex with Fe^{3+} ions, thereby preventing over-pickling. In addition, rare earth thiourea compounds are also the effective corrosion inhibitors.

3. Some phosphorus compounds, such as tributyl phosphate, can inhibit the corrosion of steel and hydrogen permeation to avoid over-pickling. It is also beneficial to the acid regeneration cycle.

Behpour et al. [2] studied 2-({-1-methyl-3-[(2-sulfoaminophenyl)imin] butylene}amino)-1-thiophenol and 2-({-1,2-diphenyl-2-[(2-phospheophenyl) imine]acetal}ammonia)-1-thiophenol for the corrosion inhibition properties of copper in 15% hydrochloric acid. The results show that both of them are mixed corrosion inhibitors. When the mass fraction of two Schiff bases is 500 μg/g, the inhibition efficiency is 95.94 and 96.75%, respectively. Bentiss et al. [3] found that the three 4-hydrogen-1,2,3-triazole derivatives have better corrosion inhibition performance on low carbon steel in hydrochloric acid solution. They are all mixed corrosion inhibitors, in which the corrosion inhibition efficiency of 3,5-bis(4-methylthiophenyl)-4-hydrogen-1,2,4-triazole can be as high as 99.6%.

2.3 Picking inhibitor in nitric acid

Boiler scale and metal oxides are highly soluble in nitric acid, which sometimes replaces hydrochloric acid. Nitric acid is a very oxidizing acid, and as most of the corrosion inhibitors are organic, thus they are prone to act as redox reactions. Therefore, there are fewer types of nitric acid pickling inhibitors. Commonly used for nitric acid pickling inhibitors is a mixture of thiourea and Na_2S and a mixture of hydrazine (C_8H_7N) and NH_4SCN or Na_2S. The corrosion inhibition mechanism of thiourea is that it can decompose nitrous acid, thereby inhibiting the dissolution of metals in nitric acid. The significant inhibition of combining thiourea and Na_2S is due to synergistic effect. There are better nitric acid pickling inhibitors as well, such as thiosulfate, aniline chloride, potassium thiocyanate, potassium dichromate, alkaloids, benzoquinone, etc.

2.4 Picking inhibitor in phosphoric acid

The development of phosphoric acid corrosion inhibitors began around the 1930s. Most of the early products were inorganic salts. In the 1940s and 1950s, organic compounds were successively developed as corrosion inhibitors for metals in phosphoric acid solutions. $Cu_3(PO_4)_2$, molybdate, tungstate, borate, nitrite and nitrate, sodium chromate, dichromate, ammonium hydrogen fluoride, basic salt of arsenic, iodine or bromine, are the inorganic corrosion inhibitors, dodecylamine or 2-aminodicyclohexylamine and potassium iodide, iodoacetic acid, composite corrosion inhibitor compounded with acridine, urotropine and ammonium thiocyanate, oxazole compounds such as triazole and benzotriazole and thiourea, sulfonated imidazoline, polyvinylpyrrolidone (PVP), and polyethyleneimine (PEI); are the organic corrosion inhibitors, which are the typical corrosion inhibitors for carbon steel; sodium molybdate, ammonium chloride, sodium silicate, potassium chromate, potassium dichromate, amine compound, aminophenol, pyridine, Tween-85, Tween-20, sodium lauryl sulfate, cetylpyridinium chloride, methylpyridine, and 8-hydroxyquinoline are successfully used for inhibiting the corrosion of aluminum and aluminum alloy; for copper and copper alloy, mercaptobenzothiazole (MBT), benzotriazole, dextrin, tannin, agar, gum arabic, pyroic acid, gelatin, cinnamic acid, and its derivatives are the effective corrosion inhibitors.

2.5 Picking inhibitor in hydrofluoric acid

Hydrofluoric acid is a weak inorganic acid that volatilizes in the air, and its vapor is highly corrosive and toxic. However, it dissolves oxides at a high rate and has the special properties of dissolving silicon scale (silicon oxide). Hydrofluoric acid pickling was first successfully used in a post-operation supercritical pressure boiler in West Germany in 1968. Since then, 40% of the boilers in West Germany have been washed with hydrofluoric acid. Although the use of hydrofluoric acid pickling has the disadvantages of unsafe operation, high price and environmental pollution, considering the characteristics of hydrofluoric acid, its application range is still extensive after the application of new corrosion inhibitors. Alkylthioureas, organic amines, benzotriazole, 2-mercaptobenzimidazole, and 2-mercaptobenzothiazole are the characteristic corrosion inhibitors for stainless steel and carbon steel.

2.6 Picking inhibitor in citric acid

When citric acid pickling is selected, since the temperature during pickling is high and the circulation speed is fast, the corrosion inhibitor must be applied to such conditions when selecting citric acid pickling. The commonly used corrosion inhibitors are urotropine, thiourea, o-xylene thiourea, ruthenium, industrial xylene thiourea, and so on.

2.7 Picking inhibitor in sulfamic acid

Sulfamic acid is a powdery and moderately acidic mineral acid. In the pickling process, it is stable and good soluble in water. It does not cause salt precipitation and precipitation and does not contain halogen ions. It is suitable for cleaning carbonates and hydroxides. However, it has poor ability to pickle iron oxide, and it decomposes above 60°C. Therefore, it is rarely used for pickling of large boilers and is generally used for pickling of copper pipes. The sulfamic acid pickling inhibitors mainly contain organic amine, dibutylthiourea, dipropynyl sulfide, propargyl alcohol, quaternary ammonium salt, ethyl thiourea, and dodecylamine. The inhibitor of inorganic compound (potassium iodide) also performs excellent performance. **Table 2** is the corrosion inhibition effects of some organic and inorganic compounds in sulfamic acid.

Compounds	Inhibitor composition	Concentration (mg/L)	Corrosion inhibition efficiency (%)
Organic amines	N-butyl amine	146	7.1
	Twelve amine	380	81.7
	Ethylene two amine	120	11.2
	Quinoline	258	23.1
	Quaternary ammonium salt	1130	91.1
	Aliphatic amine fatty acid condensate	4000	84.5
Thiourea	Thiourea	150	65.1
	N-cyclohexyl thiourea	313	64.9
	1,3-Diethyl thiourea	246	29.5
	1,3-Dihexyl thiourea	487	88.0
Alkynes	Methyl propargyl alcohol	112	81.2
	Dipropargyl sulfide	168	80.5
	Acetylene propyl disulfide	220	96.3
Inorganic salts	Sodium chloride	117	17.3
	Potassium bromide	180	14.6
	Potassium iodide	330	79.7

Table 2.
The corrosion inhibition effects of some organic and inorganic compounds in sulfamic acid.

3. Inhibitors for near-neutral solution

The neutral medium includes circulating cooling water, boiler water, heating water, washing water, oil and gas field injection water and neutral salt water, etc., and the application of corrosion inhibitor is mainly in circulating cooling water system. During the operation of the cooling water in the circulating water system, the concentration of harmful ionic dissolved in the water increases with the continuous evaporation of water, which resulting in deterioration of the quality of the cycle water, and fouling and corrosion of the heat exchanger and the cooling tower. In the open circulating cooling water system, biological slime caused by the microbial growth can also accelerate the local corrosion of metal.

The washing water and the neutral salt water systems, which are similar to the circulating water, will not be introduced in this section. The corrosion inhibitors for boiler water and water injection of oil and gas fields will be introduced in Sections 4 and 5, respectively.

3.1 Cooling water inhibitor

We can select water treatment agents according to the Ryznar index of cooling water or the possible problem. For example, zinc-phosphonate salts as water treatment agents can be chosen for serious erosion in circulating cooling water system, while polyacrylic acid, hydrolyzed polymaleic anhydride, or phosphonate can be used when a large number of scaling has occurred. As the limited compounds in the figure, it cannot be used as a standard method for choosing the right agents. However, it provides the vital reference for the design of the cooling water treatment system.

The complex corrosion inhibitors used in the neutral medium include chromate, phosphate, alkaline zinc/organic, molybdate, silicate, and organic programs. Formulations and chemical programs of some examples are discussed in this section.

3.1.1 Chromate programs

The chromate programs are based primarily on the zinc and chromate mixture, with the ratio of CrO_4^{2-}: Zn^{2+} from 6:1 to 8:1, and the chromate reserve varies from 10 to 25 ppm. If low or ultralow chrome program is employed, the ratio of CrO_4^{2-}:Zn^{2+} is typically 1:1, and the chromate reserve is from 3 to 8 ppm, respectively. In order to lower the health and environmental hazard by chromate, polyphosphate, phosphonate, or polyacrylate is often added as well. The level of zinc or other possible inhibitor is correspondingly raised for compensation.

A typical "standard" formulation for a chromate product that may be used as shown in **Table 3**, for example, in a large coastal petrochemical facility, is described here. This program formulation will tend to be dosed continuously to achieve a 35–50 ppm product reserve in the recirculating cooling water. The pH of the cooling water is typically 6.0–7.0.

3.1.2 Stabilized phosphate programs

In essence, stabilized phosphate programs involve the treatment of controlling ratios of oxygen and phosphate or phosphorus and phosphate and also the other inhibitors combining with a suitable "stabilizing" polymer. Various halogen-stabilized polymers are usually provided. Under prescribed operating conditions, this program can often provide excellent corrosion inhibition.

Chromate program formulation	w/w %
Sodium dichromate, dihydrate	50
Chromic acid	1
H_3PO_4, commercial acid	8
$ZnCl_2$	11
HEDP	14
Water	16
Total	100

Table 3.
A typical "standard" formulation for a chromate product.

In most hard waters, the typical O:P ratio is 2:1, whereas in lower hardness waters it is 3:1. Typically, the total unfiltered O-PO_4^{3-} reserve is at the concentration of 10–15 ppm. Some "stabilized phosphate" programs contain zinc, which may also be operated at higher pH levels. A formulation for a steelwork or fertilizer plant is shown in **Tables 4** and **5**, and the pH value of the cooling water is 6.5–7.0, perhaps up to pH 7.5.

This program formulation will tend to be dosed continuously to achieve a total phosphate reserve in the recirculating cooling water based on calcium hardness (as ppm $CaCC_3$) as shown in **Table 6**.

3.1.3 Alkaline zinc/organic programs

With the growing concerns of the hazard of over phosphate levels in the environment, it has resulted in the development of many programs with lower concentration of PO_4^{3-}, thus come out various combinations of zinc and polymers.

The formulation of phosphate program A	w/w %
KOH, commercial solution	40
H_3PO_4, commercial acid	15
TKPP, commercial solution	15
TTA, commercial solution	1
AA/AMPS	7
Water	22
Total	100

Table 4.
A typical formulation of phosphate program A.

The formulation of stabilizer program B	w/w %
NaOH, commercial solution	15
PCA	10
PAA	15
AA/AMPS	15
Water	45
Total	100

Table 5.
A typical formulation of phosphate program B.

Calcium hardness (ppm)	Product A required (ppm)	Product B required (ppm)	Total PO_4^{3-} in system (ppm)
300	100	30	16
600	85–90	45	14
900	70–75	60	12
1200	60–65	75	10

Table 6.
The requirement of product A and B for different concentrations of calcium hardness.

In addition, these low PO_4^{3-} programs have been designed to operate at higher pH levels than that of the stabilized phosphate programs, which reduces the risk of pH upset and iron phosphate deposition. Thus, these treatments have been called alkaline phosphate, alkaline zinc, or zinc/polymer phosphonate programs. The improved deposit control agents and phosphate/zinc stabilizers have promised these programs to operate with high levels of calcium (up to 1200 ppm) and high pH (up to 9.0).

Some key features of these programs are discussed here:

- Alkaline phosphate programs: These programs use the reverse ratios of phosphate compared with stabilized phosphate. Typically, the O/P ratio is 1:2, with about 5 ppm PO_4^{3-} at 100 ppm alkalinity in the cooling water; it is also possibly down to 2 ppm PO_4^{3-} at 300 ppm alkalinity.

- Alkaline zinc programs: These types of programs tend to combine O-PO_4^{3-} and phosphonate, together with a good scale inhibitor ($CaCC_3$)/dispersant and some zinc salt.

- Zinc/polymer/phosphonate programs: Here the PO_4^{3-} is completely removed and substituted with additional organic compounds.

- Zinc/phosphate/organic programs: These types of formulations try to incorporate the best treatments and also to operate under more difficult and stressful conditions.

Some examples of formulations were designed to operate with the cooling water products reserving of around 100 ppm as shown in **Tables** 7 and **8**, and the typical pH ranges of these programs are shown in **Table 9**. These types of formulations were applied in a wide range of general industrial field and comfort cooling systems.

3.1.4 Molybdate programs

Most formulators combine molybdate with zinc, orthophosphate, or phosphonate to reduce the cost of sodium molybdate, which maintains the excellent

Basic formulation materials	Alkaline PO_4^{3-} (%)	Alkaline Zn (%)	Zn/polymer/phosphonate (%)
H_3PO_4, commercial acid	5	10	—
H_2SO_4, commercial acid	—	—	4
TKPP	6	—	—
$ZnCl_2$	—	5	5
HEDP	5	5	10
SSMA	10	5	—
TTA, commercial solution	1	1	1
PAA	—	—	5
Water	73	74	75
Total	100	100	100

Table 7.
The typical formulation of alkaline zinc/organic program.

Basic formulation materials	Zn/phosphate organic compound (%)
H_3PO_4, commercial acid	10
$ZnCl_2$	5
PBTC	5
AA/AMPS	5
TTA, commercial solution	1
PAA	6
MA/EA/VA	6
Water	62
Total	100

Table 8.
The typical formulation of Zn/organic program.

Programs	pH range
Alkaline phosphate	8.0–9.0
Alkaline zinc	7.0–8.0
Zinc/polymer/phosphonate	7.5–8.5
Zinc/phosphate/organic	7.5–9.0

Table 9.
The typical pH ranges of these programs.

corrosion inhibition properties of molybdate. Azole and polymers are also added for this purpose. The relatively high levels of azoles in these formulations are required to make sure that the finished products are strong alkaline. Typically, the reserve of molybdenum in recirculating cooling water for such blended products is on the order of 2–5 ppm. Molybdate has a comparatively low aquatic toxicity, and it is considered as a safe alternative to replace chromate. However, in the United States (partly in response to the possible pressures on the reduction of Mo in wastewater), new formulations have been developed in recent years that perform equally well but with <1.5 ppm Mo.

Two examples of traditional molybdate formulations are shown below. Formulation A is designed for lower hardness makeup waters (i.e., 50–100 ppm or thereabout). Formulation B is designed for higher hardness makeup waters (i.e., 100–200 ppm or thereabout). Both of the formulations are designed to provide excellent corrosion inhibition and deposit control in comfort cooling systems, with a cooling water reserve of about 75–125 ppm. Formulation A will yield ~4.0 ppm Mo, and the operating pH is around 7.5–8.5, while Formulation B will yield ~2.5 ppm Mo with the operation pH of 8.0–9.0 (**Table 10**).

For those closed circuits, which in reality are semi-open systems, good corrosion control has been obtained by changing from a nitrite-based program to a molybdate program and maintaining the Mo reserve of 8–10 ppm.

3.1.5 All organic programs

These programs have become popular and diverse since the presence of improved DCAs, phosphonates, HPA, and other organic inhibitors. They have been widely available since the mid-1980s. They can successfully operate under a variety of operating conditions at high pH level, without the need of acid dosing. It means the operators can increase the cycles of concentration. In addition, this kind of program type is suitable for employing in small industrial and comfort cooling systems, where acid or continual and precise analytical control is not involved.

An example of the all organic program is given below. Typically, this program would be used at the concentration of 75–125 ppm, depending on operating conditions. The pH range is from 7.5 to 9.0, and this formulation is suitable for use with oxidizing biocides, such as bromine (**Table 11**).

There are numerous permutations of all organic formulations. Formulations containing 6–12% POCA perform good results at the presence of high chlorides. Some formulations also incorporate phosphate-polymer constituent.

3.1.6 Soft and lean water and environmental programs

Acceptable cooling water programs usually use either lean water, with very low natural hardness, or ion-exchange softened water as a makeup source. In order to minimize the deposits of calcium carbonate, it must require softened water, of which the total hardness (TH) should be <5–10 ppm, to act as the makeup water

Basic formulation materials	Formulation A (%)	Formulation B (%)
NaOH, commercial solution	7	16
MA/EA/VA	6	6
ATMP	4	6
AA/AMPS	—	6
TTA, commercial solution	4	4
PEG	2	2
Sodium molybdate solution	25	16
Water	52	44
Total	100	100

Table 10.
Two examples of traditional molybdate formulations.

Basic formulation materials	All organic formulation (%)
NaOH, commercial solution	17
POCA	6
HPA	6
MEA	1
HEDP	6
PAA	6
TTA, commercial solution	4
PEG	54
Total	100

Table 11.
An example of the all organic program.

according to some authorities. For comfort cooling systems, the presence of calcium carbonate deposits can in turn increase the risk of *Legionella* proliferation.

There is a tendency to operate cooling systems at high cycles of concentration with lean waters, perhaps 10–15 times COC, depending on the particular circumstances (including the limitation of COC for the cooling system). It is not only considering the economic benefits but also reducing the inherent highly corrosive nature in most low-calcium waters. A further benefit is that it may permit the presence of sufficient calcium to serve as a corrosion inhibitor in combination with a silicate, phosphate, or phosphonate formulation component with high COC.

Various programs exist to provide satisfactory corrosion inhibition and good deposit control. The formulations include the inhibitors noted previously. It is important that the selected type of formulation must match the operating conditions because sludge may form in the cooling system with numerous cycles. Also, if bromine is used, certain organic inhibitors may be oxidized, thus compromising the program's effectiveness with the long retention times. The formulations may also be modified to remove all metals and phosphorus-containing compounds. A typical formulation is shown in **Table 12**. This is designed to be used at 125–175 ppm reserve in a cooling water system operating at high COC.

3.1.7 Closed-loop programs

There is a wide range of metals used in the construction of closed-loop circuits, and the usual differences are the properties of the circulating water (ranging from deionized water to glycol mixtures to brine water). As a consequence, it is not possible to devise a single water treatment formulation that satisfies the requirements of all closed-loop circuits. In addition, some formulators add oxygen scavengers or indicator dyes in their blends additionally. An example of a formulation suitable for a closed-loop circuit, which is primarily for carbon steel construction, is as shown in **Table 13**. The minimal whole product reserve to be maintained in the closed-loop circuit ranges from 350 to 700 ppm (calculated as nitrite).

3.2 Hot-water inhibitor

The function of the oil well heat-tracing system is to prevent the oil temperature at the oil well exit from decreasing rapidly and to ensure that the oil production

Basic materials	Formulation (%)
NaOH, commercial solution	10
TKPP, commercial solution	5
Sodium silicate	25
EDTA, commercial solution	6
MA/EA/VA	4
PAA	5
TTA, commercial solution	3
PEG	2
Water	40
Total	100

Table 12.
A typical formulation for soft and lean water.

Basic materials	Closed-loop formulation (%)
NaOH, commercial solution	15
Borax	6
Sodium silicate	8
Sodium nitrite	30
PAA	3
TTA, commercial solution	2
Water	36
Total	100

Table 13.
A typical formulation for closed-loop program.

fluid flows normally to the gas collection station. Hot-water continuous circulation, many sources of water supply, and high operating temperature will cause severe corrosion and scaling in the heat-tracing system. In the most corrosive areas, some pipelines are even perforated once a week, which brings great hidden dangers to the normal production of oil wells and also causes a large energy waste.

Hot-water boilers and heating systems generally use alkaline chemicals for water treatment to prevent corrosion and fouling of boilers and heating systems. However, due to the increasing use of heating equipment made of copper and aluminum in heating systems, corrosion is very serious under alkaline water conditions. In particular, radiators made of aluminum are often perforated by alkali corrosion in less than one heating period. Therefore, if there is heating equipment made of copper or aluminum in the hot-water boiler and heating system, it is necessary to add additional copper and aluminum corrosion inhibitor, which greatly increases the cost of water treatment.

4. Inhibitors for alkaline solution

4.1 Boiler water inhibitors

Supplemental water for boilers should be chemically treated in advance, for lowering the possibility of corrosion and scaling to increase the heat transfer effect. Steam is often needed for power generation, thus deposition of SiO_2 on the turbine blades is inevitable if enough silicon component is contained in supplemental water. It is necessary to control the scaling by getting rid of the Ca^{2+} and Mg^{2+} with different kinds of methods, such as by ion-exchange resin and apposite precipitants. In order to control the corrosion process, the essential treatments include degassing, adding alkali and using corrosion inhibitor.

4.1.1 Extrude dissolved O_2 and CO_2

If residual dissolved O_2 in the supplemental water obeys stoichiometric ratio with the metal ions in the boiler system under the conditions of high pressure, it will cause serious pitting of the metal tubes. Degassing of the supplemental water and then adding of appropriate scavenger, such as Na_2SO_3 and N_2H_4, is the effective method for deoxygenation. The acceptable concentration of O_2 should be kept lower than 0.005 ppm. Degassing may cause the reduction of CO_2 content, especially if the supplemental water is pre-acidified to release H_2CO_3 from the dissolved carbonates. H_2CO_3 is aggressive to steel without dissolved O_2, but adding of alkali into the boiler water will mitigate the corrosion raised by CO_2.

Iron return pipe systems undergo critical corrosion; thus, if the concentration of CO_2 in the boiler water is high enough, the iron carbonate will be formed, and it changes to ferrous hydroxide and CO_2 further, which is the obvious corrosive medium. The Cu-Al cooling system also endures corrosion when both of the dissolved O_2 and CO_2 are together, but the Cu base alloys are not easily attacked when there is a lack of O_2. As CO_2 is not depleted at all during the corrosion stage, it will gather inevitably in the boiler during the process of adding supplemental water, and the concentration of CO_2 may decrease if with occasional venting.

4.1.2 Adding alkali

Addition of alkali into boiler waters is a routine operation for many boilers under high pressure in the United States and abroad. Supplemental water for a boiler with high pressure is restricted to the minimal pH of 8.5 to reduce the corrosion of iron at room temperature, and the acceptable pH value should be at the range of 9.2–9.5. For Cu alloys, the optimal pH value is from 8.5 to 9.2. As both iron and Cu alloys are usually used in boiler systems, the balanced pH value range is suggested to be between 8.8 and 9.2. Adding buffer ions, such as phosphate, will restrict the rise of pH value. Such ions also have advantage of preventing high concentrations of hydroxide in the boiler waters, which can give rise to stress corrosion cracking of any section of the boiler under high external stress. It reported that when pH value of the boiler water is 9.5–10.0, phosphate with the concentration of 5–10 ppm was more efficient than either sodium hydroxide or ammonia in slowing down the corrosion rate of boiler tubes under high-pressure conditions.

4.1.3 Adding inhibitors

It is feasible to select suitable inhibitors for inhibiting two kinds of typical corrosion in boiler systems, mainly stress corrosion cracking and return pipe systems corrosion, and the former can be restricted by supplement of phosphate. Corrosion raised by dissolved CO_2 in steam condensate can be limited by supplying volatile amines for the boiler water. Two kinds of characteristic volatile amines are chosen for neutralizing amines and film formation. The former group species contain benzylamine, cyclohexylamine, or morpholine. When one provided into the boiler water with adequate amount, it can neutralize CO_2 and alkalinize the steam condensate, thus dropping the corrosive rate of the condensate. The latter group species include volatile hexadecylamine, octadecylamine, or dioctadecylamine, which is the typical film-forming inhibitor, which prevents corrosion from constructing the stable protective film on the surface of the condenser. The film-forming amine is regarded as more apposite to the circumscription of the inhibitor, while others are only used mainly for neutralizing actually.

4.2 O_2 scavenger

O_2 scavengers are reagents which are often used to get rid of the dissolved O_2 from water through reduction reactions, thus prohibiting the corrosion ascribed to the O_2 in water. For this purpose, the desirable characters of O_2 scavengers should be (1) excellent reducing ability against O_2, (2) no violently actions of the thermal decomposition products and the final reaction products with O_2 countering the equipment.

The species and reagent names of O_2 scavengers are displayed in **Table 14** [4]. The optimal reagent is hydrazine, but it is limited by its highly toxic properties. Thus, the replacement for hydrazine was inevitable.

4.3 Corrosion inhibitors for condensate line

As mentioned before, volatile amines for neutralizing and amines for film formation are also two kinds of classic corrosion inhibitors for prohibiting the corrosion of condensate lines. Volatile amines prevent the erosion by adjusting the pH values of the condensates, while water barrier at the surface of metal can be formed attributed to the amines for film formation, which block the touching between metals and the aggressive substances, for example, O_2 and CO_2.

4.3.1 Inhibitors used in condensate lines of boilers under low pressure

Under low pressure, the main aggressive substances in the pipelines of boilers are O_2 and CO_2. The existence of slight CO_2 in the condensation product will decrease the pH, which will accelerate metal corrosion. Thus, volatile amines for neutralizing and amines for film formation are provided as the suitable corrosion inhibitors. The use of volatile amines for neutralizing displays the wonderful inhibition effect, but the efficiency is discounted at the presence of O_2. Amines for film formation can adsorb on the surface of the metal and then construct the hydrophobic protective film even on low concentration. However, once the corrosion products are covered on the surface of the metal, it will take long time to form the protective film. Therefore, the synergistic effect of using both of these two amines should be considered.

4.3.1.1 Neutralizing amines

Cyclohexyl amine ($C_6H_{13}N$), monoisopropanol amine [$NH_2CH_2CH(CH_3)OH$], morpholine (C_4H_9NO), and ammonium hydroxide ($NH_3{\cdot}H_2O$) are usually used for neutralization reaction in boiler systems. They are also dosage by dose for condensate lines. The amine is complemented to the supplemental water and then mixed with the water vapor produced from the boiler. When the water vapor cools down in the condensate line, the amine will dissolve in condensation product and neutralize CO_2 (H_2CO_3). Therefore, the pH value of the solution in the condensate line will be raised, and the crisis of metal corrosion will be eased. **Table 15** displays the needed

Species	Reagent names
N_2H_4-based O_2 scavengers	$N_2H_4{\cdot}H_2O$ $N_2H_4{\cdot}H_2SO_4$ Hydrazine phosphate
SO_3^{2-}-based O_2 scavengers	Na_2SO_3 $NaHSO_3$ $Na_2S_2O_5$
Other O_2 scavengers	Saccharides Tannins Hydrazide L-ascorbic acid Hydroquinone Alkanol amines

Table 14.
Some species of O_2 scavengers.

Neutralizing reagent	Needed concentrations of amines for the neutralization reaction for 1 mg/L CO_2 (mg/L)
$NH_3{\cdot}H_2O$	0.4
$C_6H_{13}N$	2.3
$NH_2CH_2CH(CH_3)OH$	1.8
C_4H_9NO	2.0

Table 15.
The species and amounts of amines for neutralizing CO_2.

concentrations of amines for the neutralization reaction when the concentration of CO_2 is 1 mg/L. $NH_3{\cdot}H_2O$ is efficient for the neutralization reaction for CO_2 revealed in **Table 15**; however, it is not appropriate for the copper condensate lines in boilers because it may accelerate copper corrosion.

4.3.1.2 Amines for film formation

Alkylamines (R-NH_2), where R is alkyl group which contains 10–20 carbon in the alkyl chain, are considered as the film-forming corrosion inhibitors. Octadecylamine (ODA) has 18 carbon atoms in the alkyl chain and is the typical amine for film formation. As ODA cannot dissolve in water, emulsifiers are usually needed for dispersion. The -NH_2 in ODA can adsorb on the surface of the metal and then construct single or several layer molecular adsorption coating, which is hydrophobic and can block the aggressive medium.

The influencing factors, such as the concentration of ODA, M-alkalinity of supplemental water, temperature of condensation product, and so on, will affect the corrosion inhibition effect of ODA. For instance, under the concentration of 25 mg/L ODA in the supplemental water and in order to control the corrosion rate of iron lower than 10 mg/dm^2·day, it will extend 1 day when the M-alkalinity of $CaCO_3$/L raise from 50 mg (2 days) to 100 mg (3 days). Therefore, the beginning corrosion inhibition effect of ODA is affected as the increase of the M-alkalinity in the supplemental water. However, it will show slight influence after the ODA film is formed. In addition, it has also reported that ODA displays the superior corrosion inhibition effect lower than 45°C. ODA performs the wonderful inhibition effect when the dissolved O_2 is as low as 0.2 mg O/L; however, the inhibition effect is obviously hazarded when the concentration of dissolved O_2 is as high as 3 mg O/L. Therefore, minimizing transfer of dissolved O_2 from boiler to the condenser by degassing and/or O_2 scavengers is extremely important to enhance the corrosion inhibition effect of ODA.

4.3.1.3 Combined use of neutralizing amines and amines for film formation

A high amount of neutralizing amine is needed to prohibit the corrosion behavior of condensing pipes effectively if the M-alkalinity of supplemental water is high; amines for film formation demand a relatively long period to fabricate the hydrophobic deposited film for preventing the corrosion of the metal. Therefore, the combining of these two kinds of amines must improve the corrosion inhibition enormously.

Table 16 displays the corrosion inhibition synergistic effect by combining use of two kinds of the amines. With the treatment of neutralizing amine, the corrosion rate of iron drops accompanying the raise of the pH value; with the treatment

Treatment	pH value	Amount of the amines for film formation	Corrosion rate of iron (mg/ dm^2·day)
None	5.3	0	124.5
Neutralizing amines	6.6	0	78.4
	7.0	0	54.8
Amines for film formation	5.3	20	32.1
Combining of both	6.6	20	11.2
	7.0	20	11.8

Table 16.
The corrosion inhibition synergistic effect by combining use of two kinds of the amines.

of amines for film formation, the corrosion rate cut down when the pH value is reduced further. The collaborative processing outstandingly enhances the corrosion inhibition effect. Even with dissolved O_2 in the condensing pipes, the combined treatment behaves the prior corrosion inhibition effect compared to the treatment of only with single filming amine.

4.3.2 Inhibitors used in condensate lines of boilers under moderate or high pressure

As softened water at low temperature is usually provided for boilers under moderate or high pressure, it is not easy to cause the corrosion problems in condensate lines before entering the deaerator. However, the piping and the auxiliary apparatus after the deaerator will erode sometimes as the ill control of water quality or operation of deaerator, which is due to the rise of temperature. Therefore, the pre-control of supplemental water pipeline is as important as that of the condensate line. Neutralizing amines are then selected as the corrosion inhibitors for the supplemental water and stream condensing pipe because they are fit for pH value control and stable at high temperature under moderate or high pressure.

5. Inhibition for oil and gas systems

5.1 Drilling fluid inhibitor

As the high concentration of Cl^- in the oil field sewage, and also the role of CO_2, H_2S, dissolved oxygen, oil stain and miscellaneous, it results the serious pitting problems. At present, organic amine, organic polyol phosphate complex, imidazoline, and their derivatives and long carbon chain amide carboxylate are well used in this field.

5.2 Fracturing acid inhibitor

If the gas field is dominated by carbonate fracture-porosity gas reservoirs, acid fracturing is an important stimulation method to increase the production for such gas reservoirs. Therefore, the corresponding research and application of acid chemical additives, pre-fracturing fluid chemical additives, and fracturing acid inhibitors have become important tasks in the gas field development.

The acid solution strongly corrodes the steel, and the corrosion inhibitor is the key to ensure the smooth operation of the acidification operation. According to the long-term acidification construction experience, the degree of corrosion inhibition of acid has a recognized standard according to indoor evaluation. It is generally required that the amount of corrosion of steel in an acidizing operation does not exceed 98 g/m^2; in deep well operations, it is acceptable to not exceed 245 g/m^2; pitting corrosion shall not occur. Since the 1970s, fracturing acid inhibitors have gradually developed into a variety of varieties. The main types of compounds are amines and their derivatives, pyrenes, acetylenic alcohols, etc., and some nonionic surfactants and alcohol solvents are added for the fabrication of a multicomponent formulation. In order to improve the temperature resistance, iodide or telluride is also used as a synergist, and the well temperature can be as high as 204°C to achieve the degree of corrosion inhibition required for acidification construction.

5.3 Oil and gas well inhibitor

Oil well acidification is performed through injecting hydrochloric acid, soil acid (hydrofluoric acid and hydrochloric acid mixture), or other acid solution into stratum, by means of acidic fracturing equipment. After the dissolution of the rock by the acid solution, the enlargement of the penetration channel of the rock in the reservoir, and the treatment of the blocks in the seepage channel or the cause by artificial cracks, the gas passages will be unblocked to achieve the purpose of increasing oil and gas production. For the acidification of carbonate oil and gas layers, these are the following chemical reactions.

Lime petroleum gas rock formation reacts with hydrochloric acid:

$$CaCO_3 + 2HCl \rightarrow CaCl_2 + H_2O + CO_2\uparrow \quad (1)$$

Dolomite oil and gas layer reacts with hydrochloric acid:

$$MgCa(CO_3)_2 + 4HCl \rightarrow CaCl_2 + MgCl_2 + 2H_2O + 2CO_2\uparrow \quad (2)$$

The reaction of sandstone oil and gas layer with soil acid solution:

$$CaAl_2Si_2O_8 + 16HF \rightarrow CaF_2 + 2AlF_3 + 2SiF_4 + 8H_2O \quad (3)$$

Hydrofluoric acid reacts with carbonate:

$$CaCO_3 + 2HF \rightarrow CaF_2 + 8H_2O + CO_2\uparrow \quad (4)$$

From Eqs. (3) and (4), CaF_2 is formed in the reaction of hydrofluoric acid with sandstone. When the concentration of the acid solution is lowered, it precipitates and blocks the pores. While the acid solution contains HCI, it can suppress or reduce the sinking of CaF_2 precipitate. This is why hydrofluoric acid and hydrochloric acid are mixed in the soil acid solution. The commonly used soil acid solution is 7% HCI + 3% HF or 12% HCI + 6% HF, depending on the clay composition of the sandstone reservoir containing silicate, quartz or carbonate.

From Eqs. (1) and (2), the higher the concentration of hydrochloric acid, the stronger the ability to dissolved rocks and the less volume of acid required to dissolve a certain volume of carbonate and the residual acid solution, thus the easier it is to be discharged from the formation and the favorable oil and gas output; in addition, high concentration of hydrochloric acid has a longer chemical activity time. Therefore, the concentrated hydrochloric acid solution is chemically activated before it is consumed, and its distance between the rock formations is relatively far, and the effect of acidification of the oil wells for improving the production is better. It is popular of acidizing oil wells with high concentration of hydrochloric acid and increasing oil and gas production. However, the corrosion problem caused by the acidic solution at high temperature must be solved.

From the 1960s to the 1970s, the oil wells were shallow, generally between 1000 and 2000 m, and the downhole temperature was not high. The acidizing corrosion inhibitors in oil well mainly contained formaldehyde, urotropine, arsenious acid (arsenic) and other compounds. Later, after the compound test, two or more corrosion inhibitors were compounded, such as urotropine + potassium iodide, urotropine + OP, butynediol + potassium iodide, butynediol + OP, butynediol + potassium iodide + OP, etc. The HCI concentration should be not more than 15% (10% in general) and the well temperature does not exceed 80°C. Due to the low acid concentration and temperature, the composite corrosion inhibitor can make the corrosion rate of carbon steel within the allowable conditions of construction. During this period, the acidification operation is not much, and the acid corrosion inhibitor research has just started.

After 1970s, the ultra-deep wells were above 6000 m, and the downhole temperature was as high as 180–200°C. It has been discovered that the acidification of oil wells with high concentration of hydrochloric acid can significantly increase the effect of oil and gas production. However, in such high-temperature ultra-deep wells, when the acidification is carried out with concentrated hydrochloric acid solution, the acidification inhibitor must be required to protect the mild steel in such concentrated hydrochloric acid at 200°C. Therefore, it played a driving role for the research and development of oil acidification inhibitors.

Currently, a variety of alkyl pyridine and quinoline-based benzyl quaternary ammonium salt substances can well disperse in hydrochloric acid and soil acid solution. They can become clear and transparent brown liquid and show very good corrosion inhibition for carbon steel in hydrochloric acid or soil acid solution. They have certain corrosion inhibition effect on ferric ions and are highly corrosive to H_2S. They are the concentrated hydrochloric acid acidification corrosion inhibitors with high-temperature resistance of 180–200°C. The ketone amine aldehyde condensate compounded with alkynol compound has good dissolving and dispersing property in hydrochloric acid. It has good corrosion inhibition effect on carbon steel in a hydrochloric acid solution at 105–180°C. The alkynyl oxymethylamine modified derivative also has good corrosion inhibition effect on carbon steel in hydrochloric acid or soil acid. The combination of a ketone aldehyde amine condensate, an alkyne alcohol and a cationic surfactant and a solvent to prepare a solution, is suitable for 15–28% HCl solution at 130°C, which has been applied in oil fields. Growcock and Lopp [5] developed a well-acidified corrosion inhibitor "PPO" (3-phenyl-2-propynyl alcohol), which is used in high-temperature downholes with 1–9 mol/L hydrochloric acid. The corrosion inhibition effect is as high as 99%. **Table 17** lists the types of corrosion inhibitors commonly used in oil fields.

5.4 Oil field-produced water treatment inhibitor

Metal equipment in oil field water injection systems, such as various heat exchangers, pumps, valves, oil casings, and storage tanks, are corroded by sewage, causing corrosion and perforation of equipment. It will cause serious impact on production and even force the oil well to stop production. In the construction of oil well acidification, the development of ultra-deep wells and extremely deep wells puts higher requirements on corrosion inhibition performance.

Main chemical composition	Acid-base concentration (%)	Temperature range (°C)
Pyridine slag quaternary ammonium salt	15–20 HCl	70–90
Coal tar pyridine residue extract	15–20 HCl	80–120
Coal tar pyridine residue extract adding dyeing agent	15–28 HCl	90–180
Pyridine hydrochloride residue and surfactant	15–20 HCl	80–120
Ketoamine aldehyde condensate	15–28 HCl	90–150
Imidazoline	15 HCl	90
Imidazoline plus formaldehyde	15–28 HCl	80–150
Pyridine hydrochloride slag + alkyl sulfonate	15–28 HCl	90
MBT, thiourea, OP, and other pyridine residues	7 HCl + 3 HF	30–70
Benzyl quaternary ammonium salt	15–28 HCl	90–190
Pyridine derivatives, formaldehyde, etc.	15–28 HCl	90–120
Cyclohexanone aniline	15–28 HCl	90–180
Ketoamine condensate	15–28 HCl	90–150
Acetophenone aniline, etc.	15–28 HCl	90–160
Pyridine quaternary ammonium salt, etc.	12 HCl + 6 HF	120
Quinoline quaternary ammonium salt, etc.	12 HCl + 6 HF	150
Alkylpyridine quaternary ammonium salt and other components	15–28 HCl	196
Ketoamine aldehyde condensate, alkynol, etc.	15–28 HCl	90–130
Imidazoline derivative	15 HCl	90–100
Imidazoline	15 HCl	90
Pyridine quaternary ammonium salt, etc.	15–28 HCl	90–120

Table 17.
Acidification inhibitors commonly used in oil fields and their application conditions.

It is possible that the effective corrosion inhibitor for industrial circulating cooling water may be not suitable for the treatment of oil field sewage, which containing large concentration of Cl^-. The typical oil fields in China, such as Shengli, Zhongyuan, Jianghan, Dagang, and Huabei Oilfield, are with the NaCI up to around 200,000 mg/L in the sewage, which also include CO_2, H_2S, dissolved oxygen, oil stain and miscellaneous, etc. CT2-7 corrosion inhibitor (the main component is organic amine) is mixed with HEDP and 1227, which is the promising reagent for oil field.

6. Results and discussion

Corrosion inhibitors are classified into inorganic corrosion inhibitors and organic corrosion inhibitors according to their composition. Inorganic corrosion inhibitors passivate the metal on the surface of the anode by its inorganic anion or prevent ions from the anode portion of the metal surface from entering the solution, thereby inhibiting corrosion. The organic corrosion inhibitor mainly forms a precipitation film by the reaction between a reactive group on the organic molecule and a metal ion generated during the etching process and suppresses the electrochemical processes of the anode and the cathode. They have good adsorption to the metal surface in the corrosive medium. Many corrosion inhibitors containing heteroatoms rely on functional groups to adsorb on the metal surface. The nitrogen atoms in the corrosion inhibitor become cations after quaternization and are easily adsorbed by the negatively charged metal surface to form a monomolecular protective film. The charge distribution and interfacial properties of the metal surface tend to stabilize the energy state of the metal surface. The process can increase the activation energy of the corrosion reaction, slow down the corrosion rate, and greatly inhibit the discharge of hydrogen ion, inhibit the cathode reaction, and effectively improve the corrosion inhibition efficiency of the corrosion inhibitor.

7. Summary conclusion

Corrosion inhibitors play an important role in metal protection engineering and the national economic construction. Judging from the current anticorrosion of equipment and other industrial fields, the use of corrosion inhibitors is an effective and economical anticorrosion method. The research on the theory, testing technology, and calculation method of inhibitor has made certain progress, which has promoted the development and application of new corrosion inhibitors.

In the future, the mechanism of the corrosion inhibitor and the relationship between the molecular structure of the corrosion inhibitor and the corrosion inhibition effect should be further developed. More complete and specific environmental performance assessment methods need to be proposed; neural networks, density functional theory, and other computer science, quantum chemistry, and other research methods should be better applied to the development and evaluation of new corrosion inhibitors, and the new high-efficiency corrosion inhibitors should be designed and synthesized. In addition, promote green chemistry and research on the corrosion inhibitors with low-cost and non-polluting chromium-free, zinc-free, low-phosphorus, and even phosphorus-free which is the promising development direction of water treatment corrosion inhibitor. The corrosion inhibitors can be made from natural raw materials, thereby expanding the application range. Furthermore, researchers should pay more attention to the development of multifunctional corrosion inhibitors used in different working conditions. It should focus on the development of copolymer corrosion inhibitors and other multi-purpose corrosion inhibitors that have the properties of scale inhibition, biocidal, and biodegradability.

Author details

Yun Chen* and Wenzhong Yang
School of Chemistry and Molecular Engineering, Nanjing Tech University, Nanjing, China

*Address all correspondence to: ychen@njtech.edu.cn

References

[1] Satpati AK, Ravindran PV. Electrochemical study of the inhibition of corrosion of stainless steel by 1,2,3-benzotriazole in acidic media. Materials Chemistry and Physics. 2008;**109**:352-359

[2] Behpour M, Ghoreishi SM, Salavati-Niasari M, Ebrahimi B. Evaluating two new synthesized S-N Schiff bases on the corrosion of copper in 15% hydrochloric acid. Materials Chemistry and Physics. 2008;**107**:153-157

[3] Bentiss F, Bouanis A, Mernari B, Traisnel M, Vezin H, Lagrenee M. Understanding the adsorption of 4H-1,2,4-triazole derivatives on mild steel surface in molar hydrochloric acid. Applied Surface Science. 2007;**253**:3696-3704

[4] Tetsuo H. Kurita Handbook of Water Treatment. Second English ed. Japan: Kurita Water Industries Ltd; 1999

[5] Growcock FB, Lopp VR. The inhibition of steel corrosion in hydrochloric-acid with 3-phenyl-2-propyn-1-ol. Corrosion Science. 1988;**28**:397-410

Section 3

Water Splitting

Chapter 10

On the Limits of Photocatalytic Water Splitting

Bahar Ipek and Deniz Uner

Abstract

The major drawbacks on the limited H_2 and O_2 evolution activities of one-step photocatalytic water splitting systems are given here with the emphasis on charge recombination, back-oxidation reactions, and mass transfer limitations. Suppression of these unwanted phenomena is shown to be possible with the usage of small crystal-sized photocatalysts with low defect concentrations, presence of phase junctions, selection of co-catalyst that would be active for H_2 evolution but inactive for O_2 reduction, coating of the co-catalyst or the whole photocatalyst with selectively permeable nanolayers, and usage of photocatalytic systems with high solid–liquid and liquid–gas surface areas. The mass transfer limitations are shown to be important especially in the liquid–gas interfaces for agitated and suspended systems with estimated H_2 transfer rates in the range of ~200–8000 μmol/h.

Keywords: hydrogen production, photocatalyst, water splitting, mass transfer, back-oxidation

1. Introduction

Hydrogen gas is one of the best alternatives to fossil fuels since it has a high gravimetric energy density (142 MJ/kg) and it produces zero carbon upon combustion. Hydrogen is also used as a major reactant in environmentally important reactions such as carbon dioxide hydrogenation to methanol [1] or ammonia production (Haber-Bosch reaction) [2]. For hydrogen to be used as a *clean* energy source, its production via renewable ways is of great importance. It is conventionally produced via steam reforming of methane and fossil fuels (energy intensive, ΔH^0_{rxn} = 206 kJ/mol, 700–1100°C [3]) and coal gasification, which results in significant amounts of carbon dioxide production. The renewable ways for carbon-free production include biological sources (microalgae and cyanobacteria) and electrolysis of water using wind energy and photovoltaic cells as electricity generation sources. In addition to the mentioned renewable ways, photocatalytic water splitting/oxidation is a promising alternative, in which solar energy is used as the driving force to split water molecules to hydrogen and oxygen on the surface of a catalyst. This renewable production method of hydrogen is advantageous over other renewable methods due to the free source of energy and lower cost of the photocatalysts when compared to that of photovoltaic cells or wind turbines. Solar-driven catalytic (photocatalytic) reactions are considered to be of fundamental importance to the catalysis community since the solar energy is inexhaustible; i.e., the solar energy absorbed by the lands and oceans on an hourly basis (432 EJ/h or 120,000 TW [4]) is comparable to the Earth's yearly energy consumption (reaching 575 EJ/year or 18 TW in 2017). However, the solar-to-hydrogen energy conversion efficiency value for photocatalytic water splitting systems is much lower (targeted to be 10%, currently reaching 1% [5]) than that of photovoltaic-assisted electrolysis (reaching 30% [6]) due to the major drawbacks in the one-step photocatalytic water splitting systems. Herein, we firstly introduce photocatalytic water splitting systems and give the major developments in materials such as visible light utilization and corresponding H_2 and O_2 production activity values (in Section 2). Then in Section 3, we discuss the causes of the low efficiencies in photocatalytic water splitting systems and the recent approaches in preventing energy efficiency-lowering factors such as inefficient visible light utilization, charge recombination, back-oxidation reactions, and mass transfer limitations.

2. Photocatalysis and water splitting

The first report on water splitting via harvesting photon energy is authored by Fujishima and Honda using a photoelectrochemical cell with a TiO_2 photoelectrode [7]. Following this first report suggesting the oxidation of water molecule via photo-generated holes on TiO_2 surface with the aid of small electrical voltage, photocatalytic water splitting on powder photocatalyst particles is demonstrated by other authors in the late twentieth century [8–15]. Metal-loaded semiconductors (such as Pt/TiO_2) are described as "short-circuited photoelectrochemical cells" that provide both the oxidizing centers and the reduction centers on the same catalyst (see **Figure 1**) [16].

Photocatalytic reactions are initiated by absorption of light having an energy higher than (or equal to) the bandgap of the photocatalysts that consist of semi-conductor materials. This bandgap energy should be larger than 1.23 V for overall water oxidation reaction, for which the maximum of the valence band and the minimum of the conduction band should be located at proper potentials for the oxygen and hydrogen evolution reactions to occur. To illustrate, the minimum of

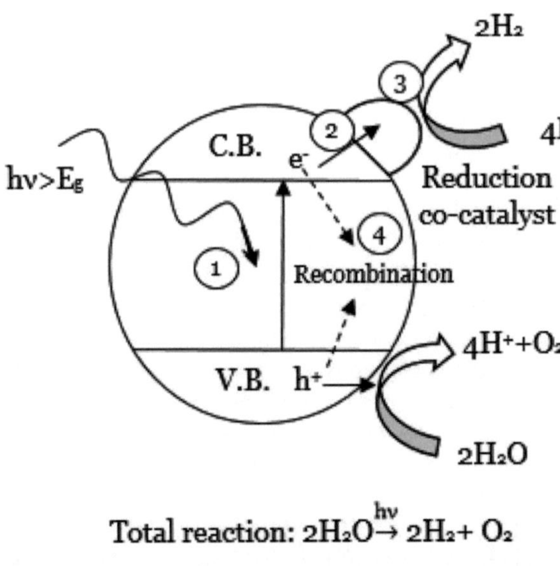

Figure 1.
Schematic representation of photocatalytic water splitting on metal-loaded semiconductor particle systems: (1) light absorption and charge excitation from valence band to conduction band, (2) transfer of the photo-generated electrons and holes to the catalyst surface, (3) surface redox reactions, and (4) charge recombination.

the conduction band energy level should be located at a more negative potential than 0 V vs. NHE, at pH = 0 for H_2 evolution (Eq. (1)), and the maximum of the valence band should be at a more positive potential than 1.23 V vs. NHE at pH = 0 for oxygen evolution reaction (Eq. (2)):

$$4H^+ + 4e^- \rightarrow 2H_2 \tag{1}$$

$$2H_2O + 4\,h^+ \rightarrow 4H^+ + O_2 \tag{2}$$

Following the light absorption, photoexcited electrons are transferred to the conduction band, while a positively charged charge carrier (hole) is generated at the valence band. These charge carriers are then transferred to the catalyst surface (step 2 in **Figure 1**) to be utilized in surface redox reactions, unless they recombine in the bulk or on the surface (step 4). Ultimately, electrons and holes reduce/oxidize the adsorbed species on the catalyst surface (step 3), the products of which should then be desorbed from the surface to complete the overall process.

2.1 Semiconductors

TiO_2, having a large bandgap (anatase: 3.2 eV), is the most commonly used photocatalyst due to its photostability, nontoxicity, and high activity (upon UV radiation $\lambda < 387$ nm). Following the report on water oxidation reaction [7], various photochemical reaction activities of TiO_2 such as carbon dioxide reduction with H_2O [17–19], alkene and alkyne hydrogenation [20, 21], CH_3Cl oxidation [22], 1-octanol degradation [23], phenol degradation [24], surfactant degradation [25], and more have been reported. Detailed reviews on TiO_2-based materials and photocatalytic performances can be found in literature [26–28].

As photostable and active TiO_2 is, UV light requirement to activate the large bandgap of TiO_2 motivated research for visible light active semiconductors as well as bandgap engineering for TiO_2 such as nonmetal ion doping (N [29], C [30], F [31], S [32]). Substitution of lattice oxygen atoms by these anions is reported to shift the valence band level upward and narrow the bandgap to as low as 2.25 eV (~550 nm) with 16.5% N doping [33].

Similar to TiO_2, oxides of other transition metals with d^0 (such as Ti^{4+}, Zr^{4+}, Nb^{5+}, Ta^{5+}, and W^{6+} [34, 35]) and d^{10} electronic configurations (such as Ga^{3+}, In^{3+}, Ge^{4+}, Sn^{4+}, and Sb^5 [36–38]) are shown to possess large bandgap energies (>3 eV) due to the maximum valence band levels consisting O_{2p} orbitals located near 3 V (vs. NHE at pH = 0). These d^0 and d^{10} metal oxide catalysts are reported to show remarkable one-step photocatalytic water splitting activity under UV light irradiation [39] reaching 71% quantum yield with photocatalysts such as Al-doped $SrTiO_3$ [40] or Zn-doped Ga_2O_3 [41]. The H_2 and O_2 evolution activity under UV radiation and the apparent quantum yields of some of these materials are given in **Table 1**. The apparent quantum yield is defined as the number of reacted electrons and holes divided by the number of incident photons on the photocatalysts. **Table 1** is not intended to cover the whole range of particulate catalysts in literature but rather to give a selection of examples. A wider selection of d^0 and d^{10} metal oxide particulate catalysts' one-step water oxidation activity and apparent quantum yields can be found in the works of Kudo et al., Chen et al., and Domen et al. [39, 42, 43].

The most remarkable upgrades in the apparent quantum yields are achieved by material engineering such as (i) doping the metal oxides/perovskites with cations having lower valences, (ii) decreasing the crystal sizes to submicron levels, and (iii) loading with H_2/O_2 evolution co-catalysts.

Semiconductor	Co-catalyst	Bandgap (eV)	H_2 activity (μmol/h)	O_2 activity (μmol/h)	AQY (%)	Reference
$La_2Ti_2O_7$:Ba	NiO_x	3.26	5000		50	[44]
$SrTiO_3$:Al	$Rh_{2-y}Cr_yO_3$	3.2	550*	280*	30 at 300 nm	[35]
$SrTiO_3$:Al (200–500 nm)	$Rh_{2-y}Cr_yO_3$	3.2	1372*	683*	56 at 365 nm	[45]
$SrTiO_3$:Al	MoO_y/$RhCrO_x$	3.2	1800*	900*	69 at 365 nm	[40]
$NaTaO_3$	NiO	4.0	3390	1580	20 at 270 nm	[46]
$NaTaO_3$:La	NiO	4.1	19,800	9700	56 at 270 nm	[34]
Ga_2O_3:Zn	NiO	4.4	4100	2200	20	[47]
Ga_2O_3:Zn	$Rh_{0.5}Cr_{1.5}O_3$	4.4	32,000	16,000	71 at 254 nm	[41]

**0.1 g of photocatalyst is used instead of 1 g.*

Table 1.
H_2 and O_2 evolution activity of d^0 and d^{10} metal oxide particulate catalysts under UV light irradiation.

2.2 Co-catalysts

An important addition to the light-harvesting semiconductors is H_2 evolution/O_2 evolution co-catalysts on the surface. The early co-catalysts that have been widely used included the noble metals and transition metal oxides such as Pt [12, 13], Rh [10], Ru [48], Au [49, 50], and NiO_x [11] that mainly promote the hydrogen evolution, and CoO_x [51] and Fe [52], Mn [52], RuO_2 [53], and IrO_2 [54] that accelerate the oxygen evolution. These metals are considered to act as charge carrier sinks that suppress electron–hole pair recombination as well as increasing the reaction kinetics by lowering the activation energy of the redox reactions. Co-catalysts are also known to inhibit photodegradation of the photocatalysts such as oxysulfides and oxynitrides by generated holes due to the effective extraction of these holes by the co-catalysts [55, 56].

Following the works of noble metal co-catalysts, Domen et al. showed water splitting activity on $SrTiO_3$ photocatalyst together with the co-catalyst NiO [57, 58], which became the choice of H_2 evolution co-catalyst for many d^0 and d^{10} metal oxides such as $La_2Ti_2O_7$:Ba [44], $La_4CaTi_5O_{17}$ [59], $Rb_4Nb_6O_{17}$ [60], $NaTaO_3$ [46], and Ga_2O_3:Zn [47]. The photocatalyst stability of NiO-loaded $K_2La_2Ti_3O_{10}$ is reported to increase by addition of a second co-catalyst, Cr, using a co-impregnation method [61]. Based on the promoting effect of Cr, a systematic study of Cr and various transition metals (such as Fe, Co, Ni, Cu, Ru, Rh, Pd, Ag, Pt) on $(Ga_{1-x}Zn_x)(N_{1-x}O_x)$ has been conducted [62], from which core-shell structures of core Rh nanoislands and shell Cr_2O_3 structures (10–30 nm in size) are found to promote H_2 and O_2 evolution reactions to significant levels [63].

2.3 Visible light utilization

The alterations to the semiconductors such as doping with low-valence cations, reducing particle sizes to submicron levels and obtaining a high degree of crystallinity help with the overall water splitting activity. However, activation of these photocatalysts uses a narrow portion of the solar spectrum (4%); i.e., the UV light sustains as a problem due to the large bandgap energies of these materials. To enable visible light utilization of the d^0- and d^{10}-type oxide semiconductors and to split water into H_2 and O_2 via one-step excitation, the valence band levels should be shifted upward without changing the conduction band potentials. One approach to do this is to use oxynitrides to make use of N_{2p} states that lie at a more negative potential than O_{2p} states. Emerging $LaMgTa_{1-x}O_{1+3x}N_{2-3x}$ [64] and $Ga_{1-x}Zn_xN_{1-x}O_x$ [65] oxynitrides are representatives of visible light active overall water splitting catalysts. Using N doping, the absorbed light wavelength can be increased up to 500 nm on solid solutions of GaN:ZnO ($Ga_{1-x}Zn_xN_{1-x}O_x$) [66] and up to 600 nm on solid solutions of $LaTaON_2$ and $LaMg_{2/3}Ta_{1/3}O_3$ ($LaMgTa_{1-x}O_{1+3x}N_{2-3x}$) [64]. Other examples include $LaSc_xTa_{1-x}O_{1+2x}N_{2-2x}$ [67] and $CaTaO_2N$ [68] in which La or Ta sites are replaced by Ca and Sc that alters O/N ratios due to charge compensation, which in turn results in valence band energy level shift.

Some examples of visible light active photocatalyst and their H_2 and O_2 evolution activity are given in **Table 2**. As it can be seen from the table, one-step water splitting quantum yields are quite lower when compared to those of the UV-activated photocatalysts (**Table 1**). The exceptions to the low activity are reported by $Rh_{2-y}Cr_yO_3$ (Rh 1.0 wt%, Cr 1.5 wt%)-loaded $(Ga_{1-x}Zn_x)(N_{1-x}O_x)$ photocatalyst [63], multiband InGaN/GaN nanowire arrays [69], and monodisperse 4 nm graphite nanoparticle-deposited C_3N_4 catalysts [70].

An alternative way to cover both oxidation and reduction reactions with semiconductors that could be activated under visible light radiation is to utilize two individual photocatalysts with an electron transfer mediator to obtain two-step excitation known as the two-step water oxidation ("Z-scheme system," see **Figure 2**). In this system, O_2 evolution photocatalysts oxidize the water molecules to O_2, while the photo-generated electron is transferred to the mediator to reduce the electron acceptor (such as Fe^{3+} ions or IO_3^- ions). Then, the reduced mediator is oxidized by donating its electron to the H_2 evolution photocatalyst. At the same time, the photo-generated electrons in the H_2 evolution photocatalyst reduce H^+s to H_2.

The semiconductors used in this two-step water splitting process should be selected based on the energy levels of their corresponding valence or conduction band maximum/minimum that would enable O_2/H_2O oxidation and H^+/H_2 reduction. As H_2 evolution and O_2 evolution reactions are realized at separate photocatalysts, these semiconductors could have bandgap energy values lower than 3 eV that would enable visible light utilization such as Pt- or RuO_2-loaded WO_3

Semiconductor	Co-catalyst	Bandgap (eV)	H_2 activity (μmol/h)	O_2 activity (μmol/h)	AQY (%)	Reference
$SrTiO_3$:Rh,Sb	IrO_2		4.4	1.9	0.1 at 420 nm	[71]
g-C_3N_4	Pt-CoO_x	2.8	~8.5	~3.5	0.3 at 405 nm	[72]
CDots-C_3N_4		2.74	46		16 at 420 nm	[70]
$Bi_{1-x}In_xV_{1-x}Mo_xO_4$	RuO_2	2.5	17		3.2 at 420 nm	[73]
$BiYWO_6$	RuO_2	2.7	4.1	1.8	0.17 at 420 nm	[74]
$LaMg_{1/3}Ta_{2/3}O_2N$	$RhCrO_x$		22	11	0.18 at 440 nm	[75]
$(Zn_{0.18}Ga_{0.82})(N_{0.82}O_{0.18})$	$Rh_{2-y}Cr_yO_3$	2.64	927	460	5.9 at 420 nm	[76]
GaN:Mg/InGaN:Mg	Rh/Cr_2O_3	2.22	38	21	12.3 at 400 nm	[69]

Table 2.
H_2 and O_2 evolution activity of one-step water splitting catalysts under visible light irradiation.

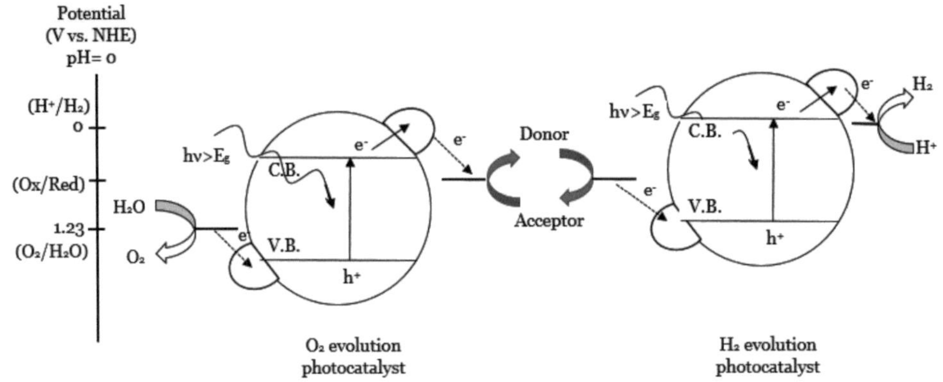

Figure 2.
Schematic diagram for photocatalytic water splitting using a two-step photoexcitation system.

H_2 photocatalyst	O_2 photocatalyst	Mediator	H_2 activity (μmol/h)	O_2 activity (μmol/h)	AQY (%)	Reference
$Pt/SrTiO_3$:Rh	$BiVO_4$	Fe^{3+}/Fe^{2+}	15	7.2	0.4 at 420 nm	[77]
$Pt/SrTiO_3$:Cr/Ta	PtO_x/WO_3	IO_3^-/I^-	~16	~8	1 at 420 nm	[78]
Pt/TaON	PtO_x/WO_3	IO_3^-/I^-	~16.5	~8	0.5 at 420 nm	[79]
Pt/ZrO_2/TaON	PtO_x/WO_3	IO_3^-/I^-	52	27	6.3 at 420 nm	[80]
$Ru/SrTiO_3$:Rh	$BiVO_4$	Fe^{3+}/Fe^{2+}	88	44	4.2 at 420 nm	[81]
$Pt/MgTa_2O_{6-x}N_y$/ TaON	PtO_x/WO_3	IO_3^-/I^-	108	55	6.8 at 420 nm	[82]

Table 3.
Z-scheme-type photocatalysts for water splitting without sacrificial agents.

($E_g \sim 2.8$ eV) or oxynitrides such as TaON ($E_g \sim 2.4$ eV) or Rh-doped $SrTiO_3$ ($E_g \sim 2.4$ eV). Examples of these materials and systems can be seen in **Table 3**. The detailed reviews on two-step photocatalytic water splitting can be found elsewhere [83].

3. Drawbacks on photocatalytic activity

There are numerous and challenging processes that need to be realized for photocatalytic evolution of H_2 and O_2 (**Table 4**) via a thermodynamically unfavorable reaction (Eq. (3)):

$$H_2O \rightarrow H_2 + 1/2\,O_2\ \Delta G^0 = 237\ \text{kJ/mol} \tag{3}$$

These processes include (i) excitation of the semiconductor photocatalyst with photon having higher energies than the bandgap energy of the material, (ii) transfer of the photo-generated electrons and holes to the reaction sites on the surface, (iii) utilization of these charge carriers in the oxidation/reduction reactions, and (iv) desorption of the products from the surface of the photocatalyst to the liquid/gas medium.

As the timescale of these processes varies, recombination of the electrons and holes in the bulk or on the surface happens more frequently than the rate of the

Process		Timescale
Light absorption and electron and hole generation	$Semiconductor \xrightarrow{h\nu} e^- + h^+$	fs
Photo-generated electron and hole transfer to the surface and trapping	$h^+_{VB} \rightarrow h^+_{trap}$ $e^-_{CB} \rightarrow e^-_{trap}$	200 fs 50 ps
Recombination of charge carriers	$e^-_{tr} + h^+_{tr} \rightarrow recombination$	>20 ns
Interfacial charge transfer	$e^-_{CB} + O_2 \rightarrow O_2^{\bullet -}$	10–100 μs
Observed O_2 evolution*	$2H_2O + 4h^+ \rightarrow O_2 + 4H^+$	37 s*
Photosynthesis of H_2O oxidation	$2H_2O + 4D^+TA \rightarrow O_2 + 4H^+ + DTA$	1.59 ms [84]

**Based on 16,000 μmol O_2/g/h O_2 evolution rate on $Rh_{0.5}Cr_{1.5}O_3$-doped Ga_2O_3:Zn upon illumination at 254 nm [41], assuming 10 m^2/g surface area and 10^{15} sites/cm^2 site density.*

Table 4.
The processes occurring in photocatalytic water splitting on TiO_2 and their timescales [27] and the references therein.

chemical oxidation/reduction reactions. Recombination is therefore considered to be one of the main reasons limiting the photocatalytic activity. Together with the recombination events, realization of back-oxidation reactions (Eq. (4)) on noble metals and the rate-limiting mass transfer events are the major drawbacks in an efficient photocatalytic process:

$$H_2 + 1/2O_2 \rightarrow H_2O \ \Delta G^0 = -237 \ kJ/mol \tag{4}$$

Natural photosynthesis yields a much higher rate of O_2 evolution (see **Table 4**) when compared to artificial water splitting due to improved charge carrier and mass transfer events. From this comparison, it is clear that the photocatalytic systems still need to be perfected to compete with the nature's intricate design.

3.1 Charge recombination

Due to the presence of the multiple processes, the overall photocatalytic reactions are extremely complicated. In order to obtain an efficient photocatalytic performance, the photo-generated charges must be transferred to the surface reaction sites as rapidly as possible while preventing recombination or trapping of these charge carriers. It is reported by Leytner and Hupp that 60% of the trapped electron–hole pairs recombine with a timescale of about 25 ns while releasing heat of 154 kJ/mol [85]. As the defects such as vacancies and dislocations are considered as recombination sites, higher crystallinity of the photocatalysts is often aimed to decrease the recombination rates. From diffusion point of view, the shorter distances for the charge carriers to the surface reaction centers are also aimed to prevent the recombination. Shorter pathways are achieved via smaller crystal/particle sizes of the photocatalysts. More than two times of increase in the H_2 and O_2 evolution rates on Al-doped $SrTiO_3$ photocatalyst (reaching an apparent quantum yield of 56% [45]) as the particle size drops from few micrometers to 200 nm is a direct evidence of the effect of the particle size. Another method for reducing the charge recombination is to make use of phase junctions. One example is the α-β-phase junction of Ga_2O_3, which results in enhanced interfacial charge transfer, charge separation, and therefore enhanced water splitting activity [86]. Loading the photocatalysts with co-catalysts such as noble metals or transition metal oxides to accelerate the reduction/oxidation reactions is a commonly employed method.

These co-catalysts are known to enhance the charge migration from the semiconductor depending on the alignment of the potentials of the semiconductor and the co-catalyst. As these co-catalysts accelerate the desired H_2 evolution and O_2 evolution reactions, they can also increase the rates of undesired secondary reactions such as hydrogen oxidation or oxygen reduction to water reactions.

3.2 Back-oxidation reactions

Introduction of one-step photocatalysts for overall water splitting combined the H_2 evolution and O_2 evolution sites on the same catalyst surface. This design of a photocatalytic system that realizes both charge trapping and reduction/oxidation reactions on the same surface not only accelerated the charge recombination but also allowed secondary reactions on these reduction/oxidation centers. When fast removal of the products, i.e., H_2 and O_2, is not provided and there are no barriers that prevent interaction of these products with highly active sites, the reaction of H_2 and O_2 on the photocatalyst surface to produce H_2O ($2H_2 + O_2 \rightarrow H_2O$) is highly probable. And back-oxidation of the produced H_2 is considered to be one of the main reasons for observed low photocatalytic water splitting activity values.

As early as 1985, Sato and coworkers realized the importance of back-oxidation of H_2 with O_2 to produce H_2O. They have realized that the metal-loaded photocatalysts, mainly Pt- or Pd-loaded TiO_2, can oxidize H_2 with O_2 easily under the same photocatalytic water oxidation conditions. They have reported first-order reaction rate constants in the range of 0.23–0.51 h^{-1} for Pt, 0.32–1.8 h^{-1} for Pd, and 0.2–0.3 h^{-1} for Rh, suggesting the least active metal for back-oxidation reaction to be Rh [10]. Later in 2000, Anpo and coworkers investigated back-oxidation reaction on Pt/TiO_2 systems under dark conditions and observed increased back-oxidation rate with increasing Pt loading (up to 0.1 wt.% [87]). While Pt is active for H_2 evolution (Eq. (1)), it is also notoriously active for dark H_2–O_2 recombination reaction (Eq. (4)) even at room temperature [88]. In order to prevent H_2–O_2 recombination reaction, the Pt surface is modified with F ions for Pt/TiO_2 catalyst, and the reaction rate decreased from 2 to 0.3 h^{-1} upon F^- modification [89]. The inhibition mechanism is suggested to be due to the occupation of the H_2 surface adsorption sites on Pt by F atoms.

Another modification to the noble metal surfaces is reported by Lercher et al., in which CO is chemisorbed on the Rh co-catalyst for GaN:ZnO semiconductor. Chemisorbed molecular layer of CO suppressed the back-oxidation reaction by selective metal poisoning of the back-oxidation sites by CO. While H_2 evolution rates of 28 μmol/h are achieved (75 mg photocatalyst, 300 W Xe lamp [90]), significant CO oxidation to CO_2 is also observed.

The back-oxidation reaction-inhibiting effects of the nanolayer coating on noble metals are shown on Rh/Cr_2O_3-loaded GaN:ZnO photocatalysts. Rh/Cr_2O_3 core-shell structure [91] is formed by photodeposition of Rh and reduction of CrO_4^{2-} by electrons coming from Rh upon radiation, resulting in few nanometer thickness of Cr_2O_3 layer (2–3 nm, see **Figure 3**). Hydrated Cr_2O_3 nanolayer is reported to selectively permeate protons for H_2 evolution reaction [92], whereas it hinders O_2 permeation from the layer inhibiting O_2 reduction reaction (Eq. (5)) on Rh sites [93]. The same effect is also valid for Cr_2O_3-coated Pt catalyst (GaN:ZnO). Back-oxidation rates on Pt-loaded GaN:ZnO photocatalyst decreased significantly from $\sim 105 \times 10^{-12}$ molecules/s to $\sim 8 \times 10^{-12}$ molecules/s, while photocatalytic H_2 evolution rate increased from $\sim 5 \times 10^{-12}$ molecules/s to $\sim 30 \times 10^{-12}$ molecules/s upon Cr_2O_3 coating [93]. Apart from the oxygen-blocking role of the Cr_2O_3 nanolayer, much lower back-oxidation rate of Rh-loaded GaN:ZnO when compared to Pt-loaded GaN:ZnO (11×10^{-12} molecules/s vs. 105×10^{-12} molecules/s)

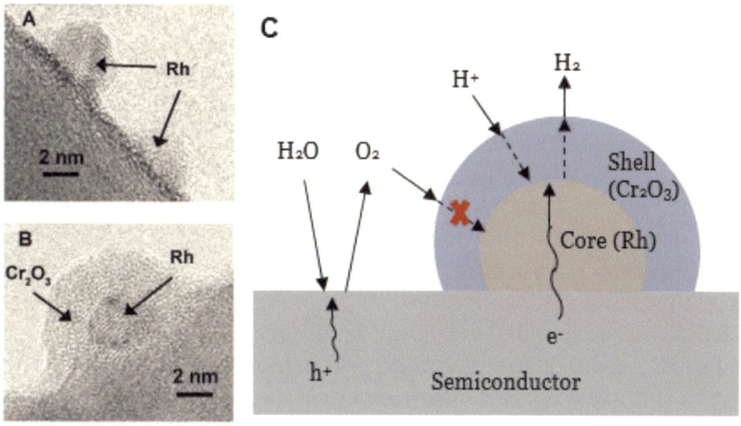

Figure 3.
HR-TEM images of GaN:ZnO photodeposited with (A) Rh and (B) Rh/Cr_2O_3. Reprinted with permission from [55]. Copyright 2007 American Chemical Society. (C) Schematic representation of O_2 and H_2 evolution reactions with inhibited O_2 permeation and O_2 reduction reaction on core-shell-type co-catalysts.

explains the significant photocatalytic H_2 evolution activity on Rh/Cr_2O_3-loaded GaN:ZnO (130×10^{-12} molecules/s). Lower back-oxidation rate of Rh-loaded GaN: ZnO could be related to the low-oxygen reduction reaction (Eq. 5) activity of Rh when compared to Pt [94]:

$$O_2 + 4H^+ + 4e^- \rightarrow 2H_2O \tag{5}$$

Similar selective permeability concept is considered to be the case for Ni/NiO core-shell structures deposited on various photocatalysts such as $SrTiO_3$ or $NaTaO_3$ [34, 57]. In these systems, in addition to the back-oxidation reaction impeding effect of NiO layer on Ni [58], low-oxygen reduction activity of NiO_x catalysts when compared to Pt can also be considered to be effective for improved water splitting activity.

Coatings of the whole photocatalyst instead of the co-catalyst by oxyhydroxides of Ti, Nb, and Ta are reported on Rh-loaded $SrTiO_3$:Sc photocatalyst. Surface nanolayer not only suppressed back-oxidation reactions but also prevented the access of sacrificial agents such as ethanol to the photocatalyst surface, resulting in nearly stoichiometric H_2/O_2 ratios [95]. Surface nanolayer coatings on the whole photocatalysts have proven to also prevent photodecomposition (N_2 evolution) of oxynitride photocatalysts while increasing the overall water splitting activity [64, 68].

Prevention of the secondary reactions such as H_2 oxidation or O_2 reduction reaction to H_2O is found to be essential for improving the overall water splitting activity and the apparent quantum yield values (reaching apparent quantum yield value of 69% under irradiation at 365 nm [40]). In addition to the reduced back-oxidation rates, complementary measures such as decreasing the charge recombination rates and enhancing the product transfer rates away from the surface (increasing the mass transfer rates) are necessary for increased photocatalytic water splitting activity.

3.3 Mass transfer limitations

Mass transfer limitations especially in the slurry photocatalytic systems can be the most overlooked problem in the photocatalytic field. To complete the photocatalytic reaction cycle, adsorption of the reactants, reduction/oxidation of the reactants, desorption of the products, and transfer of the products from the photocatalyst surface to the gas phase need to be realized. When the rates of the mass transfer of the products from the surface are slower than the reduction/oxidation rates, produced H_2 and O_2 would stay longer on the surface, resulting in promotion of back-oxidation reactions. Moreover, when the mass transfer rates are slower than the reaction kinetics, the apparent H_2 and O_2 evolution rates in the gas phase will be limited by the mass transfer rates.

Experimental evidence for mass transfer limitations in agitated systems is presented in a previous publication [96]. In a batch slurry reactor, where the catalyst particles are suspended via agitation, observed H_2 evolution rates for UV-irradiated Pt/TiO_2 photocatalyst showed improvement with increasing stirring rates up to 900 rpm (**Figure 4a**). This improvement is a direct indication of mass transfer limitations on the solid–liquid and gas–liquid interfaces as the turbulence in the liquid and therefore boundary layers are affected by increasing stirring rates. In another experiment, the effect of liquid volume is investigated by varying catalyst

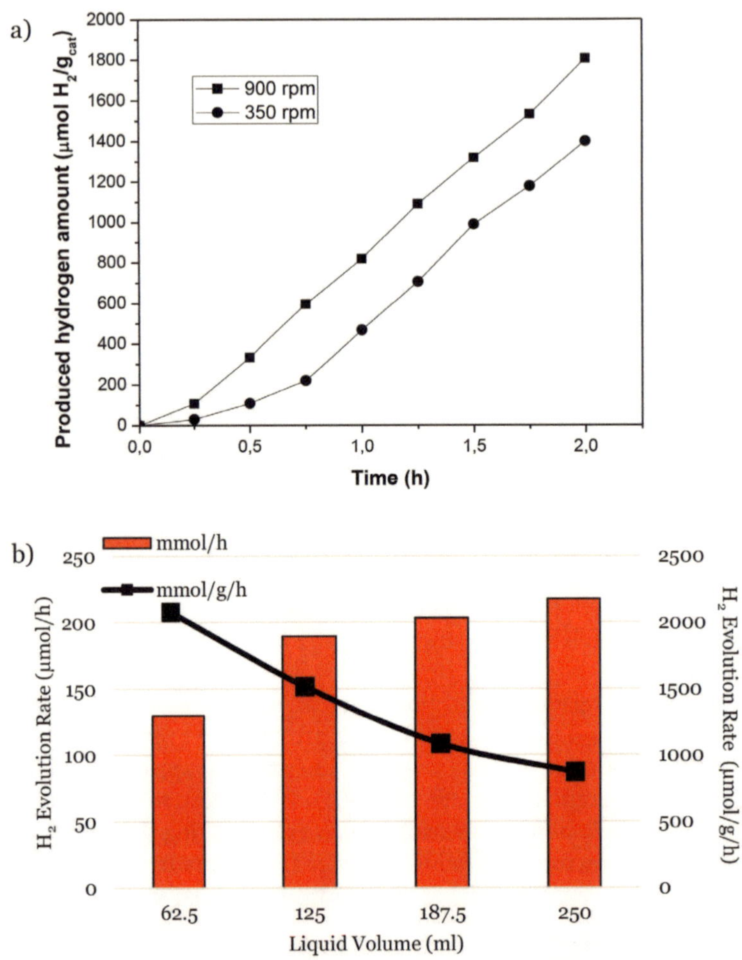

Figure 4.
(a) Effect of stirring rate on photocatalytic hydrogen evolution with methanol as sacrificial agent, with 0.5 wt% Pt/TiO_2, 250 ml deionized water, 2 ml methanol, (■) 900 rpm and (●) 350 rpm. (b) Observed hydrogen evolution rates in the gas phase with changing liquid volume, CH_3OH/H_2O:1/125 (v/v) and $C_{catalyst}$: 1 g/L for each case. Adapted from [96].

weight and liquid volumes (keeping the catalyst concentration constant). H_2 evolution rates on an hour basis (μmol H_2/h) are found the same regardless of the liquid volume (or catalyst weight) above 62.5 ml (**Figure 4b**) as the H_2 evolution rate per gram and hour basis decreased as liquid volume increased. Similar H_2 evolution rates regardless of the catalyst weight indicate significant mass transfer limitations in the liquid–gas interface.

Mass transfer limitations for different photocatalytic reaction systems are analyzed by different groups. For immobilized photocatalyst systems, the importance of internal mass transfer resistance is emphasized [97]. In another investigation, severe mass transfer limitations are observed in the product separator (liquid–gas interface) for a fluidized bed/separator system, in which modification of the liquid–gas surface area enhanced the H_2 evolution rates by 350% [98].

To prevent mass transfer limitations in the photocatalytic tests and to report actual kinetic rates; stirring rates, liquid levels, and mass transfer areas should be designed carefully. To design these parameters, approximate mass transfer rates should be known. Here, we present a sample calculation for H_2 mass transfer rate in a slurry reactor containing 0.5 g TiO_2 photocatalyst having a surface area of 40 m^2/g inside an agitated glass reactor having 200 ml liquid volume and a tank diameter of 7 cm.

Mass transfer resistances in a gas–liquid–solid multiphase photocatalytic systems involve the internal mass transfer, mass transfer from the solid catalyst particles to liquid (Eq. 6), transfer from the liquid bulk to the liquid interface (Eq. 7), and transfer from the liquid–gas interface to the gas phase (Eq. 8). Photocatalysts such as perovskites and TiO_2 are known to be nonporous (unless mesoporous versions are prepared on purpose [99, 100]) and have surface area values between 5 and 50 m^2/g. For nonporous photocatalysts, the internal mass transfer limitations can be discarded (Eq. 9). Hence, the H_2 mass transfer rate equation will have a form containing the mass transfer resistances from the solid–liquid and liquid–gas interfaces as seen in Eq. 8:

$$r_{H2,S} = k_s a_S (C_S - C_L) \tag{6}$$

$$r_{H2,L} = k_L a_L (C_L - C_{L,i}) \tag{7}$$

$$r_{H2,G} = k_G a_G (C_{G,i} - C_G) \tag{8}$$

$$r_{H2} = \frac{C_{H2,s} - HC_{H2,g}}{\left(\frac{1}{k_s a_S} + \frac{1}{k_L a_L} + \frac{H}{k_G a_G}\right)} \tag{9}$$

The mass transfer limitations coming from the solid–liquid and liquid–gas interfaces may play important role depending on the photocatalytic reactor type. The most often used photocatalytic reactor systems such as slurry reactors have solid–liquid and liquid–gas phase interfaces that suspend its catalysts by agitation using an impeller or a magnetic stirrer. The convection mass transfer coefficient for solid–liquid interface of such a system could be estimated using Eq. 10 suggested by Armenante and Kirwan for agitated tanks using Kolmogorov's theory for Reynold's number calculation to consider the effect of solid particle size [101]:

$$Sh = \frac{k_s d_p}{D_{H2-H2O}} = 2 + 0.52\,\mathrm{Re}^{0.52} Sc^{1/3} \tag{10}$$

where k_s is the convection mass transfer coefficient from solid to liquid in (m/s), d_p is the particle diameter in (m), D_{H2-H2O} is the diffusion coefficient of H_2 in liquid water (m^2/s), and Sc is the Schmidt number. A rough estimation for k_s for such system can be found in **Table 5**.

The creation of air/inert bubbles in the continuous phase (water) due to agitation could be considered as the transfer mechanism of produced H_2 from the liquid phase to the gas phase. In such systems, comparing the mass transfer resistance from liquid to interface and interface to gas, it can be assumed that nearly all of the mass transfer resistance comes from the liquid side of the interface [102], leaving Eq. 9 as Eq. 11:

$$r_{H2} = \frac{C_{H2,s} - C_{H2,Li}}{\left(\frac{1}{k_s a_S} + \frac{1}{k_L a_L}\right)} \tag{11}$$

The liquid side mass transfer coefficient for such a system could then be calculated using Calderbank and Moo-Young correlation for rising small bubbles of gas in continuous liquid phase (Eq. 12) [103]:

$$Sh = \frac{k_L d_b}{D_{H2-H2O}} = 2 + 0.31Ra^{1/3} \ where \ Ra = \frac{d_b^3(\rho_L - \rho_G)g}{\mu_L D_{H2-H2O}} \tag{12}$$

The first term on Eq. 12 is the molecular diffusion term, whereas the second term is for the rise of the bubbles due to gravitational forces independent of the agitation. With estimations on the bubble size and gas holdup of such a system (given in **Table 6**), the mass transfer coefficient and $k_L a_L$ term are calculated to be 2.5×10^{-4} m/s and 2.2×10^{-6} m^3/s, rendering liquid–gas mass transfer resistance way more important than solid–liquid resistance.

The overall mass transfer coefficient and the mass transfer rate from solid to the gas phase can be calculated with the estimated $k_s a_S$ and $k_L a_L$ values. As the concentration of H_2 in the gas phase will be negligible ($C_G \sim 0$), the liquid phase interface can also be assumed to be equal to zero ($C_{H2,Li} \sim 0$) with negligible gas phase resistance. Therefore, from Eq. 11, the rate of H_2 mass transfer can be calculated by assuming H_2 concentration at the catalyst surface and the gas holdup ratio in the liquid. The rate of H_2 mass transfer values for 200 ml of water and 0.5 g of catalyst

Parameter	Unit	Value
Particle size, d_p	μm	1
Density of water, ρ_L	kg/m^3	997 at 25°C
Viscosity of water, μ_L	Pa s	890×10^{-6} at 25°C
Kinematic viscosity, ν	m^2/s	8.93×10^{-7}
Schmidt number $\left(Sc = {}^{\nu}/_{D_{H2-H2O}}\right)$		141
Energy density, $\varepsilon = \frac{Power}{mass\ of\ liquid}$	m^2/s^3	25 for 5 W stirrer, 200 g solution
Reynold's number $\left(\mathrm{Re} = {}^{\varepsilon^{1/3} d_p{}^{4/3}}/_{\nu}\right)$		0.033
Sherwood number, $Sh = \frac{k_s d_p}{D_{H2-H2O}} = 2 + 0.52\,\mathrm{Re}^{0.52} Sc^{1/3}$		2.46
Diffusion coefficient of H_2 in water	m^2/s	$6.30{*}10^{-9}$ at 25 °C
Convection mass transfer coefficient for solid, k_s	m/s	0.015
$k_s a_S$	m^3/s	0.31

Table 5.
Convection mass transfer coefficient calculation for solid–liquid transfer and parameters used in the calculation.

are calculated in the range of ~200–8000 μmol/h (see **Figure 5**) for a surface H_2 adsorption capacity range of 50–400 μmol/g (H_2 chemisorption on 0.1% Pt/TiO_2 is reported to be ~400 μmol/g at room temperature [104]) and gas holdup ratio between 0.001 and 0.005.

The H_2 concentration on the solid surface and the gas–liquid contact area are not easy to estimate. However, for the limited gas–liquid area, the photocatalytic reaction rates above the calculated mass transfer rate will be suppressed due to the limiting mass transfer rates. Therefore, special care must be given for the UV-irradiated photocatalytic systems, in which observed H_2 and O_2 evolution rates are found to be close to the calculated mass transfer rates here (see **Table 1**).

These studies show that for each type of photocatalytic system that contains limited gas–liquid contact area or immobilized photocatalyst, mass transfer limitations should not be underestimated, and not only the materials but also the systems should be improved for better photocatalytic efficiencies.

Parameter	Unit	Value
Bubble size, d_b	μm	700
Density of water, ρ_L	kg/m^3	997 at 25°C
Density of air, ρ_G	kg/m^3	1.18 at 25°C
Viscosity of water, μ_L	Pa s	890×10^{-6} at 25°C
Raleigh number $\left(Ra = \frac{d_b^3(\rho_L - \rho_G)g}{\mu_L D_{H2-H2O}}\right)$		5.97×10^5
Sherwood number, $Sh = \frac{k_L d_b}{D_{H2-H2O}} = 2 + 0.31 Ra^{1/3}$		28.1
Diffusion coefficient of H_2 in water	m^2/s	6.3×10^{-9} at 25°C
Convection mass transfer coefficient for liquid, k_L	m/s	2.5×10^{-4}
Liquid–gas bubble contact area, $a_L = \frac{6V_G}{d_b} = \frac{6\phi}{d_b} V_L$ *where* $\phi = V_G/V_L$	m^2	0.008 (gas holdup, ϕ, assumed to be 0.005)
$k_L a_L$	m^3/s	2.2×10^{-6}

Table 6.
Convection mass transfer coefficient calculation for liquid–gas transfer and parameters used in the calculation.

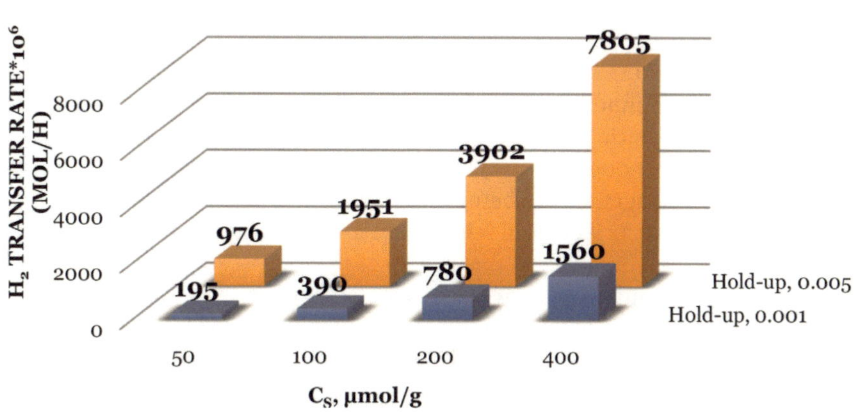

Figure 5.
Calculated H_2 mass transfer rate values (μmol H_2/h) for gas holdup values of 0.001 and 0.005 and H_2 adsorption capacity values in the range of 50–400 μmol/g. The liquid volume is taken as 200 ml and catalyst weight is taken as 0.5 g.

4. Future of photocatalytic H_2 evolution

The literature examples of photocatalytic water splitting activities show improvements in visible light utilization, charge separation, and prevention of back-oxidation reactions via fine tuning of photocatalyst materials that enabled more efficient water splitting systems. The efficiency of these systems working under sunlight is better defined with solar-to-hydrogen energy conversion efficiency (STH), i.e., hydrogen production rate times the Gibbs free energy for generating 1 mole of H_2 divided by the power of incident sunlight (Eq. 13):

$$STH = \frac{(mmol\ H_2/s) * 237 * 10^3\ J/mol}{P\left(\frac{mW}{cm^2}\right) * A(cm^2)} \times 100 \tag{13}$$

The estimated STH required for the particulate photocatalytic systems to be economically compatible with current H_2 production technologies is 10% [104]. However, the highest STH values obtained with current developed photocatalysts are in the order of 1–2% (1.8% at 400–475 nm using Rh/Cr_2O_3-loaded GaN:Mg/InGaN:Mg photocatalyst [69] and 2% at 420 nm using CDots-C_3N_4 [70]). The STH conversion efficiency depends both on the catalytic activity and the extent of the utilization of sunlight that depends on the bandgap of the semiconductor. With current photocatalysts having absorption edges around 500 nm, even 100% apparent quantum yields would not guarantee 10% STH values [106]. For a photocatalyst to show 10% STH values, it should have absorption edges at least at 600 nm with apparent quantum yields around 60%. Under the light of these calculations, it can be said that the present photocatalysts having adsorption edge values around 450 nm and quantum yields around 10% are far from being utilized in commercial systems. In order to achieve targeted STH values, the photocatalysts with lower bandgap energies such as (oxy)nitrides and (oxy)sulfides should be improved for H_2 evolution activities while ensuring their thermal stability and photostability.

Large-scale photocatalytic water splitting reactors are implemented with current low STH values as of 2015. The first example of large-scale photocatalytic water splitting utilized Pt-loaded C_3N_4 photocatalyst with sacrificial electron donor triethanolamine in a flat-panel-type photocatalytic reactor system in 2015 [107]. The solar-to-hydrogen conversion efficiency is reported to be 0.12%, for which the photocatalytic activity is monitored for 30 days. In such systems, where a sacrificial reagent such as triethanolamine or methanol is irreversibly oxidized at a more negative potential than water (thermodynamically more favorable) at the oxidation centers, the photo-generated charges can be more efficiently separated, thus increasing H_2 evolution rates. However, in those systems, hydrogen production is not solely due to the water splitting; as the carbon- and hydrogen-containing "sacrificial agents" are being oxidized at the oxidation centers, they produce hydrogen as well as aldehydes, carboxylic acid, and carbon dioxide [108].

Another large-scale photoreactor is reported by Domen et al., who used Al-doped and $RhCrO_x$-loaded $SrTiO_3$ photocatalyst sheets in their 1 × 1 m water splitting panel [45]. The achieved STH value under simulated sunlight is 0.6% at 331 K and limited to a maximum value of 1.4% due to the large bandgap energy of the photocatalyst (3.2 eV). Improved STH values (reaching 1.1%) are shown to be possible on a two-step excitation system in 2016 using photocatalyst sheets having smaller bandgap energy values such as Mo-doped $BiVO_4$ (2.4 eV) and La- and Rh-doped $SrTiO_3$ [5].

5. Conclusion

The developments in the photocatalytic water splitting reactions are explained here with the emphasis on the one-step photocatalysis systems. The early photocatalyst improvements with bandgap engineering, co-catalyst usage, and size reductions are shown to contribute to the increased visible light-driven H_2 evolution activity values. The main drawbacks in the present systems are discussed to be the charge recombination, back-oxidation reactions of the products into water, and mass transfer limitations especially in the three-phase systems. Using defect-free small crystals of photocatalysts and making use of phase junctions or metal co-catalysts are suggested to decrease charge recombination rates. Back-oxidation of H_2 into water or oxygen reduction reaction to water is expected in many noble metal-containing particulate photocatalyst systems. The prevention of these unwanted secondary reactions is shown to be possible to some extent by modification of the noble metal surfaces. Some examples of these modifications are anion coating, partial adsorption of a poison, or nanolayer coating of the co-catalyst or the whole photocatalyst. Selective permeation property of the nanolayer coatings such as Cr_2O_3 is reported to suppress the back-oxidation rates, resulting in enhanced H_2 and O_2 evolution rates. Possible mass transfer limitations, limiting the observed rates in three-phase systems, are predicted especially in the liquid–gas interfaces. The literature examples attracted attention for the liquid–gas interfaces in suspended systems and internal mass transfer limitations for the immobilized photocatalyst systems. It is concluded that, in addition to the required developments in activities with suppression of charge recombination, back-oxidation, and mass transfer limitations, future of the photocatalytic systems would necessitate active and stable photocatalysts with narrower bandgap energies (to be activated at >600 nm) for achieving targeted 10% solar-to-hydrogen energy conversion efficiency value.

Author details

Bahar Ipek* and Deniz Uner
Department of Chemical Engineering, Middle East Technical University, Ankara, Turkey

*Address all correspondence to: bipek@metu.edu.tr

References

[1] Centi G, Perathoner S. Opportunities and prospects in the chemical recycling of carbon dioxide to fuels. Catalysis Today. 2009;**148**:191-205

[2] Erisman JW, Sutton MA, Galloway J, Klimont Z, Winiwarter W. How a century of ammonia synthesis changed the world. Nature Geoscience. 2008;**1**: 636-639

[3] Aasberg-Petersen K, Dybkjær I, Ovesen CV, Schjødt NC, Sehested J, Thomsen SG. Natural gas to synthesis gas—Catalysts and catalytic processes. Journal of Natural Gas Science and Engineering. 2011;**3**:423-459

[4] Morton O. A new day dawning?: Silicon valley sunrise. Nature. 2006;**32**: 19-22

[5] Wang Q, Hisatomi T, Jia Q, Tokudome H, Zhong M, Wang C, et al. Scalable water splitting on particulate photocatalyst sheets with a solar-to-hydrogen energy conversion efficiency exceeding 1%. Nature Materials. 2016; **15**:611-615

[6] Jia J, Seitz LC, Benck JD, Huo Y, Chen Y, Ng JWD, et al. Solar water splitting by photovoltaic-electrolysis with a solar-to-hydrogen efficiency over 30%. Nature Communications. 2016;**7**:1-6

[7] Fujishima A, Honda K. Electrochemical photolysis of water at a semiconductor electrode. Nature. 1972; **238**:37-38

[8] Sato S, White JM. Photodecomposition of water over Pt/ TiO_2 catalysts. Chemical Physics Letters. 1980;**72**:83-86

[9] Wagner FT, Somorjai GA. Photocatalytic hydrogen production from water on Pt-free $SrTiO_3$ in alkali hydroxide solutions. Nature. 1980;**285**: 559-560

[10] Yamaguti K, Sato S. Photolysis of water over metallized powdered titanium dioxide. Journal of the Chemical Society, Faraday Transactions 1: Physical Chemistry in Condensed Phases. 1985;**81**:1237-1246

[11] Kudo A, Domen K, Ken-ichi M, Onishi T. Photocatalytic activities of TiO_2 loaded with NiO. Chemical Physics Letters. 1987;**133**:517-519

[12] Sayama K, Arakawa H. Effect of carbonate salt addition on the photocatalytic decomposition of liquid water over Pt-TiO_2 catalyst. Journal of the Chemical Society, Faraday Transactions. 1997;**93**:1647-1654

[13] Tabata S, Nishida H, Masaki Y, Tabata K. Stoichiometric photocatalytic decomposition of pure water in Pt/TiO_2 aqueous suspension system. Catalysis Letters. 1995;**34**:245-249

[14] Moon SC, Mametsuka H, Suzuki E, Anpo M. Stoichiometric decomposition of pure water over Pt-loaded Ti/B binary oxide under UV-radiation. Chemistry Letters. 1998;**27**:117-118

[15] Sayama K, Arakawa H. Effect of Na_2CO_3 addition on photocatalytic decomposition of liquid water over various semiconductor catalysts. Journal of Photochemistry and Photobiology A: Chemistry. 1994;**77**:243-247

[16] Bard AJ. Photoelectrochemistry. Science (80-). 1980;**207**:139-144

[17] Anpo M, Yamashita H, Ichihashi Y, Ehara S. Photocatalytic reduction of CO_2 with H_2O on various titanium oxide catalysts. Journal of Elcetroanalytical Chemistry. 1995;**396**:21-26

[18] Saladin F, Forss L, Kamber I. Photosynthesis of CH_4 at a TiO_2 surface from gaseous H_2O and CO_2. Journal of

the Chemical Society, Chemical Communications. 1995:533-534

[19] Yamashita H, Nishiguchi H, Kamada N, Anpo M, Teraoka Y, Hatano H, et al. Photocatalytic reduction of CO_2 with H_2O on TiO_2 and Cu/TiO_2 catalysts. Research on Chemical Intermediates. 1994;**20**:815-823

[20] Anpo M, Aikawa N, Kubokawa Y. Photocatalytic hydrogenation of alkynes and alkenes with water over TiO_2. Pt-loading effect on the primary processes. The Journal of Physical Chemistry. 1984;**88**:3998-4000

[21] Anpo M, Nakaya H, Kodama S, Kubokawa Y, Domen K, Onishi T. Photocatalysis over binary metal oxides. Enhancement of the photocatalytic activity of titanium dioxide in titanium-silicon oxides. The Journal of Physical Chemistry. 1986;**90**:1633-1636

[22] Choi W, Termin A, Hoffmann MR. The role of metal ion dopants in quantum-sized TiO_2: Correlation between photoreactivity and charge carrier recombination dynamics. The Journal of Physical Chemistry. 1994;**98**: 13669-13679

[23] Yamashita H, Ichihashi Y, Harada M, Stewart G, Fox MA, Anpo M. Photocatalytic degradation of 1-octanol on anchored titanium oxide and on TiO_2 powder catalysts. Journal of Catalysis. 1996;**158**:97-101

[24] Trillas M, Pujol M, Domènech X. Phenol photodegradation over titanium dioxide. Journal of Chemical Technology and Biotechnology. 1992;**55**: 85-90

[25] Hidaka H, Zhao J. Photodegradation of surfactants catalyzed by a TiO_2 semiconductor. Colloids and Surfaces. 1992;**67**:165-182

[26] Kitano M, Matsuoka M, Ueshima M, Anpo M. Recent developments in titanium oxide-based photocatalysts. Applied Catalysis A: General. 2007;**325**: 1-14

[27] Schneider J, Matsuoka M, Takeuchi M, Zhang J, Horiuchi Y, Anpo M, et al. Understanding TiO_2 photocatalysis: Mechanisms and materials. Chemical Reviews. 2014;**114**:9919-9986

[28] Gaya UI, Abdullah AH. Heterogeneous photocatalytic degradation of organic contaminants over titanium dioxide: A review of fundamentals, progress and problems. Journal of Photochemistry and Photobiology C: Photochemistry Reviews. 2008;**9**:1-12

[29] Asahi R, Morikawa T, Ohwaki T, Aoki K, Taga Y. Visible-light photocatalysis in nitrogen-doped titanium oxides. Science (80-). 2001; **293**:269-271

[30] Parayil SK, Kibombo HS, Wu CM, Peng R, Baltrusaitis J, Koodali RT. Enhanced photocatalytic water splitting activity of carbon-modified TiO_2 composite materials synthesized by a green synthetic approach. International Journal of Hydrogen Energy. 2012;**37**: 8257-8267

[31] Xing M, Qi D, Zhang J, Chen F, Tian B, Bagwas S, et al. Super-hydrophobic fluorination mesoporous MCF/TiO_2 composite as a high-performance photocatalyst. Journal of Catalysis. 2012; **294**:37-46

[32] Periyat P, Pillai SC, Mccormack DE, Colreavy J, Hinder SJ. Improved high-temperature stability and sun-light-driven photocatalytic activity of sulfur-doped anatase TiO_2. Journal of Physical Chemistry C. 2008;**112**:7644-7652

[33] Kitano M, Funatsu K, Matsuoka M, Ueshima M, Anpo M. Preparation of nitrogen-substituted TiO_2 thin film photocatalysts by the radio frequency magnetron sputtering deposition

method and their photocatalytic reactivity under visible light irradiation. The Journal of Physical Chemistry B. 2006;**110**:25266-25272

[34] Kato H, Asakura K, Kudo A. Highly efficient water splitting into H_2 and O_2 over lanthanum-doped $NaTaO_3$ photocatalysts with high crystallinity and surface nanostructure. Journal of the American Chemical Society. 2003; **125**:3082-3089

[35] Ham Y, Hisatomi T, Goto Y, Moriya Y, Sakata Y, Yamakata A, et al. Flux-mediated doping of $SrTiO_3$ photocatalysts for efficient overall water splitting. Journal of Materials Chemistry A. 2016;**4**:3027-3033

[36] Ikarashi K, Sato J, Kobayashi H, Saito N, Nishiyama H, Inoue Y. Photocatalysis for water decomposition by RuO_2-dispersed $ZnGa_2O_4$ with d10 configuration. The Journal of Physical Chemistry B. 2002;**106**:9048-9053

[37] Sato J, Saito N, Nishiyama H, Inoue Y. New photocatalyst group for water decomposition of RuO_2-loaded p-block metal (In, Sn, and Sb) oxides with d10 configuration. The Journal of Physical Chemistry. B. 2001;**105**:6061-6063

[38] Inoue Y. Photocatalytic water splitting by RuO_2-loaded metal oxides and nitrides with d0- and d10-related electronic configurations. Energy & Environmental Science. 2009;**2**:364-386

[39] Kudo A, Miseki Y. Heterogeneous photocatalyst materials for water splitting. Chemical Society Reviews. 2009;**38**:253-278

[40] Chiang TH, Lyu H, Hisatomi T, Goto Y, Takata T, Katayama M, et al. Efficient photocatalytic water splitting using Al-doped $SrTiO_3$ coloaded with molybdenum oxide and rhodium-chromium oxide. ACS Catalysis. 2018;**8**: 2782-2788

[41] Sakata Y, Hayashi T, Yasunaga R, Yanaga N, Imamura H. Remarkably high apparent quantum yield of the overall photocatalytic H_2O splitting achieved by utilizing Zn ion added Ga_2O_3 prepared using dilute $CaCl_2$ solution. Chemical Communications. Royal Society of Chemistry. 2015;**51**: 12935-12938

[42] Chen X, Shen S, Guo L, Mao SS. Semiconductor-based photocatalytic hydrogen generation. Chemical Reviews. 2010;**110**:6503-6570

[43] Chen S, Takata T, Domen K. Particulate photocatalysts for overall water splitting. Nature Reviews Materials. 2017;**2**:1-17

[44] Kim J, Hwang DW, Kim HG, Bae SW, Lee JS, Li W, et al. Highly efficient overall water splitting through optimization of preparation and operation conditions of layered perovskite photocatalysts. Topics in Catalysis. 2005;**35**:295-303

[45] Goto Y, Hisatomi T, Wang Q, Higashi T, Ishikiriyama K, Maeda T, et al. A particulate photocatalyst water-splitting panel for large-scale solar hydrogen generation. Joule. 2018;**2**:509-520

[46] Kato H, Kudo A. Water splitting into H_2 and O_2 on alkali tantalate photocatalysts $ATaO_3$ (A = Li, Na, and K). The Journal of Physical Chemistry B. 2001;**105**:4285-4292

[47] Sakata Y, Matsuda Y, Yanagida T, Hirata K, Imamura H, Teramura K. Effect of metal ion addition in a Ni supported Ga_2O_3 photocatalyst on the photocatalytic overall splitting of H2O. Catalysis Letters. 2008;**125**:22-26

[48] Inoue Y, Niiyama T, Asai Y, Sato K. Stable photocatalytic activity of $BaTi_4O_9$ combined with ruthenium oxide for decomposition of water. Journal of

Chemical Society, Chemical Communications. 1992:579-580

[49] Iwase A, Kato H, Kudo A. Nanosized Au particles as an efficient cocatalyst for photocatalytic overall water splitting. Catalysis Letters. 2006; **108**:7-10

[50] Murdoch M, Waterhouse GIN, Nadeem MA, Metson JB, Keane MA, Howe RF, et al. The effect of gold loading and particle size on photocatalytic hydrogen production from ethanol over Au/TiO_2 nanoparticles. Nature Chemistry. 2011; **3**:489-492

[51] Li R, Han H, Zhang F, Wang D, Li C. Highly efficient photocatalysts constructed by rational assembly of dual-cocatalysts separately on different facets of $BiVO_4$. Energy & Environmental Science. 2014;**7**:1369-1376

[52] Liu L, Ji Z, Zou W, Gu X, Deng Y, Gao F, et al. In situ loading transition metal oxide clusters on TiO_2 nanosheets as co-catalysts for exceptional high photoactivity. ACS Catalysis. 2013;**3**: 2052-2061

[53] Maeda K, Wang X, Nishihara Y, Lu D, Antonietti M, Domen K. Photocatalytic activities of graphitic carbon nitride powder for water reduction and oxidation under visible light. Journal of Physical Chemistry C. 2009;**113**:4940-4947

[54] Youngblood WJ, Lee S-HA, Kobayashi Y, Hernandez-Pagan EA, Hoertz PG, Moore TA, et al. Photoassisted overall water splitting in a visible light-absorbing dye-sensitized photoelectrochemical cell. Journal of the American Chemical Society. 2009;**131**: 926-927

[55] Maeda K, Domen K. New non-oxide photocatalysts designed for overall water splitting under visible light. Journal of Physical Chemistry C. 2007; **111**:7851-7861

[56] Kamata K, Maeda K, Lu D, Kako Y, Domen K. Synthesis and photocatalytic activity of gallium-zinc-indium mixed oxynitride for hydrogen and oxygen evolution under visible light. Chemical Physics Letters. 2009;**470**:90-94

[57] Domen K, Naito S, Soma M, Onishi T, Tamaru K. Photocatalytic decomposition of water vapour on an $NiO–SrTiO_3$ catalyst. Journal of the Chemical Society, Chemical Communications. 1980:543-544

[58] Domen K, Kudo A, Onishi T, Kosugi N, Kuroda H. Photocatalytic decomposition of water into H_2 and O_2 over $NiO-SrTiO_3$ powder. 1. Structure of the catalysts. Journal of Physical Chemistry A. 1986;**90**:292-295

[59] Kim HG, Hwang DW, Kim J, Kim YG, Lee JS. Highly donor-doped (1 1 0) layered perovskite materials as novel photocatalysts for overall water splitting. Chemical Communications. 1999;**2**:1077-1078

[60] Sayama K, Arakawa H, Domen K. Photocatalytic water splitting on nickel intercalated $A_4Ta_xNb_{6-x}O_{17}$ (A = K, Rb). Catalysis Today. 1996;**28**:175-182

[61] Thaminimulla CTK, Takata T, Hara M, Kondo JN, Domen K. Effect of chromium addition for photocatalytic overall water splitting on Ni-$K_2La_2Ti_3O_{10}$. Journal of Catalysis. 2000; **196**:362-365

[62] Maeda K, Teramura K, Saito N, Inoue Y, Domen K. Improvement of photocatalytic activity of $(Ga_{1-x}Zn_x)(N_{1-x}O_x)$ solid solution for overall water splitting by co-loading Cr and another transition metal. Journal of Catalysis. 2006;**243**:303-308

[63] Maeda K, Teramura K, Lu D, Saito N, Inoue Y, Domen K. Noble-

metal/Cr_2O_3 core/shell nanoparticles as a cocatalyst for photocatalytic overall water splitting. Angewandte Chemie International Edition. 2006;**45**:7806-7809

[64] Pan C, Takata T, Nakabayashi M, Matsumoto T, Shibata N, Ikuhara Y, et al. A complex perovskite-type oxynitride: The first photocatalyst for water splitting operable at up to 600 nm. Angewandte Chemie International Edition. 2015;**54**:2955-2959

[65] Ohno T, Bai L, Hisatomi T, Maeda K, Domen K. Photocatalytic water splitting using modified GaN:ZnO solid solution under visible light: Long-time operation and regeneration of activity. Journal of the American Chemical Society. 2012;**134**:8254-8259

[66] Maeda K, Teramura K, Takata T, Hara M, Saito N, Toda K, et al. Overall water splitting on $(Ga_{1-x}Zn_x)(N_{1-x}O_x)$ solid solution photocatalyst: Relationship between physical properties and photocatalytic activity. The Journal of Physical Chemistry B. 2005;**109**:20504-20510

[67] Pan C, Takata T, Kumamoto K, Khine Ma SS, Ueda K, Minegishi T, et al. Band engineering of perovskite-type transition metal oxynitrides for photocatalytic overall water splitting. Journal of Materials Chemistry A. 2016; **4**:4544-4552

[68] Xu J, Pan C, Takata T, Domen K. Photocatalytic overall water splitting on the perovskite-type transition metal oxynitride $CaTaO_2N$ under visible light irradiation. Chemical Communications. 2015;**51**:7191-7194

[69] Kibria MG, Nguyen HPT, Cui K, Zhao S, Liu D, Guo H, et al. One-step overall water splitting under visible light using multiband InGaN/GaN nanowire heterostructures. ACS Nano. 2013;**7**: 7886-7893

[70] Liu J, Liu Y, Liu N, Han Y, Zhang X, Huang H, et al. Metal-free efficient photocatalyst for stable visible water splitting via a two-electron pathway. Science (80-). 2015;**347**:970-974

[71] Asai R, Nemoto H, Jia Q, Saito K, Iwase A, Kudo A. A visible light responsive rhodium and antimony-codoped $SrTiO_3$ powdered photocatalyst loaded with an IrO_2 cocatalyst for solar water splitting. Chemical Communications. 2014;**50**:2543-2546

[72] Zhang G, Lan ZA, Lin L, Lin S, Wang X. Overall water splitting by Pt/g-C_3N_4 photocatalysts without using sacrificial agents. Chemical Science. 2016;**7**:3062-3066

[73] Jo WJ, Kang HJ, Kong K-J, Lee YS, Park H, Lee Y, et al. Phase transition-induced band edge engineering of $BiVO_4$ to split pure water under visible light. Proceedings of the National Academy of Sciences. 2015;**112**:13774-13778

[74] Liu H, Yuan J, Shangguan W, Teraoka Y. Visible-light-responding $BiYWO_3$ solid solution for stoichiometric photocatalytic water splitting. Journal of Physical Chemistry C. 2008;**112**:8521-8523

[75] Pan C, Takata T, Domen K. Overall water splitting on the transition-metal oxynitride photocatalyst $LaMg_{1/3}Ta_{2/3}O_2N$ over a large portion of the visible-light spectrum. Chemistry: A European Journal. 2016;**22**:1854-1862

[76] Maeda K, Teramura K, Domen K. Effect of post-calcination on photocatalytic activity of $(Ga_{1-x}Zn_x)(N_{1-x}O_x)$ solid solution for overall water splitting under visible light. Journal of Catalysis. 2008;**254**:198-204

[77] Kato H, Hori M, Konta R, Shimodaira Y, Kudo A. Construction of Z-scheme type heterogeneous photocatalysis systems for water

splitting into H_2 and O_2 under visible light irradiation. Chemistry Letters. 2004;**33**:1348-1349

[78] Abe R, Sayama K, Sugihara H. Development of new photocatalytic water splitting into H_2 and O_2 using two different semiconductor photocatalysts and a shuttle redox mediator IO_3^-/I^-. The Journal of Physical Chemistry. B. 2005;**109**:16052-16061

[79] Abe R, Higashi M, Domen K. Overall water splitting under visible light through a two-step photoexcitation between TaON and WO_3 in the presence of an iodate-iodide shuttle redox mediator. ChemSusChem. 2011;**4**: 228-237

[80] Maeda K, Higashi M, Lu D, Abe R, Domen K. Efficient nonsacrificial water splitting through two-step photoexcitation by visible light using a modified oxynitride as a hydrogen evolution photocatalyst. Journal of the American Chemical Society. 2010;**132**: 5858-5868

[81] Kato H, Sasaki Y, Shirakura N, Kudo A. Synthesis of highly active rhodium-doped $SrTiO_3$ powders in Z-scheme systems for visible-light-driven photocatalytic overall water splitting. Journal of Materials Chemistry A. 2013; **1**:12327-12333

[82] Chen S, Qi Y, Hisatomi T, Ding Q, Asai T, Li Z, et al. Efficient visible-light-driven Z-scheme overall water splitting using a $MgTa_2O_{6-x}Ny$/TaON heterostructure photocatalyst for H_2 evolution. Angewandte Chemie, International Edition. 2015;**54**:8498-8501

[83] Wang Y, Suzuki H, Xie J, Tomita O, Martin DJ, Higashi M, et al. Mimicking natural photosynthesis: Solar to renewable H_2 fuel synthesis by Z-scheme water splitting systems. Chemical Reviews. 2018;**118**:5201-5241

[84] Haumann M, Liebisch P, Muller C, Barra M, Grabolle M, Dau H. Photosynthetic O_2 formation tracked by time-resolved X-ray experiments. Science (80-). 2005;**310**:1019-1021

[85] Leytner S, Hupp JT. Evaluation of the energetics of electron trap states at the nanocrystalline titanium dioxide/ aqueous solution interface via time-resolved photoacoustic spectroscopy. Chemical Physics Letters. 2000;**330**:231-236

[86] Wang X, Xu Q, Li M, Shen S, Wang X, Wang Y, et al. Photocatalytic overall water splitting promoted by an α-β phase junction on Ga_2O_3. Angewandte Chemie International Edition. 2012;**51**: 13089-13092

[87] Yoshida Y, Matsuoka M, Moon SC, Mametsuka H, Suzuki E, Anpo M. Photocatalytic decomposition of liquid-water on the Pt-loaded TiO_2 catalysts: Effects of the oxidation states of Pt species on the photocatalytic reactivity and the rate of the back reaction. Research on Chemical Intermediates. 2000;**26**:567-574

[88] Sanap KK, Varma S, Dalavi D, Patil PS, Waghmode SB, Bharadwaj SR. Variation in noble metal morphology and its impact on functioning of hydrogen mitigation catalyst. International Journal of Hydrogen Energy. 2011;**36**:10455-10467

[89] Wang M, Li Z, Wu Y, Ma J, Lu G. Inhibition of hydrogen and oxygen reverse recombination reaction over Pt/ TiO_2 by F− ions and its impact on the photocatalytic hydrogen formation. Journal of Catalysis. 2017;**353**:162-170

[90] Berto TF, Sanwald KE, Byers JP, Browning ND, Gutiérrez OY, Lercher JA. Enabling overall water splitting on photocatalysts by CO-covered noble metal co-catalysts. Journal of Physical Chemistry Letters. 2016;**7**:4358-4362

[91] Maeda K, Teramura K, Lu D, Saito N, Inoue Y, Domen K. Roles of Rh/ Cr_2O_3 (core/shell) nanoparticles photodeposited on visible-light-responsive $(Ga_{1-x}Zn_x)(N_{1-x}O_x)$ solid solutions in photocatalytic overall water splitting. Journal of Physical Chemistry C. 2007;**111**:7554-7560

[92] Yoshida M, Takanabe K, Maeda K, Ishikawa A, Kubota J, Sakata Y, et al. Role and function of noble-metal/Cr-layer core/shell structure cocatalysts for photocatalytic overall water splitting studied by model electrodes. Journal of Physical Chemistry C. 2009;**113**:10151-10157

[93] Dionigi F, Vesborg PCK, Pedersen T, Hansen O, Dahl S, Xiong A, et al. Suppression of the water splitting back reaction on GaN:ZnO photocatalysts loaded with core/shell cocatalysts, investigated using a μ-reactor. Journal of Catalysis. 2012;**292**:26-31

[94] Nørskov JK, Rossmeisl J, Logadottir A, Lindqvist L, Kitchin JR, Bligaard T, et al. Origin of the overpotential for oxygen reduction at a fuel-cell cathode. The Journal of Physical Chemistry B. 2004;**108**:17886-17892

[95] Takata T, Pan C, Nakabayashi M, Shibata N, Domen K. Fabrication of a core-shell-type photocatalyst via photodeposition of group IV and V transition metal oxyhydroxides: An effective surface modification method for overall water splitting. Journal of the American Chemical Society. 2015;**137**: 9627-9634

[96] Ipek B, Uner D. Artificial photosynthesis from a chemical engineering perspective. In: Najafpour M, editor. Artificial Photosynthesis. IntechOpen; 2012. pp. 13-36

[97] Chen D, Li F, Ray AK. External and internal mass transfer effect on photocatalytic degradation. Catalysis Today. 2001;**66**:475-485

[98] Reilly K, Wilkinson DP, Taghipour F. Photocatalytic water splitting in a fluidized bed system: Computational modeling and experimental studies. Applied Energy. 2018;**222**:423-436

[99] Niu B, Wang X, Wu K, He X, Zhang R. Mesoporous titanium dioxide: Synthesis and applications in photocatalysis, energy and biology. Materials (Basel). 2018;**11**:1-23

[100] Sangle AL, Singh S, Jian J, Bajpe SR, Wang H, Khare N, et al. Very high surface area mesoporous thin films of $SrTiO_3$ grown by pulsed laser deposition and application to efficient photoelectrochemical water splitting. Nano Letters. 2016;**16**:7338-7345

[101] Armenante PM, Kirwan DJ. Mass transfer to microparticles in agitated systems. Chemical Engineering Science. 1989;**44**:2781-2796

[102] Escudero JC, Simarro R, Cervera-March S, Gimenez J. Rate-controlling steps in a three-phase (solid-liquid-gas) photoreactor: A phenomenological approach applied to hydrogen photoproduction using Pt-TiO_2 aqueous suspensions. Chemical Engineering Science. 1989;**44**:583-593

[103] Calderbank PH, Moo-Young MB. The continuous phase heat and mass-transfer properties of dispersions. Chemical Engineering Science. 1961;**16**: 39-54

[104] Uner D, Tapan NA, Özen I, Üner M. Oxygen adsorption on Pt/TiO_2 catalysts. Applied Catalysis A: General. 2003;**251**:225-234

[105] Pinaud BA, Benck JD, Seitz LC, Forman AJ, Chen Z, Deutsch TG, et al. Technical and economic feasibility of centralized facilities for solar hydrogen production via photocatalysis and photoelectrochemistry. Energy & Environmental Science. 2013;**6**:1983-2002

[106] Hisatomi T, Takanabe K, Domen K. Photocatalytic water-splitting reaction from catalytic and kinetic perspectives. Catalysis Letters. 2015;**145**: 95-108

[107] Schröder M, Kailasam K, Borgmeyer J, Neumann M, Thomas A, Schomäcker R, et al. Hydrogen evolution reaction in a large-scale reactor using a carbon nitride photocatalyst under natural sunlight irradiation. Energy Technology. 2015;**3**: 1014-1017

[108] Schneider J, Bahnemann DW. Undesired role of sacrificial reagents in photocatalysis. Journal of Physical Chemistry Letters. 2013;**4**:3479-3483

Chapter 11

Water Splitting Electrocatalysis within Layered Inorganic Nanomaterials

Mario V. Ramos-Garcés, Joel Sanchez, Isabel Barraza Alvarez, Yanyu Wu, Dino Villagrán, Thomas F. Jaramillo and Jorge L. Colón

Abstract

The conversion of solar energy into chemical fuel is one of the "Holy Grails" of twenty-first century chemistry. Solar energy can be used to split water into oxygen and protons, which are then used to make hydrogen fuel. Nature is able to catalyze both the oxygen evolution reaction (OER) and the hydrogen evolution reaction (HER) required for the conversion of solar energy into chemical fuel through the employment of enzymes that are composed of inexpensive transition metals. Instead of using expensive catalysts such as platinum, cheaper alternatives (such as cobalt, iron, or nickel) would provide the opportunity to make solar energy competitive with fossil fuels. However, obtaining efficient catalysts based on earth-abundant materials is still a daunting task. In this chapter, we review the advancements made with zirconium phosphate (ZrP) as a support for earth-abundant transition metals for the OER. Our studies have found that ZrP is a suitable support for transition metals as it provides an accessible surface where the OER can occur. Further findings have also shown that exfoliation of ZrP increases the availability of sites where active species can be adsorbed and performance is improved with this strategy.

Keywords: water splitting, electrocatalysis, zirconium phosphate, inorganic nanomaterials, oxygen evolution

1. Introduction

Global energy consumption is projected to increase drastically in the coming decades [1]. To meet this demand, it is estimated that there are 1000–2000 years of fossil fuel resources [2]. Nonetheless, while fossil fuels could meet this huge demand of energy, CO_2 emissions from these resources would contribute to the recognized danger of climate change by increasing anthropogenic carbon emissions to the atmosphere. This motivates the development of sustainable energy production technologies, including fuel production, using solar energy in a process that has been called artificial photosynthesis. However, there are large scientific and technical challenges involved in these schemes.

One promising scheme for this purpose is the use of hydrogen as a fuel. Hydrogen has the largest energy density over any other fuel and it is the most abundant molecule in the universe. Hydrogen's energy density is 120 MJ/kg, more than twice than that of natural gas and almost three times higher than petroleum [3]. The problem with hydrogen is that even though it is very abundant, it is hard to obtain in pure form since it readily reacts with other substances and it is mostly found in compounds. Currently, ~96% of hydrogen is produced from fossil fuels with the steam methane reforming process [4]. Thus, methods for producing hydrogen from hydrogen-containing resources like biomass and water need to be more environmentally friendly and economical in order to substitute current methods of hydrogen production [4]. The high interest of hydrogen as a fuel arises because this gas is highly flammable, burns cleanly, and the cost of solar-based electricity is falling rapidly, including that used for hydrogen production [5]. The product of hydrogen combustion is water and energy, making this process extremely clean:

$$2H_2(g) + O_2(g) \rightarrow 2H_2O(g)\ \Delta H = -286\ kJ/mol$$

Out of all energy resources, solar energy is the most abundant, but it is an intermittent resource [6]. Therefore, to effectively use solar energy, we must convert and store it. One way to store this energy is in the form of chemical fuels, such as hydrogen. The idea is to split water in its components (hydrogen and oxygen) with the help of solar energy since 4.92 eV is stored when two water molecules are split [7]. This approach to store energy in the form of chemical bonds (a process that mimics the natural photosynthetic process) is called artificial photosynthesis. An example of artificial photosynthesis is the process occurring in a solar fuel cell. In such cells water is split using sunlight as the energy source. This reaction involves two separate redox reactions, one being the oxidation of water to produce oxygen and protons (a 4-electron process) and the other one is the reduction of protons to form dihydrogen:

$$\text{Water oxidation:} 2H_2O \rightarrow O_2 + 4H^+ + 4e^- \text{ (Oxygen evolution reaction, OER).}$$

$$\text{Proton reduction:} 2H^+ + 2e^- \rightarrow H_2 \text{ (Hydrogen evolution reaction, HER).}$$

Electrochemical water splitting can be achieved by using devices that can harvest the sun's energy. The two main configurations of these devices consist of (1) a photovoltaic (PV) device connected to a separate electrolyzer with catalysts that drive the necessary half reactions (PV/electrolysis) and (2) a fully integrated system where the catalysts are deposited on top of the light absorbing materials (photoelectrochemical, PEC devices) [8]. The efficiency of these devices is calculated based on the solar-to-hydrogen (STH) or solar-to-fuel (STF) efficiency, which is defined as the amount of chemical energy produced in the form of fuel divided by the solar energy input, with no external bias applied [9]. High STH efficiencies are desired as it has been proved that it is the factor with the biggest impact on the final cost of the fuel produced on any of these systems [8]. Although, other factors such as stability and material cost are also important for the final cost of the fuel.

Theoretical efficiencies calculated using combinations of published catalysts for the OER and the HER in a PEC device show that the STH efficiencies are far lower than the maximum thermodynamically achievable efficiency of 41% [8]. This highlights the need to develop more active catalysts, especially for the OER as it is the main cause of energy loss in the form of kinetic overpotentials during fuel production. Furthermore, to bring these technologies towards economical implementation, it is of much importance to continually improve device performance. Besides, benchmarking studies have shown that catalyst stability is also a major issue as the reactions are mostly carried in harsh chemical conditions, especially in very high or low pH [10–12]. Recently, density functional theory (DFT) calculations have shown that performing the OER in a confined nanoscopic environment improves the electrochemistry of the reaction by lowering the overpotential and increasing the catalytic efficiency by 10% [13]. These theoretical results were modeled on a layered RuO_2 system and attributed the improvement in activity to interactions of intermediates with the opposite surface of the metal oxide. There is evidence that encapsulation of catalysts can lead to improvements on selectivity and activity for a variety of reactions, including water oxidation [14–17]. This motivated us to use the layered compound zirconium phosphate (ZrP) as a support for active OER catalysts to mimic an environment that theoretical works have modeled. We want to target the issues presented by OER catalysts by developing catalytic systems based on ZrP nanomaterials with the goal of optimizing efficiencies of future solar water splitting devices.

1.1 Zirconium phosphates

Zirconium phosphates are part of the group of water-insoluble phosphates of tetravalent metals containing layered structures. Zirconium bis(monohydrogen orthophosphate) monohydrate ($Zr(O_3POH)_2{\cdot}H_2O$, α-ZrP) is the most extensively studied phase of ZrP. α-ZrP has an interlayer distance of 7.6 Å with a layer thickness of 6.6 Å (**Figure 1a** and **b**) [18]. α-ZrP has a structure in which the zirconium atoms in each layer align nearly to a plane with bridging phosphate groups located alternately above and below the metal atom plane [19]. Three oxygen atoms of the phosphate group bond to three different Zr^{4+} and each Zr^{4+} ion coordinates with oxygens from six different phosphate groups [19]. The fourth oxygen from the phosphate group, which has a proton, points towards the interlayer region and the surface of the nanoparticles. This proton can be exchanged with cations or molecules. The structure of α-ZrP contains a zeolitic cavity in the interlayer region with a diameter of 2.61 Å that is occupied by a water molecule [20, 21].

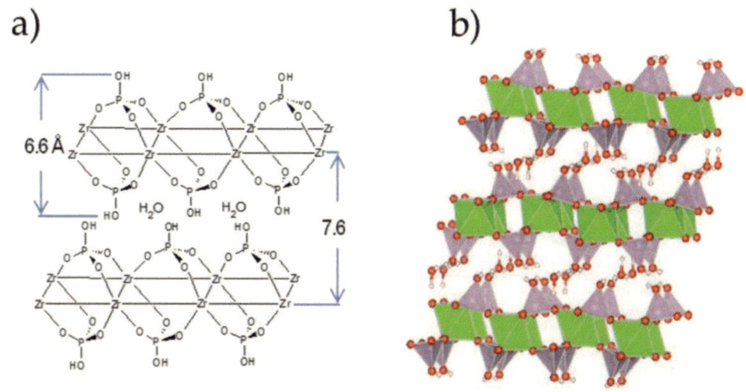

Figure 1.
(a) The structure of α-ZrP. (b) Polyhedral model of the structure of α-ZrP.

1.2 Intercalation of guest species into ZrP

Intercalation is defined as the reversible insertion of guest species into a lamellar host structure with maintenance of the structure features of the host [22]. For α-ZrP, the direct intercalation of small cations is possible if they are smaller than 2.61 Å, but for larger cations and molecules intercalation is not significant and/or these species are exchanged at very slow rates [23–26]. To circumvent this problem, α-ZrP pre-intercalated phases with sodium ions or n-butylammonium (both produce expanded phases) are commonly used as precursors to intercalate the intended guest species. One problem that arises with this method is that the pre-intercalated species do not necessarily exchange completely with the intended guest, thus becoming a contaminant in the intercalation product.

Martí and Colón developed a new direct intercalation method that does not require a pre-intercalation step using a highly hydrated phase of zirconium phosphate, θ-ZrP [27]. θ-ZrP maintains the α-ZrP-type layered structure (**Figure 2**) but has an interlayer distance of 10.4 Å and has six water molecules per formula unit, in contrast with α-ZrP that only has one [28]. Zirconium bis(monohydrogen orthophosphate) hexahydrate (θ-ZrP) converts back to α-ZrP when it dehydrates. X-ray powder diffraction (XRPD) can be used to distinguished between both ZrP phases. When θ-ZrP is dried, producing α-ZrP, the first diffraction peak at $2\theta = 8.6°$ which corresponds to the 002-plane reflection of ZrP and that of the interlayer distance, shifts towards 11.6°; the angle corresponding to an interlayer distance of 7.6 Å, characteristic of α-ZrP (**Figure 3**). For this reason, if a dry intercalation product is analyzed by XRPD and the first diffraction peak corresponds to a distance greater than 7.6 Å, this indicates that the intercalation reaction was successful [29]. One of three possible patterns can be observed by XRPD for intercalation products of ZrP; either (i) a pattern with a peak corresponding to a larger interlayer spacing at lower 2θ values than 11.6° indicates that the intercalant was introduced into the interlayer, (ii) a pattern with two distinct peaks, one at $2\theta = 11.6°$ and one that appears at lower

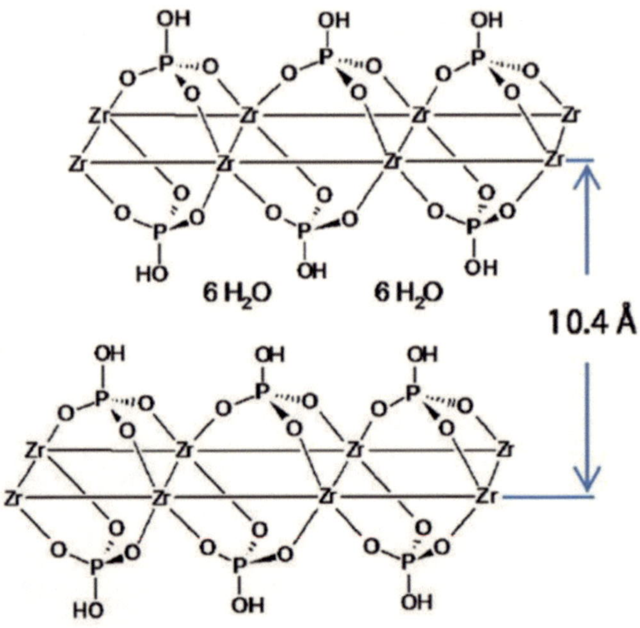

Figure 2.
The structure of θ-ZrP.

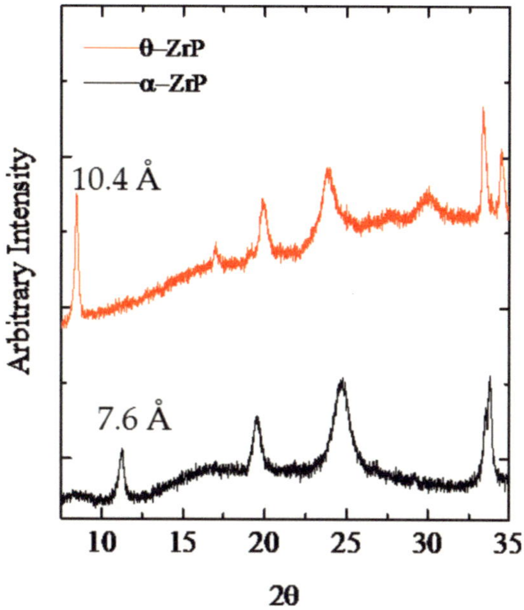

Figure 3.
XRPD patterns of α-ZrP (black) and θ-ZrP (red).

2θ values than 11.6° indicates that a mixed phase is present [30], and (iii) a pattern with no change in the reference peak, indicating that the intercalant species did not intercalate and is adsorbed on the outer surface of the layered structure or that is not present at all.

ZrP has been used for the intercalation of several photo-, bio- and redox-active compounds for a wide variety of applications including artificial photosynthesis, amperometric biosensors, and drug delivery [27, 30–42]. Even though ZrP has previously been studied for catalysis [43–47], membrane composites for proton exchange water electrolyzers [48–52], and as additive for catalytic layers for OER in order to protect metal oxide catalysts [53], our work is, to the best of our knowledge, the first time ZrP is used as an inorganic support for catalysts for the OER.

1.3 Chemical exfoliation of ZrP nanoparticles

The process of separating the layers of a layered material is known as exfoliation. This process has been extensively studied for a myriad of layered materials and the two-dimensional materials (2D) that result have been shown to have several advantages over their bulk systems [54]. *α*-ZrP has been successfully exfoliated through a variety of methods [55–58], and its nanosheets used for different applications [59–62]. The main strategy for ZrP exfoliation consists on the intercalation of small amines with positive charges that can easily displace the protons from the phosphate groups in an acid-base reaction and enter the interlayer space. If a high enough concentration of these amines is used, an amine double layer will form in the interlayer space, leading to exfoliation due to cation-cation repulsions (**Figure 4**) [56].

One of the most highly used amines for the exfoliation of ZrP is tetra-*n*-butylammonium hydroxide (TBA^+OH^-). If TBA^+OH^- is used, the exfoliated material will consist of single nanosheets of ZrP suspended with TBA^+ attached to them. This reaction is temperature sensitive as it has been found that that hydrolysis of the ZrP edges occurs due to the OH^- ions. However, the rate of hydrolysis of ZrP during exfoliation with TBA^+OH^- at 0 °C is essentially zero [58]. If the exfoliated material

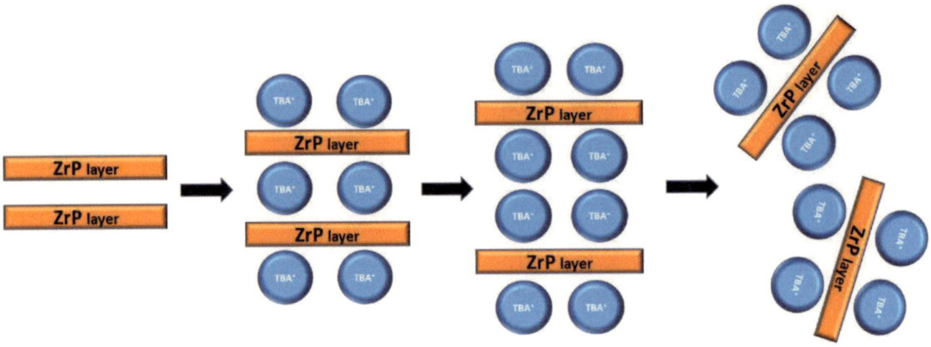

Figure 4.
Schematic drawing of the ZrP exfoliation process with TBA^+.

is dried, restacking of the layers occurs with a new expanded phase of 16.8 Å corresponding to TBA^+ intercalated in ZrP [63]. The TBA^+ cations can be displaced with another cationic species if the latter is put in contact with a suspension of the exfoliated ZrP nanoparticles. Hence, if the desired material is the exfoliated nanosheets with their phosphate groups protonated, then a follow up reaction with an acid can be performed [60].

2. Metal-modified ZrP based electrocatalysts for the OER

To facilitate the economic viability of water splitting, the efficiency of electrolyzers must be improved by addressing the overpotential losses associated with the sluggish OER kinetics. To this end, recent studies have focused on developing catalysts materials using earth-abundant transition metals [64]. Significant research has been devoted to improving OER electrocatalysts by using a wide variety of strategies that either increase the intrinsic activity of active sites or by increasing the number of them [65]. One general strategy that has been effective is to support active materials onto supports that engender improved performance [65–68]. ZrP properties make it a potential candidate as a support for active OER catalysts. Its ability to confine catalysts, high thermal stability, stability under a wide range of pH values, and its overall robustness are all desired for an ideal support. In our work, we intercalated the earth-abundant transition metal cations Fe^{2+}, Fe^{3+}, Co^{2+}, and Ni^{2+} into ZrP and assessed these composite materials as OER electrocatalysts [69].

2.1 Metal-intercalated and surface adsorbed ZrP systems

To intercalate the desired transition metals, a suspension of θ-ZrP must be mixed with a solution of the metal salt precursor and left stirring for 5 days so that ion-exchange reaches equilibrium. To optimize metal loading for improved catalysis performance, we synthetized these composite materials with several synthesis metal salt:ZrP molar ratios (10:1, 5:1, 3:1, 1:1, 1:3, 1:5, 1:10, and 1:20 M:ZrP). A stepwise process is expected as a function of intercalant solution molarity; the intercalation reaction initiates from the edges of the particle and proceeds by diffusion of the metal cations towards the interior of the interlayer sheets [70]. The XRPD patterns (**Figure 5a**) for all four metal samples show that the first diffraction peak of ZrP is shifted to lower 2θ angles, indicating larger interlayer distances and successful intercalation. Increasing the M:ZrP molar ratios results in peak broadening and shifting in all samples indicating a more mixed phase is present and that the layered structure has not achieved its maximum cation loading within the interlayer. However, at the highest loadings (i.e., 1:1–10:1 molar ratios), the original peak at 2θ = 11.6° disappeared, and a new peak emerged at significantly lower values of 2θ, reaching a final value indicative of the maximum interlayer distance for that particular metal cation intercalated within ZrP. As expected, +2 cations produced intercalated products with the first diffraction peak at lower angles than those produced by +3 cations. Compared to α-ZrP, the maximum interlayer distance increase observed for +2 cations was 2 Å, while for +3 cations it was 0.6 Å (**Figure 5b**). This difference in the increase in interlayer distance between the divalent and trivalent metal cations can be attributed to the difference electrostatic forces within the layers, consistent with Coulomb's Law. Trivalent cations produced a smaller increase because of a stronger electrostatic attraction between the metal cation and the negatively charged ZrP layers.

Ion-exchange in ZrP occurs at the Brönsted acid groups (P-OH) which are also present at the surface of the nanoparticles. Hence, there is no way of preventing that the metal cations get adsorbed to the surface. To obtain more insights into the nature of the activity of the samples, a metal-modified ZrP system in which the metals are only adsorbed onto the surface of the nanoparticles was also prepared. To prepare these samples, α-ZrP must be used as the ZrP source as the metal cations are large enough to not intercalate into the interlayer. XRPD data shows that the interlayer distance of these dry sample remains that of α-ZrP, 7.6 Å (**Figure 5b**). The presence of the metals in these systems was confirmed by high resolution X-ray photoelectron spectroscopy (XPS). XPS was also used to determine the atomic concentration on both metal-modified ZrP systems, intercalated and adsorbed [69]. Due to the uptake of metal cations within the much larger area of the interlayers of ZrP rather than solely on the surface in the adsorbed case, XPS high resolution scans show that intercalated ZrP systems have higher atomic metal content when compared to adsorbed systems at similar M:ZrP ratios.

Another useful tool to characterize ZrP systems is Fourier transform infrared spectroscopy (FT-IR). α-ZrP has four characteristic bands associated

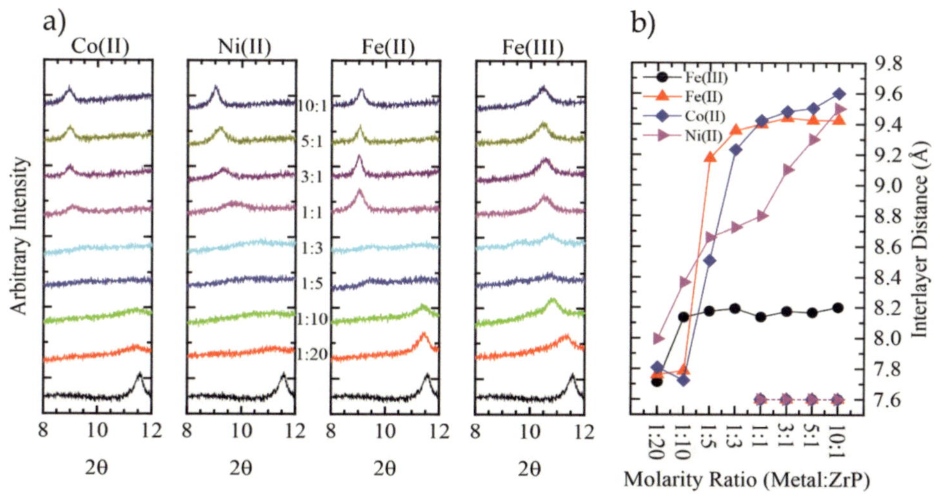

Figure 5.
(a) XRPD patterns for Fe(II), Fe(III), Co(II), and Ni(II)-intercalated ZrP at (from top to bottom) 10:1, 5:1, 3:1, 1:1, 1:3, 1:5, 1:10, and 1:20 M:ZrP molar ratios. The bottom diffraction pattern in all frames is that of pure α-ZrP; (b) interlayer distance as a function of M:ZrP molar ratio for the various metal-intercalated ZrP materials. Metal-adsorbed systems are represented as dashed lines which have an exact interlayer spacing as pure α-ZrP indicating that metal intercalation did not occur. Taken from reference [69].

with lattice water molecules. These bands appear at ~3600, ~3500, ~3140, and ~1600 cm^{-1} [71]. When intercalation occurs, the intercalant species will displace interlayer water molecules. For this reason, bands associated with these water vibrational modes showed reduced relative intensity in the intercalated materials (**Figure 6a**). In contrast, metal-adsorbed samples showed very similar spectra to that of α-ZrP, with the water bands still present, indicating negligible intercalation (**Figure 6b**). The characteristic orthophosphate group vibrations of ZrP are observed in the region of ~1100–950 cm^{-1} (**Figure 6a**). Intercalated samples show a diminished relative intensity of the shoulder at the left part of the orthophosphate group vibrations at ~1050 cm^{-1} that is attributed to the vibration of the exchangeable proton of the phosphate group, which is lost when the proton is exchanged by intercalation via ion exchange with other species. For metal-adsorbed samples, this vibration is still present indicating once again that no intercalation is observed.

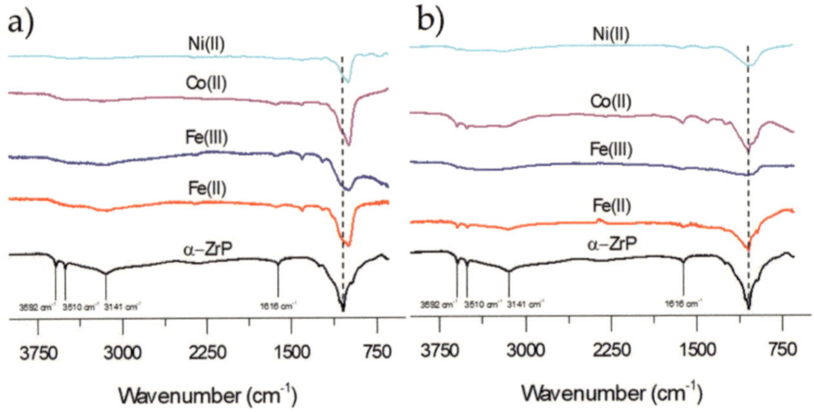

Figure 6.
FTIR spectra of (a) intercalated and (b) adsorbed Fe(II), Fe(III), Co(II), and Ni(II) at 10:1 M:ZrP ratio. Taken from Ref. [69].

2.2 OER electrochemical performance of metal-intercalated and surface adsorbed ZrP systems

To determine the activity of our metal-modified ZrP products towards the OER, cyclic voltammetry experiments were conducted using a Rotating Disk Electrode (RDE) assembly in alkaline electrolyte (0.1 M KOH). The methodology employed was according to the benchmarking protocols suggested for OER electrocatalysts [10–12]. The primary figure of merit from this data is the overpotential necessary to achieve 10 mA/cm^2 ($\eta_{j\,=\,10\,mA/cm^2}$). The overpotential measured at 10 mA/cm^2 is the potential difference between the potential to achieve 10 mA/cm^2 and the thermodynamic potential of water oxidation (1.23 V vs. RHE). All samples were active for the OER, requiring between 0.5 and 0.7 V of overpotential to reach 10 mA/cm^2, depending on the choice of metal cation, the M:ZrP molar ratio used during synthesis, and whether the metal was intercalated into or adsorbed onto ZrP. In general, lower overpotentials are observed for the higher M:ZrP molar ratios, ascribed to higher metal loadings. Also, OER activities for the metal-adsorbed ZrP catalysts are greater than or equal to those of their metal-intercalated counterparts at the same loading, as seen by their lower overpotentials (**Figure 7**). This is

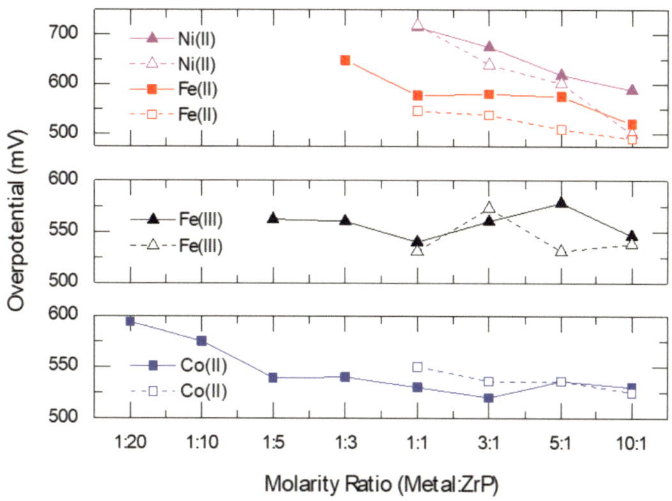

Figure 7.
Electrochemical performance comparison of all four metal systems for adsorbed and intercalated species at 10 mA/cm² except for Ni(II) which was compared at 3 mA/cm². Solid and dashed lines represent intercalated and adsorbed metal ZrP systems, respectively. Taken from reference [69].

somewhat surprising as XPS showed that higher metal loadings were achieved in the intercalated systems. This suggests that the OER is dominated by catalysis on the outer surface of the ZrP supported metal-based systems rather than within the layers, which may be limited by mass transport. These results serve as a basis for developing improved OER catalyst systems.

2.3 Metal-modified exfoliated ZrP

Our previous finding suggests that ZrP can serve as a support for transition metal-based OER catalysts and that the reaction occurs preferentially on the surface of the layered ZrP nanoparticles rather than the interlayer space [69]. Based on these results, we expected that exposing surface sites through exfoliation of ZrP could improve these catalytic systems. With the goal of developing more active materials, we prepared exfoliated ZrP nanosheets and modified these exfoliated nanoparticles with Co^{2+} and Ni^{2+} [72]. These systems underwent reaction at the same molar ratio than that of the best performing metal-adsorbed ZrP system (10:1 M:ZrP).

ZrP exfoliation was carried out by adding an excess of TBA^+OH^- in an ice bath followed by an acid wash with HCl. To modify the exfoliated ZrP with the transition metals, an aqueous suspension of the nanosheets is put in contact with an aqueous solution of the metal salt precursor. The XRPD pattern of exfoliated ZrP shows the extreme broadening characteristic of successful exfoliation (**Figure 8**). The diffractograms of Co and Ni-modified ZrP nanosheets are very similar to that of exfoliated ZrP confirming that no further restacking occurs after metal modification (**Figure 8**). Transmission electron microscopy (TEM) also confirms this as the ZrP nanosheets show a fainter contrast when compared with α-ZrP nanoparticles which is consistent with its thinner nature, since in TEM areas that contain heavy atoms or are thick appear darker (**Figure 9a–d**). After exfoliation, the nanosheets retain the hexagonal shape of α-ZrP and no hydrated zirconia nanoparticles are observed decorating the edges of the sheets, indicating that the hydrolysis prone edges were preserved by temperature control during the exfoliation reaction and that the structure of the layers did not change [58]. This was also confirmed by XPS as the P/Zr ratio after exfoliation remains constant at ~2.

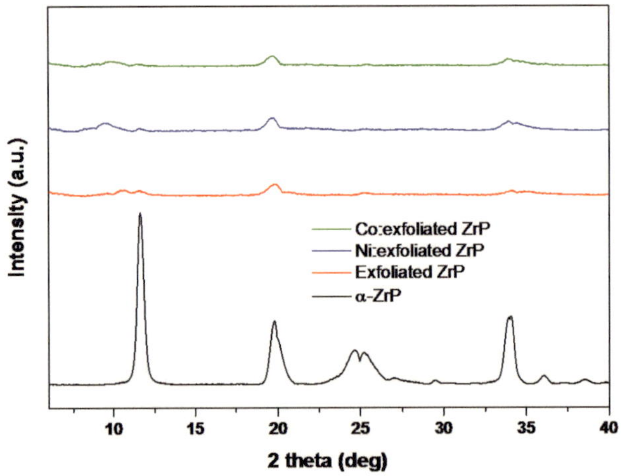

Figure 8.
XRPD patterns of α-ZrP, exfoliated ZrP, and metal-modified exfoliated ZrP samples. Reprinted with permission from [72].

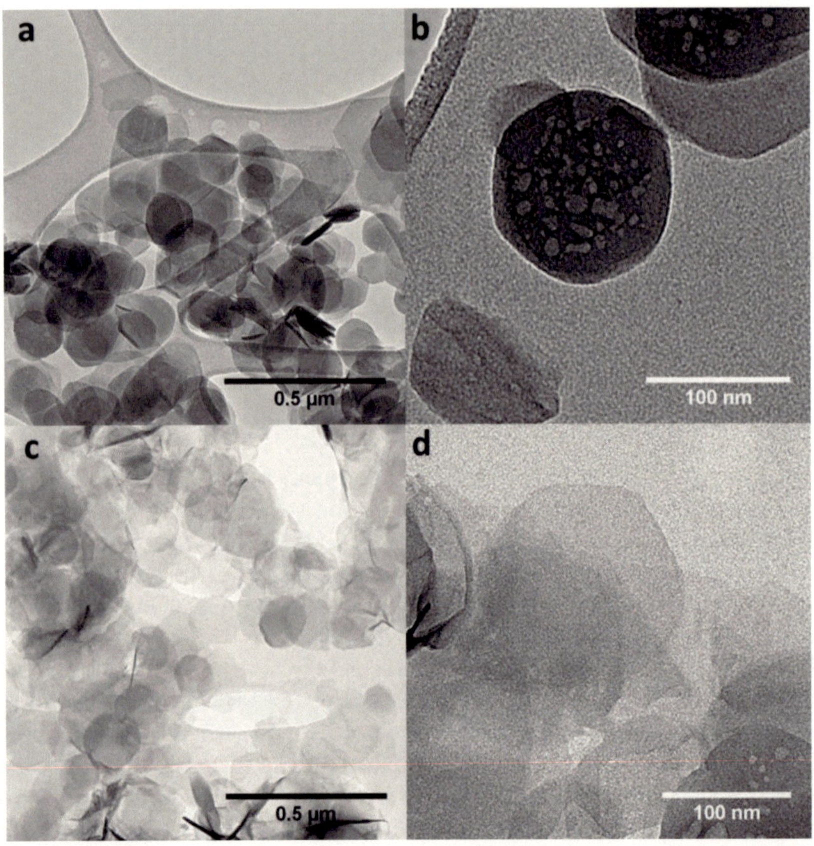

Figure 9.
(a, b) TEM images of α-ZrP nanoparticles. Scale bar: 0.5 and 100 nm, respectively. (c, d) TEM images of exfoliated ZrP. Scale bar: 0.5 µm and 100 nm, respectively. Reprinted with permission from [72].

2.4 OER electrochemical performance of metal-modified exfoliated ZrP electrocatalysts

Linear sweep voltammetry (LSV) was used to assess the activity of these exfoliated materials (**Figure 10**) [72]. OER catalytic currents for the exfoliated materials were shifted to lower potentials when compared to their surface adsorbed counterparts. The overpotential necessary to reach a current density of 10 mA/cm^2 for the Co-modified exfoliated nanosheets was 0.450 V, an improvement of 41 mV over the surface adsorbed Co material. For the Ni-modified the overpotential necessary to reach a current density of 3 mA/cm^2 is 0.410 V, an improvement of 181 mV over the surface adsorbed Ni material.

To elucidate the nature of the increased activity of the exfoliated materials, we determined the intrinsic activity of each catalytic site in both types of systems [72]. To construct a mass normalized current plot, we performed inductively plasma-mass spectrometry (ICP-MS) measurements on our samples to quantify the amount of nickel and cobalt metal content in the exfoliated and bulk materials. ICP-MS measurements show that the exfoliated samples are substantially better at adsorbing Co and Ni cations, leading to higher loadings than non-exfoliated ZrP. For our mass normalized plots, we assumed that all metal content quantified by ICP-MS

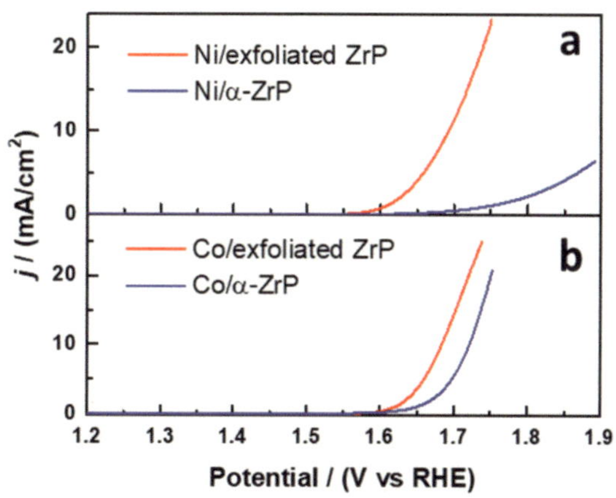

Figure 10.
Linear sweep voltammograms of (a) Ni(II)/ZrP systems and (b) Co(II)/ZrP systems. Reprinted with permission from Ref. [72].

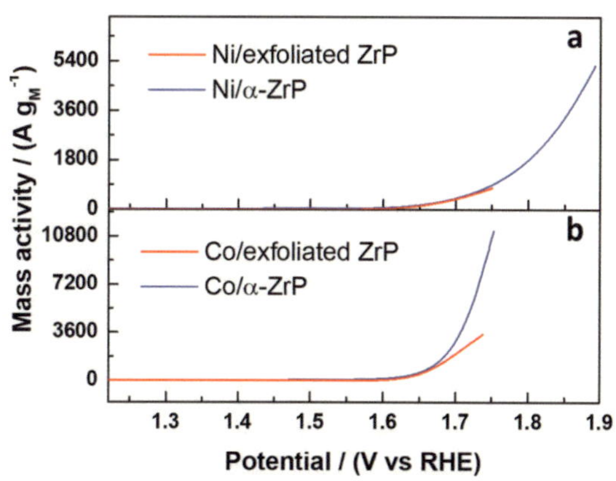

Figure 11.
Mass normalized catalytic currents for (a) Ni(II)/ZrP systems and (b) Co(II)/ZrP systems. Reprinted with permission from Ref. [72].

in the materials were active and accessible to perform OER. **Figure 11a** and **b** show the LSVs where the OER currents are normalized by the mass of the metal content. These exfoliated systems maintain reasonably high intrinsic activity values that, when coupled to a significant greater number of active sites leads to higher geometric activity. We concluded that the enhancement in activity is due to the fact that the inner layer surfaces are now more electrochemically accessible [72]. Through exfoliation the number of ion-exchange sites increases which increases the number of catalytic species that are distributed on the surface of zirconium phosphate, therefore giving rise to the improved performance.

Acknowledgements

We would like to acknowledge the NSF Center for Chemical Innovation in Solar Fuels CHE-1305124 for funding these research efforts. M.V.R.-G. was also supported by the NSF-PREM Center for Interfacial Electrochemistry of Energy Materials (CiE2M) grant DMR-1827622.

Conflict of interest

The authors declare no conflict of interest.

Notes

M.V.R.-G. and J.L.C wrote the chapter. J.S., I.B.A., Y.W., D.V., and T.F.J. revised the work. All authors helped in the conception of the work, acquisition, analysis, and interpretation of the data.

Author details

Mario V. Ramos-Garcés[1], Joel Sanchez[2], Isabel Barraza Alvarez[3], Yanyu Wu[3], Dino Villagrán[3], Thomas F. Jaramillo[2] and Jorge L. Colón[1*]

1 University of Puerto Rico at Río Piedras, San Juan, Puerto Rico, United States

2 Stanford University, Stanford, United States

3 University of Texas at El Paso, El Paso, United States

*Address all correspondence to: jorge.colon10@upr.edu

References

[1] Gray HB. Powering the planet with solar fuel. Nature Chemistry. 2009;**1**:7

[2] Lewis NS, Nocera DG. Powering the planet: Chemical challenges in solar energy utilization. Proceedings of the National Academy of Sciences of the United States of America. 2006;**103**:15729-15735

[3] Thomas G. Overview of Storage Development DOE Hydrogen Program [Internet]. Available from: http://www1.eere.energy.gov/hydrogenandfuelcells/pdfs/storage

[4] Cheng Y, Jiang SP. Advances in electrocatalysts for oxygen evolution reaction of water electrolysis-from metal oxides to carbon nanotubes. Progress in Natural Science: Materials International. 2015;**25**:545-553

[5] Bockris JO'M. Hydrogen no longer a high cost solution to global warming: New ideas. International Journal of Hydrogen Energy. 2008;**33**:2129-2131

[6] Solar Fuels. Lewis Research Group [Internet]. Available from: http://nsl.caltech.edu/home/solar-fuels

[7] Dempsey JL, Brunschwig BS, Winkler JR, Gray HB. Hydrogen evolution catalyzed by Cobaloximes. Accounts of Chemical Research. 2009;**42**:1995-2004

[8] Montoya JH, Seitz LC, Chakthranont P, Vojvodic A, Jaramillo TF, Nørskov JK. Materials for solar fuels and chemicals. Nature Materials. 2017;**16**:70-81

[9] Chen Z, Jaramillo TF, Deutsch TG, Kleiman-Shwarsctein A, Forman AJ, Gaillard R, et al. Accelerating materials development for photoelectrochemical hydrogen production: Standards for methods, definitions, and reporting protocols. Journal of Materials Research. 2010;**25**:3-16

[10] McCrory CCL, Jung S, Ferrer IM, Chatman SM, Peters JC, Jaramillo TF. Benchmarking hydrogen evolving and oxygen evolving electrocatalysts for solar water splitting devices. Journal of the American Chemical Society. 2015;**137**:4347-4357

[11] McCrory CCL, Jung S, Peters JC, Jaramillo TF. Benchmarking heterogeneous electrocatalysts for the oxygen evolution reaction. Journal of the American Chemical Society. 2013;**135**:16977-16987

[12] Jung S, McCrory CCL, Ferrer IM, Peters JC, Jaramillo TF. Benchmarking nanoparticulate metal oxide electrocatalysts for the alkaline water oxidation. Journal of Materials Chemistry A. 2016;**4**:3068-3076

[13] Doyle AD, Montoya JH, Vojvodic A. Improving oxygen electrochemistry through nanoscopic confinement. ChemCatChem. 2015;**7**:738-742

[14] Das SK, Dutta PK. Synthesis and characterization of a ruthenium oxide-zeolite Y catalyst for photochemical oxidation of water to dioxygen. Microporous and Mesoporous Materials. 1998;**22**:475-483

[15] Zhan B-Z, White MA, Sham T-K, Pincock JA, Doucet RJ, Ramana Rao KV, et al. Zeolite-confined nano-RuO_2: A green, selective, and efficient catalyst for aerobic alcohol oxidation. Journal of the American Chemical Society. 2003;**125**:2195-2199

[16] Zhan B-Z, Iglesia E. RuO_2 clusters within LTA zeolite cages: Consequences of encapsulation on catalytic reactivity and selectivity. Angewandte Chemie. 2007;**119**:3771-3774

[17] Gong M, Li H, Liang Y, Wu JZ, Zhou J, Wang J. An advanced Ni-Fe layered double hydroxide electrocatalyst for water oxidation. Journal of the American Chemical Society. 2013;**135**:8452-8455

[18] Troup JM, Clearfield A. Mechanism of ion exchange in zirconium phosphates. 20. Refinement of the crystal structure of α-zirconium phosphate. Inorganic Chemistry. 1977;**16**:3311-3314

[19] Clearfield A, Díaz A. Zirconium phosphate nanoparticles and their extraordinary properties. In: Brunet E, Colón JL, Clearfield A, editors. Tailored Organic-Inorganic Materials. New Jersey: Wiley; 2015. pp. 1-44

[20] Clearfield A, Blessing RH, Stynes JA. New crystalline phases of zirconium phosphate possessing ion-exchange properties. Journal of Inorganic and Nuclear Chemistry. 1968;**30**:2249-2258

[21] Clearfield A, Duax WL, Medina AS, Smith GD, Thomas JR. On the mechanism of ion exchange in crystalline zirconium phosphates. Sodium ion exchange of α-zirconium phosphate. The Journal of Physical Chemistry A. 1969;**73**:3424-3430

[22] Stanley Whittingham M. Intercalation chemistry: An introduction. In: Whittingham M, Jacobson AJ, editors. Intercalation Chemistry. New York: Academic Press; 1982. pp. 1-18

[23] Alberti G, Costantino U, Gupta JP. Crystalline insoluble acid salts of tetravalent metals—XIX: Na^+-catalyzed H^+-Mg^{2+} and H^+-Cs^{2+} ion exchanges on α-zirconium phosphate. Journal of Inorganic and Nuclear Chemistry. 1974;**36**:2109-2114

[24] Alberti G, Costantino U, Gupta JP. Crystalline insoluble acid salts of tetravalent metals—XXII: Effect of small amounts of Na^+ on the ion exchange of alkaline earth metal ions on crystalline $Zr(HPO_4)_2{\bullet}H_2O$. Journal of Inorganic and Nuclear Chemistry. 1976;**38**:1729-1732

[25] Alberti G, Constantino U. Recent progress in the intercalation chemistry of layered α-zirconium phosphate and its derivatives, and future perspectives for their use in catalysis. Journal of Molecular Catalysis. 1984;**27**:235-250

[26] Clearfield A, Duax WL, Medina AS, Smith GD, Thomas JR. On the mechanism of ion exchange in crystalline zirconium phosphates. I. Sodium ion exchange of α-zirconium phosphate. The Journal of Physical Chemistry. 1969;**78**:3424-3430

[27] Martí AA, Colón JL. Direct ion exchange of tris(2,2′-bipyridine) ruthenium(II) into an α-zirconium phosphate framework. Inorganic Chemistry. 2003;**42**:2830-2832

[28] Alberti G, Constantino U, Gill JS. Crystalline insoluble acid salts of tetravalent metals—XXIII: Preparation and main ion exchange properties of highly hydrated zirconium bis monohydrogen orthophosphates. Journal of Inorganic and Nuclear Chemistry. 1976;**38**:1733-1738

[29] Colón JL, Casañas B. Drug carriers based on zirconium phosphate nanoparticles. In: Brunet E, Colón JL, Clearfield A, editors. Tailored Organic-Inorganic Materials. New Jersey: Wiley; 2015. pp. 395-437

[30] Díaz A, David A, Pérez R, González ML, Báez A, Wark SE, et al. Nanoencapsulation of insulin into zirconium phosphate for oral delivery applications. Biomacromolecules. 2010;**11**:2465-2470

[31] Casañas-Montes B, Díaz A, Barbosa C, Ramos C, Collazo C, Meléndez E, et al. Molybdocene

dichloride intercalation into zirconium phosphate nanoparticles. Journal of Organometallic Chemistry. 2015;**791**:34-40

[32] Martí AA, Paralitici G, Maldonado L, Colón JL. Photophysical characterization of methyl viologen ion-exchanged within α-zirconium phosphate framework. Inorganica Chimica Acta. 2007;**360**:1535-1542

[33] Martí AA, Rivera N, Soto K, Maldonado L, Colón JL. Intercalation of Re(phen)(CO)$_3$Cl into zirconium phosphate: A water insoluble complex immobilized in a highly polar rigid matrix. Dalton Transactions. 2007;**17**:1713-1718

[34] Santiago M, Declet-Flores C, Díaz A, Velez MM, Bosques MZ, Sanakis Y, et al. Layered inorganic materials as redox agents: Ferrocenium-intercalated zirconium phosphate. Langmuir. 2007;**23**:7810-7817

[35] Díaz A, Saxena V, González J, David A, Casañas B, Batteas J, et al. Zirconium phosphate nano-platelets: A platform for drug delivery in cancer therapy. Chemical Communications. 2012;**48**:1754-1756

[36] Díaz A, González ML, Pérez RJ, David A, Mukherjee A, Báez A, et al. Direct intercalation of cisplatin into zirconium phosphate nanoplatelets for potential cancer nanotherapy. Nanoscale. 2013;**5**:11456-11463

[37] Santiago MB, Velez MM, Borrero S, Diaz A, Casillas CA, Hoffman C, et al. NADH electrooxidation using bis(1,10-phenanthroline-5,6-dione)(2,2′-bipyridine)ruthenium(II)-exchanged zirconium phosphate modified carbon paste electrodes. Electroanalysis. 2006;**18**:559-572

[38] Santiago MB, Daniel G, David A, Casañas B, Hernández G, Guadalupe A, et al. Effect of enzyme and cofactor immobilization on the response of ethanol amperometric biosensors modified with layered zirconium phosphate. Electroanalysis. 2010;**22**:1097-1105

[39] Rivera EJ, Barbosa C, Torres R, Grove L, Taylor S, Connick WB, et al. Vapochromic and vapoluminescent response of materials based on platinum(II) complexes intercalated into layered zirconium phosphate. Journal of Materials Chemistry. 2011;**21**:15899-15902

[40] Bermúdez RA, Colón Y, Tejada GA, Colón JL. Intercalation and photophysical characterization of 1-pyrenemethylamine into zirconium phosphate layered materials. Langmuir. 2005;**21**:890-895

[41] Rivera EJ, Figueroa C, Colón JL, Grove L, Connick WB. Room-temperature emission from platinum(II) complexes intercalated into zirconium phosphate-layered materials. Inorganic Chemistry. 2007;**46**:8569-8576

[42] Rivera EJ, Barbosa C, Torres R, Rivera H, Fachini ER, Green TW, et al. Luminescence rigidochromism and redox chemistry of pyrazolate-bridged binuclear platinum (II) diamine complex intercalated into zirconium phosphate layers. Inorganic Chemistry. 2012;**51**:2777-2784

[43] Clearfield A. Group IV phosphates as catalysts and catalyst supports. Journal of Molecular Catalysis. 1984;**27**:251-262

[44] Clearfield A, Thakur DS. Zirconium and titanium phosphates as catalysts: A review. Applied Catalysis. 1986;**26**:1-26

[45] Colón JL, Thakur DS, Yang C-Y, Clearfield A, Martin CR. X-ray phoelectron spectroscopy and catalytic activity of α-zirconium phosphate and zirconium phosphate

sulfophenylphosphonate. Journal of Catalysis. 1990;**124**:148-159

[46] Niño ME, Giraldo SA, Páez-Mozo EA. Olefin oxidation with dioxygen catalyzed by porphyrins and phthalocyanines intercalated in α-zirconium phosphate. Journal of Molecular Catalysis A: Chemical. 2001;**175**:139-151

[47] Hajipour AR, Karimi H. Zirconium phosphate nanoparticles as a remarkable solid acid catalyst for selective solvent-free alkylation of phenol. Chinese Journal of Catalysis. 2014;**35**:1136-1147

[48] Costamagna P, Yang C, Bocarsly AB, Srinivasan S. Nafion® 115/zirconium phosphate composite membranes for operation of PEMFCs above 100°C. Electrochimica Acta. 2002;**47**:1023-1033

[49] Alberti G, Casciola M. Composite membranes for medium-temperature PEM fuel cells. Annual Review of Materials Research. 2003;**33**:129-154

[50] Yang C, Srinivasan S, Bocarsly AB, Tulyani S, Benziger JB. A comparison of physical properties and fuel cell performance of Nafion and zirconium phosphate/Nafion composite membranes. Journal of Membrane Science. 2004;**237**:145-161

[51] Hou H, Sun G, Wu Z, Jin W, Xin Q. Zirconium phosphate/Nafion115 composite membrane for high-concentration DMFC. International Journal of Hydrogen Energy. 2008;**33**:3402-3409

[52] Rodgers MP, Shi Z, Holdcroft S. Ex situ characterization of composite Nafion membranes containing zirconium hydrogen phosphate. Fuel Cells. 2009;**9**:534-546

[53] Zlotorowicz A, Sunde S, Seland F. Zirconium hydrogen phosphate as an additive in electrocatalytic layers for the oxygen evolution reaction in PEM water electrolysis. International Journal of Hydrogen Energy. 2015;**40**:9982-9988

[54] Cai X, Luo Y, Liu B, Cheng H-M. Preparation of 2D material dispersions and their applications. Chemical Society Reviews. 2018;**47**:6224-6266

[55] Chen L, Sun D, Li J, Zhu G. Exfoliation of layered zirconium phosphate nanoplatelets by melt compounding. Materials and Design. 2017;**122**:247-254

[56] Sun L, Boo WJ, Sun D, Clearfield A, Sue H-J. Preparation of exfoliated epoxy/α-zirconium phosphate nanocomposites containing high aspect ratio nanoplatelets. Chemistry of Materials. 2007;**19**:1749-1754

[57] Xia F, Yong H, Han X, Sun D. Small molecule-assisted exfoliation of layered zirconium phosphate nanoplatelets by ionic liquids. Nanoscale Research Letters. 2016;**11**:348

[58] Kaschak DM, Johnson SA, Hooks DE, Kim H-N, Ward MD, Mallouk TE. Chemistry on the edge: A microscopic analysis of the intercalation, exfoliation, edge functionalization, and monolayer surface tiling reactions of α-zirconium phosphate. Journal of the American Chemical Society. 1998;**120**:10887-10894

[59] Casciola M, Alberti G, Donnadio A, Pica M, Marmottini F, Bottino A, et al. Gels of zirconium phosphate in organic solvents and their use for the preparation of polymeric nanocomposites. Journal of Materials Chemistry. 2005;**15**:4262-4267

[60] Zhou Y, Huang R, Ding F, Brittain AD, Liu J, Zhang M, et al. Sulfonic acid-functionalized α-zirconium phosphate single-layer nanosheets as a strong

solid acid for heterogeneous catalysis applications. ACS Applied Materials & Interfaces. 2014;**6**:7417-7425

[61] Zhou Y, Noshadi I, Ding H, Liu J, Parnas RS, Clearfield A, et al. Solid acid catalyst based on single-layer α-zirconium phosphate nanosheets for biodiesel production via esterification. Catalysts. 2018;**8**:17

[62] Wang L, Xu W-H, Yang R, Zhou T, Hou D, Zheng X, et al. Electrochemical and density functional theory investigation on high selectivity and sensitivity of exfoliated nano-zirconium phosphate toward lead. Analytical Chemistry. 2013;**85**:3984-3990

[63] Kim HN, Keller SW, Mallouk TE, Schmitt J, Decher G. Characterization of zirconium phosphate polycation thin films grown by sequential adsorption reactions. Chemistry of Materials. 1997;**9**:1414-1421

[64] Hunter BM, Gray HB, Müller AM. Earth-abundant heterogeneous water oxidation catalysts. Chemical Reviews. 2016;**116**:14120-14136

[65] Seh ZW, Kibsgaard K, Dickens CF, Chorkendorff I, Nørskov JK, Jaramillo TF. Combining theory and experiment in electrocatalysis: Insights into material design. Science. 2017;**355**:6321

[66] Weng B, Xu F, Wang C, Meng W, Grice CR, Yan Y. A layered $Na_{1-x}Ni_yFe_{1-y}O_2$ double oxide oxygen evolution reaction electrocatalyst for highly efficient water-splitting. Energy & Environmental Science. 2017;**10**:121-128

[67] Lu Z, Wang H, Kong D, Yan K, Hsu P-C, Zheng G, et al. Electrochemical tuning of layered lithium transition metal oxides for improvement of oxygen evolution reaction. Nature Communications. 2014;**5**:4345

[68] Oh H-S, Nong HN, Reier T, Bergmann A, Gliech M, Ferreira de Araújo J, et al. Electrochemical catalyst-support effects and their stabilizing role for IrO_x nanoparticle catalysts during the oxygen evolution reaction. Journal of the American Chemical Society. 2016;**138**:12552-12563

[69] Sanchez J, Ramos-Garcés MV, Narkeviciute I, Colón JL, Jaramillo TF. Transition metal-modified zirconium phosphate electrocatalysts for the oxygen evolution reaction. Catalysts. 2017;7:132

[70] Mosby BM, Díaz A, Bakhmutov V, Clearfield A. Surface functionalization of zirconium phosphate nanoplatelets for the design of polymer fillers. ACS Applied Materials & Interfaces. 2014;**6**:585-592

[71] Horsley SE, Nowell DV, Stewart DT. The infrared and Raman spectra of α-zirconium phosphate. Spectrochimica Acta, Part A: Molecular and Biomolecular Spectroscopy. 1974;**30**:535-541

[72] Ramos-Garcés MV, Sanchez J, Del Toro-Pedrosa DE, Barraza Alvarez I, Wu Y, Valle E, et al. Transition metal-modified exfoliated zirconium phosphate electrocatalyst for the oxygen evolution reaction. ACS Applied Energy Materials. 2019;**2**:3561-3567. DOI: 10.1021/acsaem.9b00299

Chapter 12

Strategies in Absorbing Materials Productivity (H_2O) of Renewable Energy Utilization by a Solar Still to Enhancement of Water Flowing over Glass Cover with the Influence of PCM and Nanoparticles

S. Shanmugan

Abstract

The solar thermal applications existing to investigative relationships of absorbing materials of water flowing over glass cover through the influence of PCM and nanoparticles for the enhancement of a single-slope single-basin solar still are presented and discussed. The results are compared with and without PCM and nanoparticles summer days for a conventional solar still. Numerically designed and experimental annotations have been written for the investigative solutions for the temperature of flowing water, glass cover, absorbing materials (FWCW and FWJW) and PCM and nanoparticles basin liner, respectively. The 24 h distillate manufacture rate of the solar still has been enhanced to usage of drip button through pure saline water to absorptive influence of FWCW capability is 70.02% and during (24 hours) daily distillate harvest of FWCW is 9.429 kg/m^2 day, water flowing glass cover influence is 13.37%, respectively. A solar still analysis of Fourier coefficients with (6 to −6) harmonics Fourier series has been used for enhancement, and it is found to be a good representation of the observed variation. It is a good treaty among theoretical and experimental annotations of the structure.

Keywords: single-basin solar still, PCM and nanoparticles, fin wick, drip button, water flowing glass cover, Fourier series

1. Introduction

Water is mislaid every day through numerous physical processes and desired by each active cell; nearly all process that grosses place within the body is reliant on water. A solar energy using in application of thermal process to solar desalination is bid in which evaporative cooling to produce an enhanced for harvests to grow in clean water. Numerical modeling of water flowing over the glass cover in a solar still is actual substantial to design a frugally optimal still and to enhance the manufacture performance for a given cost. Prakash and Kavatherkar [1] have exposed the enactment of the regenerative still and its diurnal harvest is around 7.5 L/m^2, associated to conventional solar stills. Singh and Tiwari [2] have approved out a passive regenerative solar still and the system is good agreement between theoretical and experimental results. An analytical expression for the thermal efficiency of an active regenerative solar still was observed by Singh and Tiwari [3]. It found that to be heat transfer unit and the collector to overall thermal efficiency is 50%. The yield of concentrator assisted regenerative solar still is much higher than any other passive/active, regenerative/non regenerative stills and the overall efficiency increases with an increase in the flow rate of the cold water over the glass cover was suggested by Kumar and Sinha [4]. Prasad et al. [5] reported that the regenerative active solar still harvests the thermal enactment increased. Suneja and Tiwari [6] have proposed that for a particular flow rate of water over the glass cover, the evaporative heat transfer coefficients decreases in increasing the water depth of the basin where as radiative and convective heat transfer coefficients does not very much. Zurigat and Abu-Arabi [7] analyzed a double-glass cover cooling desalination unit and inferred that the arrangement is double-fold which lowered the glass temperature and preheated the entering brine. Ultimately the stills efficiency was increased by over 25% than conventional single-basin single-glass solar stills. Zurigat and Abu-Arabi [8] had shown the performance of a regenerative solar still consisting of two basin effects (first effect and second effect) and the distillate yield is founded to be 20% higher than that of conventional solar stills. Janarthanan et al. [9] proposed the performance of a tilted wick-type solar still and concluded that the glass cover temperature decrease, water flow over the glass cover and the flow rate of 1.5 m/s has increased the production significantly. Murugavel et al. [10] had reviewed the productivity of single-basin solar still with different materials in the basin and inferred that rubber material in the basin improved the absorption, storage and evaporation effects. Boutebila [11] had shown that, the initial film thickness, plate inclination, the length of the still and the radiation reaching the flux plate are the factors affecting the still performance. Zeroual et al. [12] had investigated a double slope solar still with two effects and insisted that the productivity increased by 11.82% by cooling the condenser using flowing water over it (first effect) and 2.94% by shading the north wall from 12 to 14 hours (second effect). The performance of inverted absorber solar still by Dev et al. [13] and found that to be thermal efficiency of inverted absorber solar still is thrice than that of the normal solar still. Khalifa [14] had found that the effect of condensing cover tilt angle of simple solar still on the productivity in different seasons and latitudes. It has been found that the tilt angle should be large in winter and small in summer. Kumar and Dwivedi [15] anticipated the reformed single-slope single-basin active solar still with enhanced condensation procedure. It is found that the yield increment of 14.5% associated with the ordinary design. Shanmugan et al. [16] thermal model industrialized for an energy and exergy analysis of a single-slope single-basin solar still. Rahmani et al. [17] was urbanized for a natural circulation in a solar still and established the distillate yield of 3.72 L/m^2/day by 45.15% efficiency. Ibrahim and Elshamarka [18] was amended basin type solar still and accomplished that the maximum freshwater productivity of 2.93 L/m^2/day. Single-slope solar still was urbanized to integration for the systems to progress the efficiency.

Sahota and Tiwari [19] were advanced for a double slope solar still used (Al_2O_3) nanoparticles in the basin. They clinched without and with nanofluids for three different concentrations are 0.04, 0.08 and 0.12%. Al_2O_3 nanoparticles efficiency in the system is 0.12%. It is the charity for the system in 35 and 80 kg base fluid equaled to the efficiency of 12.2 and 8.4% in the system. Shanmugan et al. [20] was premeditated in the experimental analysis of a double Slope—Tribasin solar still. It established that the associate with and without nanofluids enhanced the performance of first, second and third basin contributes were 35.71, 35.7 and 28.5%.

Sellamia et al. [21] have enriched in solar still exploration of altered thickness besmirched layers of sponge absorber in more energy saving to recover presentation. It is charity in altered sponge liner thickness like that on 0.5, 1.0 and 1.5 cm and harvest sponge liner are 58, 23.03 and 30%, respectively.

Sharshir et al. [22] was twisted in belongings of flake graphite nanoparticles, PCM and the film cooling on the solar still enactment. It is amended of a solar still in FGN and PCM, film cooling high yield on 73.8% and matched to conventional still, the upshot of intensifications on 13% and water depth are 2–0.5 cm. The classified a heat transfer if the structure is extensive time and extra energy rise to temperature formed at pick time.

Many researchers were designed of a single-slope single-basin solar still and have been emarginated to water flowing the glass cover with enormous scope of application developing PCM and nanoparticles by a system. It has twisted internal heat transfer mode scope of advance application in rules of functions. The internal heat transfer increase of a basin solar still is with new impression of harvest in expenditure of human nature foundation.

2. Materials and methods

2.1 Investigation of water flowing glass cover to evolvements absorbing materials by a solar still

An experimental analysis of a single-slope single-basin solar still with water flowing over the glass cover to performance of absorbing materials have been offered in **Figure 1**. The solar still comprises of outward and inward attachment made of plywood through element of 1.3 × 1.3 and 1.25 × 1.25 m. The fissure between the attachments is occupied with glass wool having the thermal conductivity of 0.0038 W/mK. The stature of the back wall is 0.03 m and front wall of 0.10 m. The glass cover of thinness 4 mm is secondhand as the condensing surface assistance to water flowing and the slope of the glass cover are immovable as 11° which is equivalent to the leeway of the position (Chennai). It is accomplished vapor tight through the benefit of metal putty. The j-shaped drainage channel is immobile nigh the front wall to accumulate the concentrate harvest and the output

Figure 1.
Experimental analysis of water flowing influence with fin absorbing materials.

slobbered miserable to the gaging jar. The investigational engaged classical has been established for water flowing through a solar still follow the schematic plan **Figure 2**.

2.2 Heat transfer coefficient and productivity of a still

It have been finished in a basin area of copper leaf and black paint coating in a basin area, dye miscellaneous Al_2O_3 Nanoparticles coating through inner area with absorbing wick materials to absorb surplus solar radiation. The solar radiation communicated through the water flowing transfer to glass cover and fascinated by a wick and fin wick materials apparent—Al_2O_3 to comportment by the copper coil and then shadows the phase change materials—$C_{18}H_{36}O_2$. The solar radiation is immersed through Al_2O_3 Nano particles soars esoteric the absorbing materials area intensification further heat transfer chic inside nigh to visible and IR spectrum. The saline water from side to side of an exceptional establishment has been finalized to pour saline water drop by drop over the absorbing material kept in the basin. The drip heat transfer pipes occupied coating Nano particles miscellaneous black paint to the lengthways dripping association is finished of drip button stationary at consistent intermissions of 0.10 m and heat pipe static in amidst the gap is 0.10 m horizontally in the basin.

The PCMs energy spread stabilities are melting in a wide temperature range [23]. Various evaluations are escorted [24, 25] and a full association of the hottest progresses in PCMs with their thermo physical properties can be found that to be literature. **Figure 3** current these progresses of the absorbing materials setting in basin to improve heat, based on the above reviews, with polymeric and solid-solid PCMs involved.

Figure 3, new solar still have been manufactured in the basin area static a copper coil arrangement and is immovable drip heat pipes. The influence of the solar still is immovable in 10 drip heat pipes and buttons are each a distance of 0.10 × 0.10 m by a placed horizontally in the basin with south to north orientation. It have been

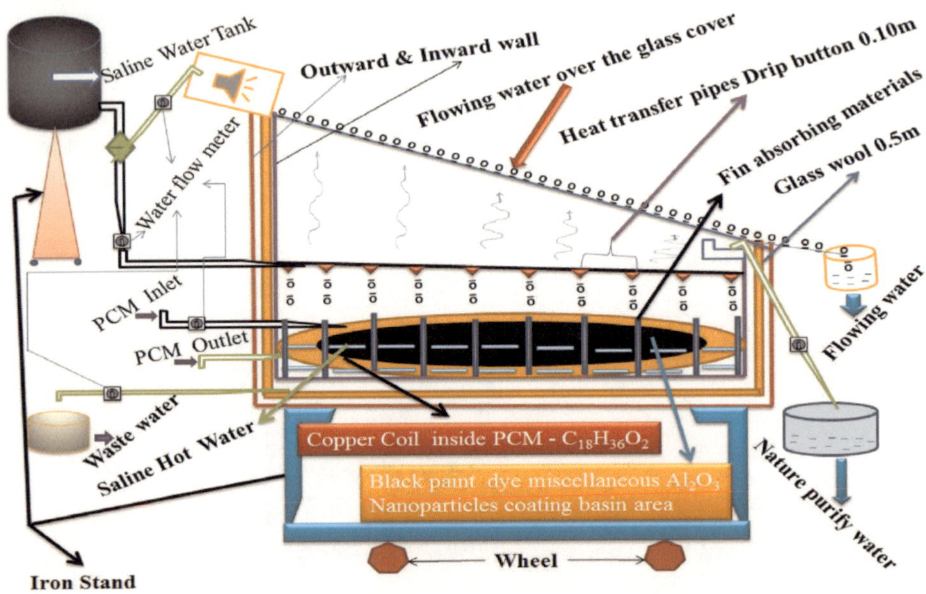

Figure 2.
Sectional view of the solar still with flowing water influence of absorbing materials.

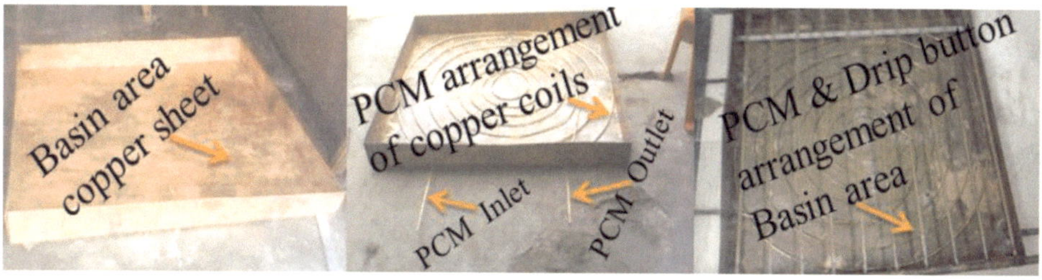

Figure 3.
Basin area arrangement of copper coil in fixed drip button heat pipes.

prepared in the basin vulnerable copper coil in 1 diameter, intervals of 0.10 m and total coil 10 m fixed in the basin. The coil occupied mad black painted and mixed Al_2O_3—nano particles more absorb solar radiation and heat extraction supplementary progress the still is the charity for $C_{18}H_{36}O_2$—PCM for 8 kg inside the coil secure basin area. The PCM melting point increase has been debauched evaporation of the solar still. The saline water tank is providing with a gate valve and is associated with the fjord of the dripping organization.

A provisional organization of water flowing over the glass cover of a single-slope single-basin solar still have been finished with the help of PVC pipe of ½ inch diameter. The length of the pipe is taken precisely equal to the width of the still so that the water would not flow in the lengthwise direction. A number of holes are made in the pipe by equal spacing to maintain the uniform flow over the glass cover. The PVC pipe is clipped at the top of the glass cover. The water at the lower end of the glass cover was collected in a small plastic bucket. Due to the water flow over the glass cover, most of the heat was utilized for evaporation during the day, uptown mid night in this process. There existed fast evaporation due to the large temperature difference between the glass cover and the water surface as expected shown that the **Figure 4** water flowing the experimental glass cover cooling effect.

The water temperature in the absorbing materials is to condensing water flowing cover temperature quantity by fixing copper-constantan thermocouples which have

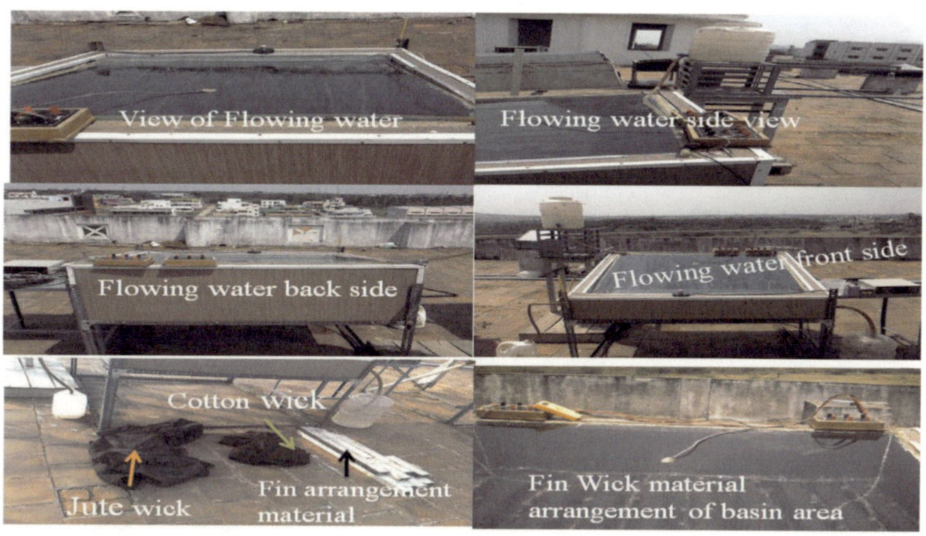

Figure 4.
View of experimental works of water flowing over glass cover used by a solar still with absorbing materials.

been calibrated originally. Solar radiation intensity and ambient temperature have been restrained with solar radiation monitor and digital thermometer. Experimental work analysis of a solar still have been carried out from 6 to 6 am of 24 hours duration with water flowing over the glass cover by absorbing materials deliverable during 2017 at Research Center of Physics, Veltech Multitech Engineering College, Avadi, Chennai—600062 [Latitude 13.1067°N, 80.0970°E], Tamilnadu, India.

2.3 Investigation of thermal modeling by a solar still with effect of absorbing materials

The solar radiation is absorbing to water flowing by glass cover transported over the absorbing materials with working of inside heat ability accessible in the organization as shown in **Figures 1** and **4**. The novel design made of fin wick absorbing materials following the flowchart **Figure 5**.

Figure 6 the water flowing over the glass cover influence of absorbing materials surface to harvest by a solar still with succeeding suppositions have been prepared to different parameters script the energy equilibrium equations.

1. The solar still (PCM and nanoparticles) performance of the water flowing over the glass cover during absorbing materials surface of heat ability have been organized full tight of insulation in the scheme and glass cover is negligible.

2. There is no hotness escape of vapor surface in the scheme.

3. A single-slope single-basin solar still is water flowing over the glass cover, absorbing materials surface and distillate water fragment.

4. Vapor pressure of water have been made full tight of experimental assumed to be linear with temperature work proof of ($P = R_1T + R_2$) single-slope single-basin solar still is given as:

Flowing water with absorbing materials by a solar still

$$bl_{fw}\rho c_{fw}\frac{dT_{fw}}{dt}dx + m_{fw}c_{fw}\frac{dT_{fw}}{dx}dx = \alpha_{fw}H_s bdx + h_4\left(T_g - T_{fw}\right)bdx - h_2\left(T_{fw} - T_a\right)bdx \quad (1)$$

Glass cover with absorbing materials by a solar still

$$H_s\alpha_g + h_1\left(T_{Fw} - T_g\right) = h_4\left(T_g - T_{fw}\right) \quad (2)$$

Basin liner with absorbing materials by a solar still

$$\alpha_{(b+PCM+Nanoparticles)}H_s = h_3\left(T_{b+PCM+Nanoparticles} - T_{Fw}\right) + h_{b+PCM+Nanoparticles}\left(T_{b+PCM+Nanoparticles} - T_a\right) \quad (3)$$

Water mass with absorbing materials by a solar still

$$\alpha_{Fw}H_s + h_3\left(T_{b+PCM+Nanoparticles} - T_{Fw}\right) = M_{Fw}C_{Fw}\frac{dT_{Fw}}{dt} + h_1\left(T_{Fw} - T_g\right) \quad (4)$$

where $h_1, h_2, h_3, h_4, h_{b+PCM+Nanoparticles}$ are demarcated in the Appendix

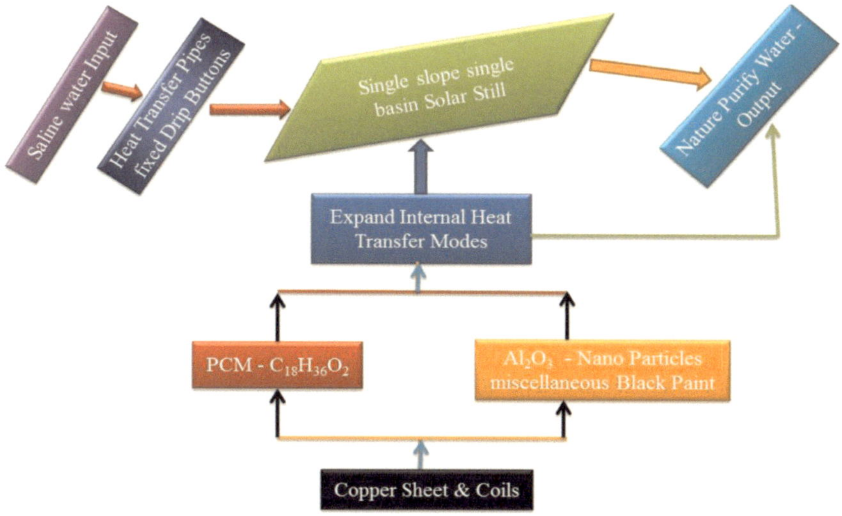

Figure 5.
A flowchart to influence a solar still.

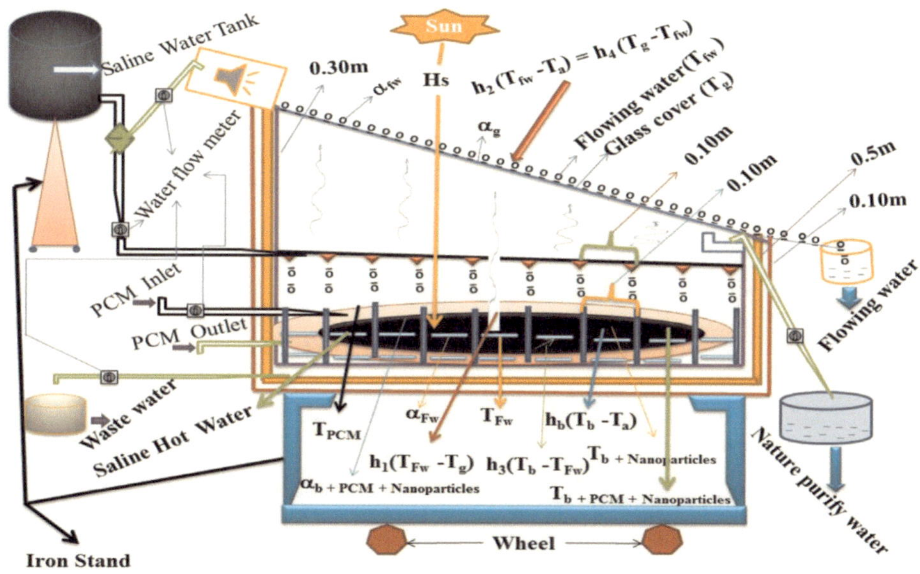

Figure 6.
A solar still with different parameters script form of energy equilibrium equations.

Equation (1), after eliminating T_g from Eq. (2), can be redrafted as

$$\frac{dT_{fw}}{dx} + a_1 T_{fw} = a_2 T_{Fw} + a_3 \tag{5}$$

where

$$a_1 = \frac{(\mathrm{h}_4 + \mathrm{h}_2)\mathrm{b}}{m_{fw C_{fw}}} + \frac{\mathrm{h}_4^2 \mathrm{b}}{(\mathrm{h}_4 + \mathrm{h}_2) m_{fw C_{fw}}}$$

$$a_2 = \frac{(\mathrm{h_1 h_4})\mathrm{b}}{(\mathrm{h_4}+\mathrm{h_2})m_{fwC_{fw}}}$$

$$a_3 = \frac{\alpha_{\mathrm{fw}}\mathrm{H_s b}}{m_{fwC_{fw}}} + \frac{\alpha_{\mathrm{g}}\mathrm{H_s h_4 b}}{(\mathrm{h_4}+\mathrm{h_1})m_{fwC_{fw}}} + \frac{(\mathrm{h_4}+\mathrm{h_2})\mathrm{b}}{m_{fwC_{fw}}}$$

The explanation of the Eq. (5) can be written as

$$T_{fw} = \frac{(a_2 T_{Fw} + a_3)}{a_1} + Ce^{-a_1 t} \quad (6)$$

The explanation of Eq. (6) exposed to the initial conditions by a still

$$T_{fw} = T_{fw0} \quad (7)$$

for all value of t at $x = 0$
Substituting the equation for c in Eq. (6), we get

$$T_{fw} = \frac{(a_2 T_{Fw} + a_3)}{a_1}[1 - e^{-a_1 t}] + T_a e^{-a_1 t} \quad (8)$$

Equation (8) is the obligatory explicit expressions for the temperatures with fin wick materials and flowing over the glass cover by a still, respectively.

Solar radiation and ambient temperature are episodic in environment these can be Fourier series in the form

$$f(t) = a_0 + \sum_{n=1}^{\infty}(A_n \cos n\omega t + B_n \sin n\omega t) \quad (9)$$

The flowing water over glass cover with influence of fin wick materials by a solar still of variation with time can be articulated by

$$f(t) = a_0 + \sum_{n=1}^{\infty} A_n \exp(in\omega t)$$

Since, solar radiation and ambient temperature are periodic nature can be Fourier analyzed in the form of the solar still

$$H_s(t) = H_0 + \sum_{n=1}^{\infty} \mathrm{H}_{sn} \exp(in\omega t) \quad (10)$$

$$T_a(t) = T_{a0} + \sum_{n=1}^{\infty} T_{an} \exp(in\omega t) \quad (11)$$

$$\mathrm{T_g}(t) = T_{g0} + \sum_{n=1}^{\infty} T_{gn} \exp(in\omega t) \quad (12)$$

$$T_{fw}(t) = T_{fw0} + \sum_{n=1}^{\infty} \mathrm{T_{fwn}} \exp(in\omega t) \quad (13)$$

and

$$T_{Fw}(t) = T_{Fwo} + \sum_{n=1}^{\infty} \mathrm{T_{Fwn}} \exp(in\omega t) \quad (14)$$

where constants $T_{g0}, T_{fw0}, T_{Fwo}, T_{gn}, \mathrm{T_{fwn}}, \mathrm{T_{Fwn}}$ are to be determined by substituting for $H_s(t), T_a(t), \mathrm{T_g}(t), T_{fw}(t) and\ T_{Fw}(t)$ with help of Eqs. (10)–(14) in Eqs. (2), (4) and (8) can be explained by integration with the initial condition $T_{Fw} = T_{Fw0}$ at $t = 0$ as

$$T_{Fw}(t) = \frac{b_0}{a_4}\left(1 - e^{-a_4 t}\right) + T_{Fw0}e^{-a_4 t} + \sum_{n=1}^{\infty} \frac{b_n}{i\omega n + a_4}\left(e^{in\omega t} - e^{-a_4 t}\right) \tag{15}$$

where b_0 is the time independent component of $b\ (t)$; b_0 is the coefficient of the time dependent component of b (t), and they are in the form

$$b_0 = a_5 H_{s0} + a_6 T_{a0}$$

$$b_n = a_5 H_{sn} + a_6 T_{an}$$

where

$$a_4 = \frac{\mathrm{h_1} + \mathrm{h_2}}{m_{Fw}C_{Fw}} - \frac{\mathrm{h_3^2}}{\left(\mathrm{h_3} + \mathrm{h}_{b+PCM+Nanoparticles}\right)\mathrm{m_{Fw}C_{Fw}}} - \frac{\mathrm{h_1^2 h_1 h_4}}{(\mathrm{h_4} + \mathrm{h_1})\mathrm{m_{Fw}C_{Fw}}}\left(\frac{a_2}{a_1}\right)\left(1 - e^{-a_1 t}\right)$$

$$a_5 = \frac{\alpha_{\mathrm{Fw}}}{m_{Fw}C_{Fw}} + \frac{\mathrm{h_3}\alpha_{\mathrm{g}}}{\left(\mathrm{h_3} + \mathrm{h}_{b+PCM+Nanoparticles}\right)\mathrm{m_{Fw}C_{Fw}}} + \left(\frac{\mathrm{h_1}\alpha_{\mathrm{g}}}{(\mathrm{h_4} + \mathrm{h_1})\mathrm{m_{Fw}C_{Fw}}}\right)\left(\frac{b\alpha_{fw}}{a_1 m_{fw}C_{fw}}\right)\left(1 - e^{-a_1 t}\right)$$

$$a_6 = \frac{\mathrm{h_3 h}_{b+PCM+Nanoparticles}}{\left(\mathrm{h_3} + \mathrm{h}_{b+PCM+Nanoparticles}\right)} + \frac{\mathrm{h_4 h_1}}{\mathrm{h_4} + \mathrm{h_1}}e^{-a_1 t} + \frac{\mathrm{h_1 h_4}}{(\mathrm{h_4} + \mathrm{h_1})}\left(\frac{\mathrm{h_2}b}{\mathrm{m_{fw}C_{fw}}}\right)\left(\frac{a_3}{a_1}\right)\left(1 - e^{-a_1 t}\right)$$

The values of $T_{Fw}\ (t)$ are designed at intervals of ½ hour starting from sun rise by Equation (15), and $T_g\ (t)$ is calculated from the following relation

$$T_{gn} = \frac{\mathrm{H_s}\alpha_{\mathrm{g}} + \mathrm{h_1 T_{Fwn}} + \mathrm{h_4 T_{fw}}}{\mathrm{h_4} + \mathrm{h_1}} \tag{16}$$

During off-sunshine hours, solar intensity and ambient temperature terms will evaporate, then the analysis will be the same.

The instantaneous hourly distillate output per unit with absorbing materials area of the still is premeditated by

$$m_e = \frac{\mathrm{h_{ewg}}\left(\mathrm{T_{Fw}(t)} - \mathrm{T_g(t)}\right)}{\mathrm{L}} \times 3600 \tag{17}$$

The efficiency of the proposed structure may be articulated as

$$\eta\% = \frac{M_e L}{A_{b+PCM+Nanoparticles} \int Hs\Delta t} \times 100 \tag{18}$$

where Δt mentions to the time interval over which the solar intensity is measured.

3. Result and discussion

Numerical calculations have been made in instruction to escalate the fin with cotton wick (FWCW) and fin with jute wick (FWJW) in the basin surface area. Expending of a still is hourly variations of the data in summer days through the solar intensity and ambient temperature measurement of monitors for two of the typical days in 2017 at Chennai weather conditions, the Fourier coefficients of solar intensity and ambient temperature have been evaluated to (6 to −6) harmonics of the Fourier series. The water flowing over the glass cover with influence of FWCW and FWJW of the solar still following parameters have been evaluated the instantaneous thermal efficiency of the suggested solar still.

$A_g = 1.69\ \text{m}^2$, $\tau_g = 0.75$, $A_w = 1.69\ \text{m}^2$, $M_w = M_{fw} = 12$ kg, $\varepsilon_g = 0.88$, $\sigma = 5.66 \times 10^{-8}\ \text{W/m}^2\text{K}^4$, $\alpha_g = 0.05$, $C_w = 4190$ J/kg, $V = 1.4$ m/K = 0.038 W/mK, $h_1 = 22.52\ \text{W/m}^2\ °\text{C}$; $h_2 = 15.64\ \text{W/m}^2\ °\text{C}$; $h_3 = 137.05\ \text{W/m}^2\ °\text{C}$; $h_4 = 135.5\ \text{W/m}^2\ °\text{C}$; $h_b = 0.7686\ \text{W/m}^2\ °\text{C}$; $h_{ew} = 14.01\ \text{W/m}^2\ °\text{C}$; $h_i = 6.27\ \text{W/m}^2\ °\text{C}$; $K_i = 0.04\ \text{W/m°C}$; $L = 2372.52$ kJ/kg; $l_i = 0.05$ m; $\omega = 7.2722 \times 10^{-5}\ \text{s}^{-1}$; $\tau_w = 0.1$; $\tau_{fw} = 0.0$; $\tau_b = 0.6$; $\alpha_{fw} = \alpha_w = \alpha_b = 0.88$; $\rho = 1000.0\ \text{kg/m}^3$; $C_w = C_{fw} = 4190$ J/kg °C.

The solar still have been implemented for high thermal energy accumulation through benefit of PCM and nanoparticles immersed by solar intensity and ambient temperature for 2 days with absorbing wick materials, i.e., FWCW and FWJW in summer typical days have been shown in the **Figure 7**. Expending of the absorbing materials has been implemented for Fourier coefficients in H_s and T_a in 2017 at Chennai in Tamilnadu, India. It is traceable that the (6 to −6) harmonics used in design is adequate for the convergence of the Fourier series. The hourly variation of solar still have same trend for all the days and solar radiation appears to be supreme among 12–2 pm.

Numerically designed and experimental annotations of with and without PCM and nanoparticles using in the basin liner to more increases of FWCW, FWJW temperature, glass cover temperature and flowing over the water temperature of the optional solar still have been expected in **Figures 8A** and **9**. It is clear that the merchandise of FWCW, FWJW temperature through experimental and theoretical consequences have the equivalent tendency. The innovative solar still has been manufactured through a thermal application of internal heat transfer modes to construction of numerous parameters enhancement occupation to originate of basin liner stable of PCM temperature (copper coils) and black paint dye miscellaneous nanoparticles (basin) temperature and more engross solar radiation in the organi-

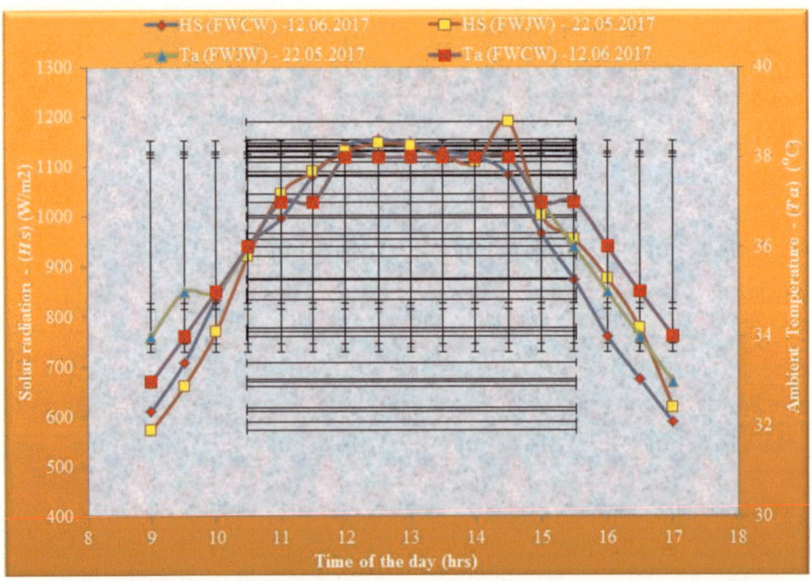

Figure 7.
Hourly variations of solar radiation and ambient temperature absorb by a solar still.

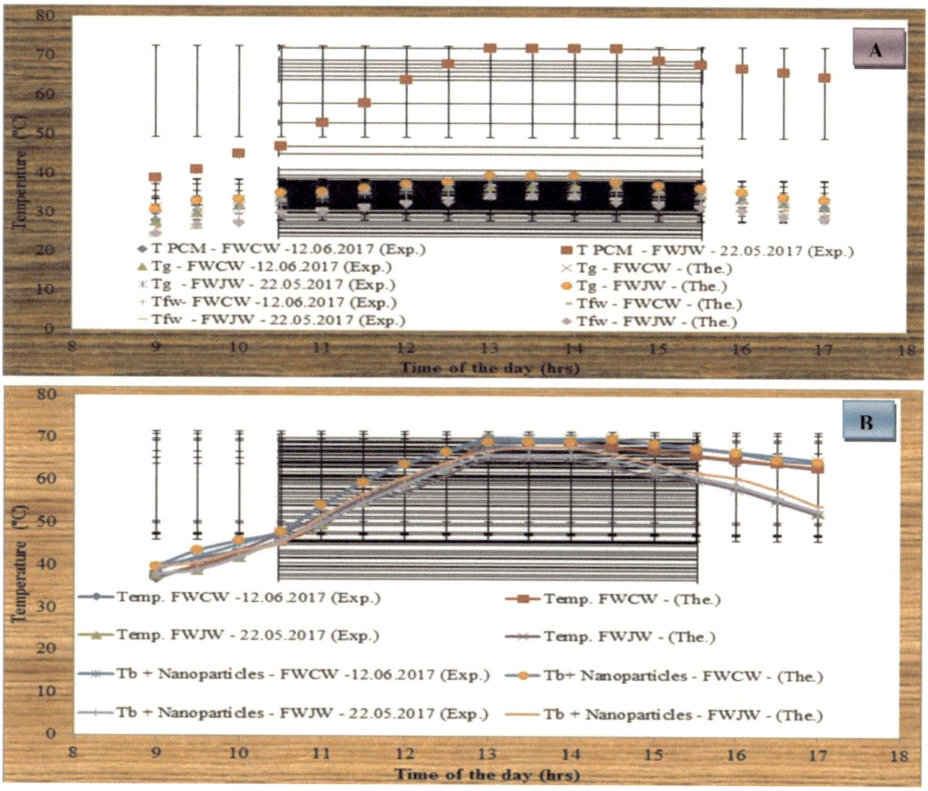

Figure 8.
(A and B) PCM and nanoparticles influence of hourly variations with temperature for FWCW, FWJW, glass cover, flowing water, and basin (nanoparticles) temperature of the solar still.

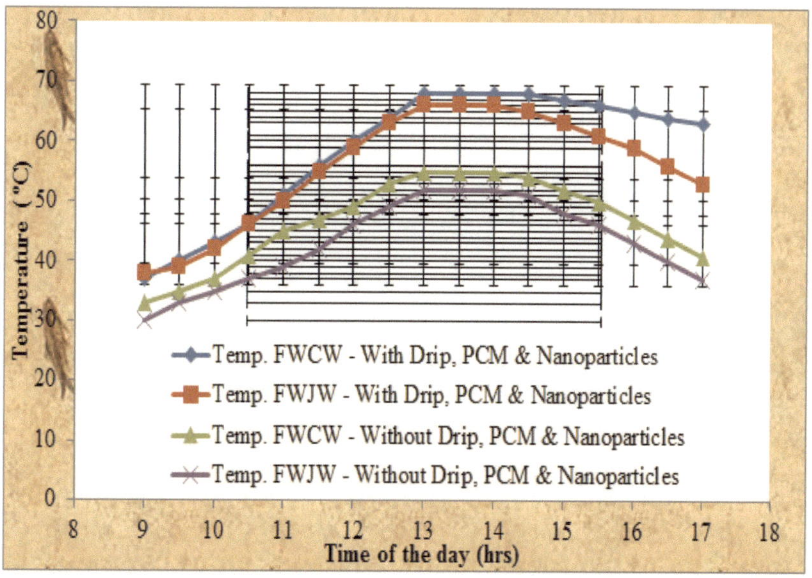

Figure 9.
Hourly variation of production rate of absorbing materials in comparison with and without drip, PCM and nanoparticles of the solar still.

zation to hourly variation of temperature to established in exposed **Figure 8B**, through of PCM and nanoparticles using in the basin liner.

The hourly variation to manufacture rate of FWCW and FWJW temperature, glass cover temperature, flowing over the water temperature for the basin liner to use in PCM and nanoparticles preparation of absorbing materials through dripping pure saline water to continue least water depth has been compared to with and without PCM and nanoparticles of expressions in **Figure 10**. From the diagram it is clear that with PCM and nanoparticles of saline water in the absorbing materials illustrations augmented rate of evaporation due to the large temperature difference between absorbing materials and glass cover temperature. The temperature

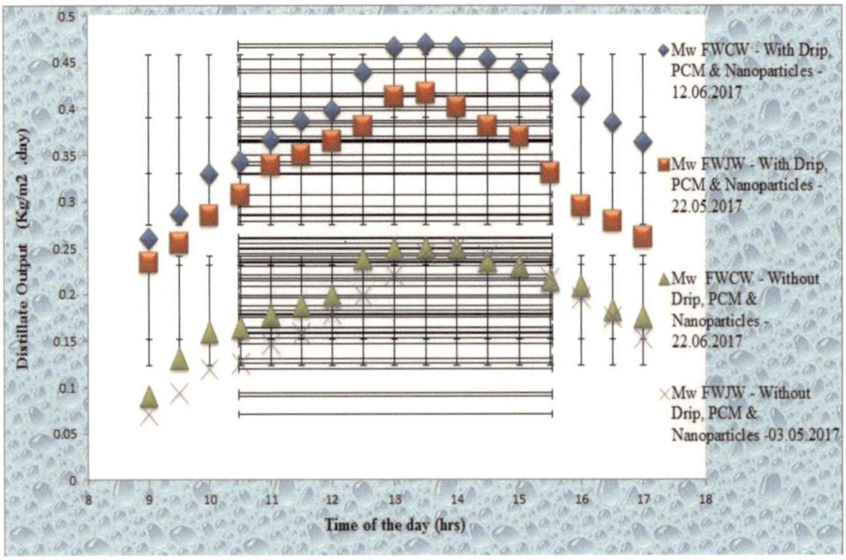

Figure 10.
Hourly variation of distillate production rate of with and without PCM and nanoparticles testing with absorbing materials.

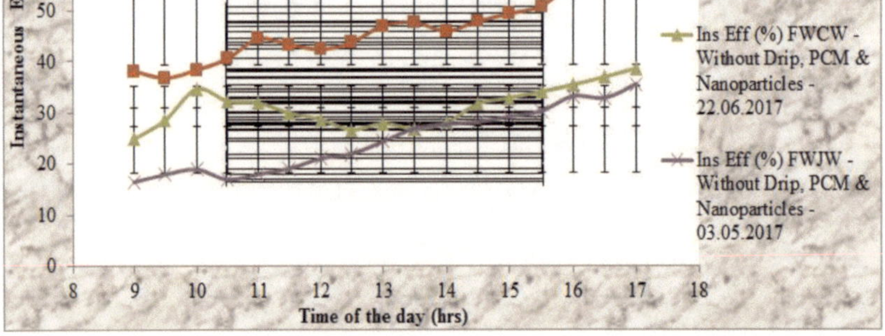

Figure 11.
Hourly variations of absorbing materials of instantaneous energy efficiency with and without PCM and nanoparticles of the solar still.

variance among the absorbing materials and glass cover temperature without PCM and nanoparticles are small due to huge thermal ability and the amount of evaporation is moderate. The supreme FWCW and FWJW distillate harvest of the structure is 0.469 and 0.415 kg/m^2 30 minutes during 12.30–2 pm and without PCM and nanoparticles 0.250 and 0.244 kg/m^2 30 minutes. The whole FWCW and FWJW distillate harvest during 9 am–17 pm is found to 6.699 and 5.659 kg/m^2 and without PCM and nanoparticles 3.343 and 3.014 kg/m^2. Hence, peak luminous during of the solar still through PCM and nanoparticles except to FWCW and FWJW has been distributed with a progressive sunshine distillate harvest as well as nocturnal output in 2.730 and 2.130 kg/m^2 nightly concentrate produce. Over 24 hours cycle, the overall FWCW and FWJW manufacture of the anticipated structure is found to be 9.429 and 7.789 kg/m^2 day and without PCM and nanoparticles is found to be 5.234 and 4.434 kg/m^2 day. Numerical results from FWCW, FWJW, glass cover, flowing over the water, basin liner temperature are in neighboring agreement through the experimental observations.

The absorbing materials instantaneous efficiency of the suggested structure has been established for exposed in the **Figure 11**. The absorptive of the wick surface are noteworthy working parameter of the solar still and should be inactive optimal to provide enhanced efficiency. The water flowing over the glass cover to influence of instantaneous overall efficiency in FWCW, FWJW with and without PCM and nanoparticles to varies of absorbing materials from 70.02 to 46.69% and 31.19 to 24.62%, respectively. It has clearly reflected that the instantaneous distillate harvest with PCM and nanoparticles to water flowing over the glass cover influence solar still is found to be 25% higher than the without PCM and nanoparticles of the solar still. Furthermore the theoretical results authenticated through the experimental explanations are rather good as there is no aberration in the tendency.

Numerical results from FWCW, FWJW, glass cover, flowing over the water, basin liner temperature in neighboring agreement through of the solar still have been observed energy to form into heat to force of a higher efficiency and compare to the upsurge in saturated vapor pressure is substantiated by Huang et al. and Shanmugan et al. [26, 27] augmentation is pretentious for a nanoparticle and found that to be hydrophobic nanoparticles and PCM assistance to the fast evaporation to vapor heaviness of productivity water to the enhancement.

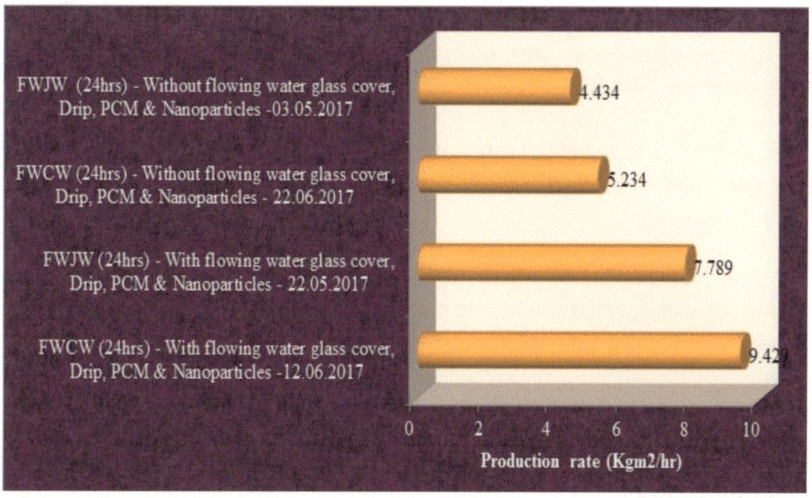

Figure 12.
Modulation of absorbing materials to absorb energy to a solar still for 24 hours output with and without PCM and nanoparticles.

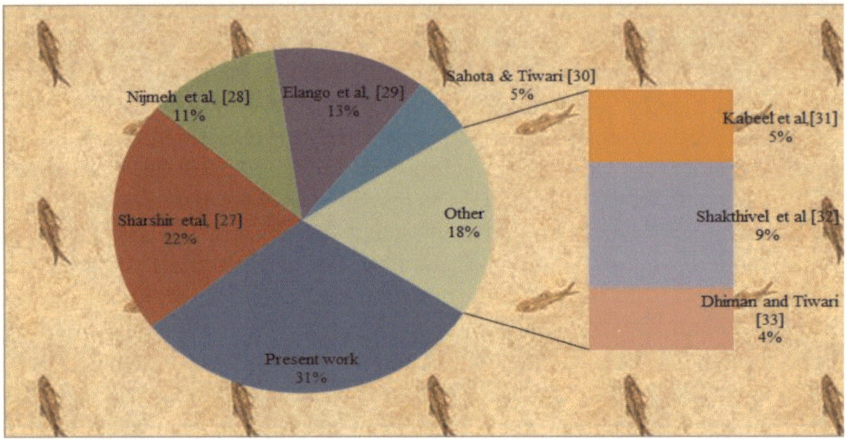

Figure 13.
Comparison of flowing water over the glass, drip, PCM and nanoparticles of overall (absorbing materials) efficiency of the solar still.

Figure 12, modulation of flowing water over the glass to high enhancement of FWCM in basin solar still have been compared to with and without PCM and nanoparticles total productivity (24 hours) on 9.429 kg/m^2 day. **Figure 13** is the photograph of comparison of flowing water over the glass, drip, PCM and nanoparticles of overall high efficiency of the solar still.

3.1 Analysis of water quality by a solar still

A new solar still have been established in natural purification (H_2O) standard water in produce and contaminants are removed. The aim of this reduction in elevations of pH, EC, TDS and total hardness is due to solar desalination process.

4. Conclusion

The solar thermal applications exhibiting and investigative relationships for the instantaneous energy productivity is done and substantiated. The subsequent assumptions have been haggard in the existing investigation of absorbing materials to water flowing glass cover with influence of PCM and nanoparticles for enhancement of a solar still [28–33].

i. The influence of flowing water over the glass cover has been absorbed role on the enactment and is good treaty among theoretical and experimental annotations of the structure.

ii. During the effective of hours in the structure is improvement of PCM and nanoparticles for different heat movements of storage density and the isothermal nature extension of intact productivity of the solar still.

iii. The daily distillate manufacture rate of the still has been enhanced to usage of drip button through pure saline water to absorptive influence of FWCW capability is 70.02%. FWCW surface should be upheld by several eliminations of salt deposition.

iv. The significance absolute produce superior of the solar still is entire daily distillate harvest in through 24 hours for FWCW is 9.429 kg/m^2 day, respectively.

v. The investigation of absorbing materials to water flowing glass cover with influence of PCM and nanoparticles for enhancement of a solar still analyses of solar intensity and ambient temperature with 6 to −6 harmonics are used and found to be a good representation of the observed variation.

Nomenclature

b	breadth of the solar still (m)
$C_{Fw} - c_{fw}$	specific heat of fin wick and flowing water (J/kg °C)
H_s	solar radiation (W/m^2)
$h_{b+PCM+Nanoparticles}$	overall bottom heat loss coefficient from basin liner improve heat use of PCM and nanoparticles to ambient through bottom insulation (W/m^2 °C)
h_1	total heat transfer coefficient from fin wick water surface to glass cover (W/m^2 °C)
h_2	convective and radiative heat transfer coefficient from water flow cooling glass cover to ambient (W/m^2 °C)
h_3	convective heat transfer coefficient from PCM and nanoparticles by basin liner to water mass (W/m^2 °C)
h_4	convective heat transfer coefficient from glass cover to flowing water (W/m^2 °C)
l_w	the thickness of flowing water over the glass covers (m)
M_{Fw}	fin wick mass in the basin surface area (kg)
m_{fw}	mass of flowing water rate (kg/m^{2}h)
$\dot{Q}$	heat flux of the still (W/m^2)
$\dot{q}_{ew}$	evaporative heat transfer rate (W/m^2)
R_1 and R_2	two constants obtained from saturation vapor data (°C)
T_a	temperature of the ambient (°C)
$T_{b+PCM+Nanoparticles}$	temperature of the PCM and nanoparticles basin surface area (°C)
T_g	temperature of the glass covers (°C)
T_{fw}	flowing water temperature (°C)
T_{Fw}	temperature of the fin wick water surface (°C)
T_s	surface temperature of the sun (°C)

Greek letters

$\alpha_{b+PCM+Nanoparticles}$	energy absorptivity of PCM and nanoparticles basin surface area
α_g	energy absorptivity of glass cover
α_{Fw}	energy absorptivity of fin wick water mass
α_{fw}	energy absorptivity of flowing water
η	energy efficiency of the still

Abbreviation

FWCW	fin with cotton wick
FWJW	fin with jute wick

Appendix

The experimental work analysis of a solar still have been approved out from 6 to 6 am of 24 hours duration the collective evaporative, convective and radiative heat transfer coefficient with water flowing over the glass cover by absorbing materials deliverable during 2017 at Research Center of Physics, Veltech Multitech Engineering College, Avadi, Chennai—600062 [Latitude 13.1067°N, 80.0970°E], Tamilnadu, India to the atmosphere can be written as

$$h_4 = 0.016 \times h_2 \times (P_{fw} - \gamma P_a) \quad (19)$$

$$P_{fw} = R_1 T_{fw} + R_2$$

$$\gamma P_a = R_1 T_a + R_2$$

$$h_1 = h_{cwg} + h_{ewg} + h_{rwg} \quad (20)$$

$$h_{cwg} = 0.884 \times \left[(T_{Fw} - T_g) + \frac{(P_{Fw} - P_g)(T_{Fw} + 273)}{268900 - P_{Fw}} \right]^{1/3}$$

$$h_{ewg} = 0.016273 \times h_{cwg} \left(\frac{P_{Fw} - P_g}{T_{Fw} - T_g} \right)$$

$$h_{rwg} = \frac{\varepsilon\sigma\left[(T_{Fw} + 273)^4 - (T_g + 273)^4\right]}{(T_{Fw} - T_g)}$$

$$h_2 = h_{cwa} + h_{rwa} \quad (21)$$

$$h_{ca} = 5.7 + 3.8\ \mathrm{V}$$

$$h_{cwa} = 0.884 \times \left[(T_{Fw} - T_a) + \frac{(P_{Fw} - P_a)(T_{Fw} + 273)}{268900 - P_{Fw}} \right]^{1/3}$$

$$h_{rwa} = \frac{\varepsilon\sigma\left[(T_{Fw} + 273)^4 - (T_a + 273)^4\right]}{(T_{Fw} - T_a)}$$

$$m_{fw} = b l_w$$

where l_w is the thickness of flowing water over the glass cover

$$h_3 = h_{cbw} + h_{rbw} \quad (22)$$

$$h_{cbw} = 0.884 \times \left[(T_{b+PCM+Nanoparticles} - T_w) + \frac{(P_{b+PCM+Nanoparticles} - P_w)(T_{b+PCM+Nanoparticles} + 273)}{268900 - P_b} \right]^{1/3}$$

$$h_{rbw} = \frac{\varepsilon\sigma\left[(T_{b+PCM+Nanoparticles} + 273)^4 - (T_w + 273)^4\right]}{(T_{b+PCM+Nanoparticles} - T_w)}$$

$$h_4 = h_{cgw} + h_{rgw} \quad (23)$$

$$h_{cgw} = 0.884 \times \left[(T_g - T_w) + \frac{(P_g - P_w)(T_g + 273)}{268900 - P_g} \right]^{1/3}$$

$$h_{rgw} = \frac{\varepsilon\sigma\left[(T_g + 273)^4 - (T_w + 273)^4\right]}{(T_g - T_w)}$$

$$h_3 = h_{cbw} + h_{rbw} \tag{24}$$

$$h_{cbw} = 0.884 \times \left[\left(T_{b+PCM+Nanoparticles} - T_w \right) + \frac{\left(P_{b+PCM+Nanoparticles} - P_w \right)\left(T_{b+PCM+Nanoparticles} + 273 \right)}{268900 - P_{b+PCM+Nanoparticles}} \right]^{1/3}$$

$$h_{rbw} = \frac{\varepsilon\sigma\left[\left(T_{b+PCM+Nanoparticles} + 273 \right)^4 - \left(T_w + 273 \right)^4 \right]}{\left(T_{b+PCM+Nanoparticles} - T_w \right)}$$

where the water, flowing water temperature at $x = 0$; and glass cover is instantaneous glass temperature for each hour.

Author details

S. Shanmugan
Research Center of Physics, Koneru Lakshmaiah Education Foundation (KLEF—KLU), Guntur District, Andhra Pradesh, India

*Address all correspondence to: s.shanmugam1982@gmail.com

References

[1] Prakash J, Kavatherkar AK. Performance prediction of a regenerative solar still. Solar and Wind Technology. 1986;**3**(2):119-125

[2] Singh AK, Tiwari GN. Experimental validation of passive regenerative solar still. International Journal of Energy Research. 1992;**16**(6):497-506

[3] Singh AK, Tiwari GN. Design parameters of an active regenerative solar still: An experimental study. International Journal of Energy Research. 1993;**17**(5):365-375

[4] Kumar S, Sinha S. Transient model and comparative study of concentrator coupled regenerative solar still in forced circulation mode. Energy Conversion and Management. 1996;**37**(5):629-636

[5] Prasad B, Bhagat N, Tiwari GN. Enhancement in daily yield due to regenerative effect in solar distillation. International Journal of Ambient Energy. 1997;**18**(2):83-92

[6] Suneja S, Tiwari GN. Effect of water flow on internal heat transfer solar distillation. Energy Conversion and Management. 1999;**40**:509-518

[7] Zurigat YH, Abu-Arabi MK. Modeling and performance analysis of a solar desalination unit with double-glass cover cooling. Desalination. 2001;**138**: 145

[8] Zurigat YH, Abu-Arabi MK. Modeling and performance analysis of a regenerative solar desalination unit. Applied Thermal Engineering. 2004; **24**(7):1061-1072

[9] Janarthanan B, Chandrasekaran J, Kumar S. Performance of floating cum-tilted wick type solar still with the effect of water flowing over the glass cover. Desalination. 2006;**190**:51-62

[10] Murugavel KK, Chockalingam KKSK, Srithar K. Progresses in improving the effectiveness of the single basin passive solar still. Desalination. 2008;**220**(1-3): 677-680

[11] Boutebila H. A theoretical model of a free flow solution over an inclined long flat plate solar still. Desalination. 2009;**249**:1249-1258

[12] Zeroual M, Bouguettaia H, Bechki D, Boughali S, Bouchekima B, Mahcene H. Experimental investigation on a double slope solar still with partially cooled condenser in the region of Ouargla (Algeria). Energy Procedia. 2011;**6**:736-742

[13] Dev R, Abdul-Wahab SA, Tiwari GN. Performance study of the inverted absorber solar still with water depth and total dissolved solid. Applied Energy. 2011;**88**:252-264

[14] Khalifa AJN. On the effect of cover tilt angle of the simple solar still on its productivity in different seasons and latitudes. Energy Conversion and Management. 2011;**52**:431-436

[15] Kumar SS, Dwivedi VK. Experimental study on modified single slope single basin active solar still. Desalination. 2015;**367**:69-75

[16] Shanmugan S, Manikandan V, Shanmugasundaram K, Janarathanan B, Chandrasekaran J. Energy and exergy analysis of single slope single basin solar still. International Journal of Ambient Energy. 2012;**33**:142-151

[17] Rahmani A, Boutria A, Hadef A. An experimental approach to improve the basin type solar still using an integrated natural circulation loop. Energy Conversion and Management. 2015;**93**: 298-308

[18] Ibrahim AGM, Elshamarka SE. Performance study of a modified basin type solar still. Solar Energy. 2015;**118**: 397-409

[19] Sahota L, Tiwari GN. Effect of Al_2O_3 nanoparticles on the performance of passive double slope solar still. Solar Energy. 2016;**130**:260-272

[20] Shanmugan S, Raj K, Arunnarayanan SR. Design and performance analysis of an innovative V-shape double slope Tribasin solar nano still. International Journal of Applied Engineering Research. 2015; **10**(83):261-266

[21] Sellamia MH, Belkisa T, Aliouara ML, Meddoura SD, Bouguettaiab H, Loudiyic K. Improvement of solar still performance by covering absorber with blackened layers of sponge. Groundwater for Sustainable Development. 2017;**5**: 111-117

[22] Sharshir SW, Peng G, Wu L, Essa FA, Kabeel AE, Yang N. The effects of flake graphite nanoparticles PCM and film cooling on the solar still performance. Applied Energy. 2017;**191**: 358-366

[23] Zhou D, Zhao CY, Tian Y. Review on thermal energy storage with phase change materials (PCMs) in building applications. Applied Energy. 2012;**92**: 593-605

[24] Pielichowska K, Pielichowski K. Phase change materials for thermal energy storage. Progress in Materials Science. 2014;**65**:67-123

[25] Shalabyet SM. An experimental investigation of a V-corrugated absorber single-basin solar still using PCM. Desalination. 2016;**398**:247-255

[26] Huang Z, Li X, Yuan H, Feng Y, Zhang X. Hydrophobically modified nanoparticle suspensions to enhance water evaporation rate. Applied Physics Letters. 2016;**109**:161-702

[27] Shanmugan S, Palani S, Janarthanan B. Productivity enhancement of solar still by PCM and nanoparticles miscellaneous basin absorbing materials. Desalination. 2017. DOI: 10.1016/j.desal.2017.11.045

[28] Sharshir SW, Peng G, Wu L, Essa FA, Kabeel AE, Yang N. The effects of flake graphite nanoparticles, phase change material and film cooling on the solar still performance. Applied Energy. 2017;**191**:358-366

[29] Nijmeh S, Odeh S, Akash B. Experimental and theoretical study of a single- basin solar still in Jordan. International Communications in Heat and Mass Transfer. 2005;**32**:565-572

[30] Elango T, Kannan A, Murugavel KK. Performance study on single basin single slope solar still with different water nanofluids. Desalination. 2015;**360**:45-51

[31] Kabeel AE, Omara ZM, Essa FA. Enhancement of modified solar still integrated with external condenser using nanofluids: An experimental approach. Energy Conversion and Management. 2014;**78**:493-508

[32] Sakthivel S, Shanmugasundaram S, Alwarsamy T. An experimental study on a regenerative solar still with energy storage medium—Jute cloth. Desalination. 2010;**264**(1-2):24-31

[33] Dhiman NK, Tiwari GN. Effect of water flowing over the glass cover of a multi-wick solar still. Energy Conversion and Management. 1990; **30**(3):245-250